W9-ABH-513

From the s[...] [...]nbers to the concise
presentation [...] [...]eparation of roots to
the iterative solution of the frequency equation, this classic text
offers a complete, basic survey of the theory of equations with
equal emphasis on the numerical methods of solution. The complete development of each theory and method is at your fingertips.

J. V. U[...] [...]tics at
Stanfo[...] [...]e the
Introd[...] [...]acks
#6672[...] [...]er he
compl[...]

DATE DUE

Theory of Equations
J. V. Uspensky

McGRAW-HILL BOOK COMPANY, INC.
New York   Toronto   London

THEORY OF EQUATIONS

6 7 8 9 0 MPC 75 74 73 72 71

ISBN 07-066736-5 pbk
ISBN 07-066735-7 cl

PREFACE

This book was written as a textbook to be used in the standard American university and college courses devoted to the theory of equations. As such it is elementary in character and, with few exceptions, contains only material ordinarily included in texts of this kind. But the presentation is made so explicit that the book can be studied by students without a teacher's help.

Everything that is stated in the text is presented with full development, and nowhere is reference made to results that are beyond the scope of this book. This accounts for the fact that the book, though containing chiefly the same matters as other currently used texts, surpasses them in size. A few topics that might be omitted without harm are marked by black stars. Numerous problems are added after each significant section. For the most part they are simple exercises, but the more difficult among them are marked by asterisks.

In four chapters the exposition differs considerably from custom. In the chapter on complex numbers the superficial approach so common in many books is replaced by a simple and yet thorough presentation of the theory of complex numbers. The author's experience shows that students, almost without exception, follow this presentation without difficulty.

In the chapter on separation of roots the author gives a very efficient method for separating of real roots, much superior in practice to that based on Sturm's Theorem. He believes that no other book mentions this method, which he invented many years ago and has been teaching to his students for a number of years.

In the chapter on numerical computation of roots Horner's method is presented in the original form, including the process of contraction, which unfortunately has disappeared from American texts. Also, a thorough examination is made of the error caused by contraction.

Determinants are introduced not by formal definition, as is usual, but by their characteristic properties, following Weierstrass. The advantage of this is apparent, for example, in the proof of the theorem of multiplication of determinants. Some elementary notions about the algebra of matrices are also developed in this chapter.

Certain matters because of their intrinsic difficulty are referred to the appendixes. Appendix I deals with the fundamental theorem of algebra.

The author chose as the most intuitive proof, and therefore most suitable for beginners, the fourth proof by Gauss.

Appendix II gives the proof of a theorem of Vincent on which is based the method of separation of roots mentioned above.

Appendixes III and IV were added on the advice of Professor S. P. Timoshenko as likely to interest engineering students. Appendix III is devoted to a simple derivation of criteria for an equation to have all roots with negative real part. Appendix IV deals with iterative solution of the frequency equation.

Appendix V gives an explanation of Graeffe's method for computing roots and is of particular value in the calculation of the imaginary roots of an equation.

J. V. Uspensky

Stanford University, Calif.
December, 1946

The author's explanation to the publisher concerning the purpose of this book has been made the Preface, for it seemed to fulfill the requirements and it expressed his thoughts.

To Max. A. Heaslet of the National Advisory Committee for Aeronautics and Carl Douglas Olds of San Jose State College for the assistance they freely offered and gave in the editing and proofreading of this text by their former professor, I acknowledge great indebtedness. They took on this responsibility, which normally falls to the author himself, when they already carried heavy loads of their own, the death of the author having occurred just after the finished manuscript had gone to the publishers.

L. Z. U.

May, 1948

CONTENTS

CHAPTER I

COMPLEX NUMBERS

1. What Are Complex Numbers? In elementary courses of algebra the letters employed ordinarily represent real numbers, that is, positive and negative integers and fractions, including 0, which are called *rational numbers*, and *irrational numbers* such as $\sqrt{2}$, $\sqrt[3]{3}$, etc. Only occasionally, in connection with the solution of quadratic equations, is some mention made of imaginary or complex numbers. For instance, applying the ordinary formula for the roots of a quadratic equation to the equation

$$x^2 + x + 3 = 0,$$

students are told that it has the roots

$$\frac{-1 + \sqrt{-11}}{2}, \qquad \frac{-1 - \sqrt{-11}}{2}$$

where the symbol $\sqrt{-11}$ is an imaginary quantity since negative numbers cannot have real square roots. Students are taught how to perform operations with such imaginary numbers in a purely formal way, but no adequate explanation is given of what lies behind blind operations on symbols that in themselves have no meaning. Perhaps there is some justification for such a procedure on the ground that at the age when first they encounter "imaginary" numbers students have not yet developed a sufficient faculty for abstraction to understand what they are really dealing with, and that all they can hope to accomplish is the acquisition of some skill in formal manipulations. But when the time arrives to begin more serious study of that part of algebra that is called the theory of equations, it becomes necessary to return to a reexamination of imaginary or complex numbers in order to lay down a firm foundation on which all the subsequent developments will solidly rest.

In what follows, letters a, b, c, . . . , etc., (with the sole exception of the letter i, which will be used in a special sense) will serve to designate real numbers. An ordered pair of real numbers

$$(a,b)$$

of which a is the *first element* and b is the *second element* will be considered as a new entity or a new object of mathematical investigation and henceforth will be called a *complex number*. In order to be able to make ordered pairs or complex numbers an object of mathematical investigation we must extend to them the notion of equality, and we must also define the meaning of the four fundamental operations

addition
subtraction
multiplication
division

that will be performed on them.

2. Definition of Equality. Two complex numbers (a,b) and (c,d) are *equal* if and only if $a = c$ and $b = d$. Complex numbers that do not satisfy this condition of equality are called *unequal*. To denote equality the ordinary sign = is used. Thus, the equality

$$(a,b) = (c,d)$$

means

$$a = c, \qquad b = d.$$

According to this definition

$$(2,\sqrt{12}) = (\tfrac{1}{2}\sqrt{7 + 4\sqrt{3}} + \tfrac{1}{2}\sqrt{7 - 4\sqrt{3}}, 2\sqrt{3})$$

since

$$2 = \tfrac{1}{2}\sqrt{7 + 4\sqrt{3}} + \tfrac{1}{2}\sqrt{7 - 4\sqrt{3}} \qquad \text{(why?)}$$

and

$$\sqrt{12} = 2\sqrt{3}.$$

On the contrary, complex numbers $(1, -1)$ and $(-1, 1)$ are unequal, and this is indicated by writing

$$(1, -1) \neq (-1, 1).$$

3. Definitions of Addition and Multiplication. Of the four fundamental operations, addition and multiplication are called "direct operations" and by means of them the "inverse operations," subtraction and division, are defined. For the addition and multiplication of complex numbers the following definitions are adopted:

Definition of Addition. The sum of two complex numbers (a,b) and (c,d) is the complex number $(a + c, b + d)$ obtained by adding, respectively, the first and the second elements of the two given pairs. To indicate the addition the ordinary sign of addition is used, so that the content of the definition may be conveniently expressed thus:

$$(a,b) + (c,d) = (a + c, b + d).$$

For example,

$$(1, -1) + (2, 1) = (3, 0),$$
$$(0, 1) + (1, 0) = (1, 1),$$
$$(3, 2) + (-3, -2) = (0, 0).$$

Definition of Multiplication. The product of two complex numbers (a,b) and (c,d) is the complex number $(ac - bd, ad + bc)$. The multiplication is indicated by placing the multiplication sign or $\times$ between the factors; sometimes, when there is no danger of misunderstanding, the sign of multiplication may be omitted. The content of the definition is thus conveniently expressed by writing

$$(a,b) \cdot (c,d) = (ac - bd, ad + bc),$$

or

$$(a,b)(c,d) = (ac - bd, ad + bc).$$

According to this definition we have, for example,

$$(2,3) \cdot (1,2) = (-4,7),$$
$$(1,-1) \cdot (1,1) = (2,0),$$
$$(0,1) \cdot (0,1) = (-1,0).$$

4. Fundamental Laws for Addition and Multiplication. While the definition adopted for addition of complex numbers is readily accepted by students as natural, they are puzzled by the apparently artificial character of the definition of multiplication and always ask for reasons for adopting such a definition. Since complex numbers as ordered pairs are new objects for which the notions of equality, addition, and multiplication are not defined at the outset, it is our privilege to define these notions as we please, striving only to choose definitions in such a manner that all the fundamental laws of algebraic operations on real numbers retain their validity for complex numbers and that, moreover, the complex numbers subjected to such laws of operations exactly fill the place of the meaningless "imaginary" numbers. The fundamental laws for addition and multiplication of real numbers are the following:

1. $a + b = b + a$. Commutative law for addition.
2. $(a + b) + c = a + (b + c)$. Associative law for addition.
3. $ab = ba$. Commutative law for multiplication.
4. $(ab)c = a(bc)$. Associative law for multiplication.
5. $(a + b)c = ac + bc$. Distributive law.

It is an easy matter to verify that these laws retain their validity for complex numbers under the adopted definitions of equality, addition, and multiplication. This straightforward verification will be left to the student.

5. Subtraction and Division. Once we have the definitions of equality, addition, and multiplication, we can define subtraction and division in exactly the same way as for real numbers. To subtract b from a means to find a number x such that

$$b + x = a.$$

Such a number—the difference of a and b—is unique. The same definition can be extended to complex numbers.

Definition of Subtraction. To subtract the complex number (c,d) from (a,b) means to find a complex number (x,y) so that

$$(c,d) + (x,y) = (a,b).$$

Since by the definition of addition

$$(c,d) + (x,y) = (c+x, d+y),$$

the unknown numbers x and y must be determined from the equations

$$c + x = a, \qquad d + y = b$$

which have the unique solution

$$x = a - c, \qquad y = b - d.$$

Thus, the difference of (a,b) and (c,d) is a uniquely determined complex number

$$(a,b) - (c,d) = (a-c, b-d).$$

In particular, we have

$$(a,b) - (a,b) = (0,0)$$

or

$$(a,b) + (0,0) = (a,b)$$

so that the complex number $(0,0)$ plays the same role as 0 does for real numbers. To define the division of complex numbers we can again imitate the definition of division of real numbers. To divide a by a number b different from 0 means to find a number x such that

$$bx = a.$$

By analogy we say:

Definition of Division. To divide the complex number (a,b) by (c,d), different from $(0,0)$, means to find such a complex number (x,y) that

$$(c,d)(x,y) = (a,b).$$

Since

$$(c,d)(x,y) = (cx - dy, dx + cy),$$

the unknown numbers x and y must be found by solving the system of equations

$$cx - dy = a, \qquad dx + cy = b.$$

By eliminating first y and then x we find

$$(c^2 + d^2)x = ac + bd,$$
$$(c^2 + d^2)y = bc - ad.$$

By hypothesis c and d are not equal to 0 simultaneously, and consequently $c^2 + d^2 > 0$. Hence, x and y have completely determined values:

$$x = \frac{ac + bd}{c^2 + d^2}, \qquad y = \frac{bc - ad}{c^2 + d^2}$$

which, as can be found by direct substitution, actually satisfy the proposed system. Consequently, division by $(c,d) \neq (0,0)$ leads to a completely determined quotient, which, retaining the usual notation, will be

$$(a,b) : (c,d) = \left(\frac{ac + bd}{c^2 + d^2}, \frac{bc - ad}{c^2 + d^2}\right)$$

or

$$\frac{(a,b)}{(c,d)} = \left(\frac{ac + bd}{c^2 + d^2}, \frac{bc - ad}{c^2 + d^2}\right).$$

6. Normal Form of Complex Numbers. Every complex number can be presented in a certain *normal form*. In the first place,

$$(a,b) = (a,0) + (0,b).$$

Again using the rule for multiplication, one easily verifies that

$$(0,b) = (b,0)(0,1)$$

and so

$$(a,b) = (a,0) + (b,0)(0,1),$$

which means that each complex number can be expressed through special complex numbers of the type $(a,0)$ with the second element 0, and one particular complex number $(0,1)$ that henceforth will be denoted by a single letter i, the initial letter of the word imaginary. When the fundamental operations are applied to complex numbers of the type $(a,0)$, the following results are found:

$$(a,0) + (b,0) = (a + b,0),$$
$$(a,0) - (b,0) = (a - b,0),$$
$$(a,0) \cdot (b,0) = (ab,0),$$
$$(a,0) : (b,0) = \left(\frac{a}{b},0\right),$$

whence the following noteworthy conclusion can be drawn: If complex numbers with the second element 0 are subjected to the operations of addition, subtraction, multiplication, and division (operations called *rational operations*), each repeated any number of times, the resulting complex number will again have its second element 0 while the first element results from performing the prescribed operations on the first elements of the complex numbers involved. And this means that complex numbers with the second elements 0 behave, with respect to rational operations, exactly as their first elements, which are real numbers. In dealing solely with such complex numbers we may therefore identify them, without fear of confusion, with real numbers—their first elements. Accordingly, we shall agree from now on to denote a complex number of the type $(a,0)$ simply by a. In this manner, one and the same symbol a will have two meanings: one as a symbol of a real number and another as a symbol of a complex number $(a,0)$. As long as we have a formula involving only rational operations on such symbols, the formula will be true whether we interpret symbols in one way or the other. For instance, in the identity

$$a^2 - b^2 = (a + b)(a - b)$$

the symbols a, b, $a + b$, $a - b$ may be interpreted as symbols of real numbers or as symbols of the complex numbers $(a,0)$, $(b,0)$, $(a + b,0)$, $(a - b,0)$, and the identity will be true in both cases. According to the adopted convention every complex number (a,b) can be presented in the following *normal form:*

$$(a,b) = a + bi$$

where i stands for $(0,1)$ and a,b are the complex numbers $(a,0)$, $(b,0)$. By the rule of multiplication we have

$$(0,1)^2 = (-1,0)$$

or, with the conventional notations adopted,

$$i^2 = -1.$$

Again, observing that the fundamental laws of operations holding good for real numbers remain true for complex numbers, we can draw the conclusion that in performing fundamental operations on complex numbers presented in the normal form we can treat them as algebraic binomials, taking care always to replace i^2 when it appears by -1. It is customary to use the abbreviated notation bi for complex numbers of the type $0 + bi$, and in case $b = \pm 1$ simply to write $a \pm i$ instead of $a + 1i$ or $a - 1i$.

A few examples will show the advantages of operating on complex numbers presented in the normal form.

Example 1. To find $(1 + i)^3$. We have

$$(1 + i)^2 = 1 + 2i + i^2 = 2i$$

and

$$(1 + i)^3 = (1 + i)^2(1 + i) = 2i(1 + i) = -2 + 2i.$$

The same example can be worked out also as follows: We have

$$(1 + i)^3 = 1 + 3i + 3i^2 + i^3.$$

But
$$i^2 = -1, \qquad i^3 = i^2 \cdot i = -i$$
and so
$$(1 + i)^3 = 1 + 3i - 3 - i = -2 + 2i.$$

Example 2. Find the cube of the complex number
$$\omega = -\frac{1}{2} + i\frac{\sqrt{3}}{2}.$$

In the first place,
$$\omega^2 = \frac{1}{4} + \frac{3}{4}i^2 - i\frac{\sqrt{3}}{2} = \frac{1}{4} - \frac{3}{4} - i\frac{\sqrt{3}}{2} = -\frac{1}{2} - i\frac{\sqrt{3}}{2},$$

and further
$$\omega^3 = \omega \cdot \omega^2 = \left(-\frac{1}{2} + i\frac{\sqrt{3}}{2}\right)\left(-\frac{1}{2} - i\frac{\sqrt{3}}{2}\right) = \left(-\frac{1}{2}\right)^2 - \frac{3}{4}i^2 = 1.$$

Exmple 3. Reduce the complex number
$$\frac{(3 + 2i)^2(1 - 3i)}{(3 + i)^2(1 + 2i)} + \frac{1 + i}{1 - i}$$
to the normal form. We have
$$(3 + 2i)^2 = 9 + 12i + 4i^2 = 5 + 12i,$$
$$(5 + 12i)(1 - 3i) = 5 - 36i^2 - 3i = 41 - 3i,$$
$$(3 + i)^2 = 9 + 6i + i^2 = 8 + 6i,$$
$$(8 + 6i)(1 + 2i) = 8 + 12i^2 + 22i = -4 + 22i.$$

To compute the quotient
$$\frac{41 - 3i}{-4 + 22i}$$
we may, without changing it, multiply both numerator and denominator by
$$-4 - 22i;$$
this gives
$$\frac{41 - 3i}{-4 + 22i} = \frac{(41 - 3i)(-4 - 22i)}{(-4)^2 + 22^2} = \frac{-230 - 890i}{500} = -\frac{23}{50} - \frac{89}{50}i.$$

Again,
$$\frac{1 + i}{1 - i} = \frac{(1 + i)^2}{2} = i,$$
and so the final answer is
$$-{}^{23}\!/_{50} - {}^{39}\!/_{50}i.$$

Problems

Reduce to normal form:

1. $7 - i + (-6 + 3i) - (4 + 3i)$.

2. $(2 - 3i)i$.

3. $(2 + i)(1 + 2i)$.

4. $\dfrac{1}{i}$.

5. $\dfrac{1 + i}{-i}$.

6. $\dfrac{1 + i}{i} + \dfrac{i}{1 - i}$.

7. $\dfrac{(2 + i)(1 - 2i)}{3 - i}$.

8. $\dfrac{(4 + 3i)(1 - 2i)}{7 - i}$.

9. $\dfrac{(1+i)^3}{1-i}.$ **10.** $\left(\dfrac{1+i}{\sqrt{2}}\right)^4.$

11. $\left(\dfrac{1}{2} + i\dfrac{\sqrt{3}}{2}\right)^3.$ **12.** $\dfrac{i}{1+i} + \dfrac{i}{1+i} + \dfrac{i}{1+i}.$

13. Find real values x and y that satisfy the equation

$$(1+i)(x+2y) - (3-2i)(x-y) = 8+3i$$

14. Find the real root of the equation

$$(1+i)x^3 + (1+2i)x^2 - (1+4i)x - 1 + i = 0.$$

15. Find the real root of the equation

$$(1+i)x^3 + (1+2i)x^2 - (1+i)x - 1 - 2i = 0.$$

7. Real and Imaginary Parts. Conjugate Numbers. Absolute Value or Modulus. In a complex number $a + bi$ presented in the normal form a is called the *real part* and b (not $bi!$) the *imaginary part*. Real and imaginary parts are usually denoted as follows:

$$a = R(a+bi),$$
$$b = I(a+bi),$$

where R and I are the initial letters of the words real and imaginary. Complex numbers with imaginary part 0 are called real numbers on account of their close resemblance to ordinary real numbers, and numbers of the form bi with real part 0 are called pure imaginaries. In general, complex numbers with an imaginary part not 0 are called imaginary numbers merely to conform to usage and historical tradition, since complex numbers as ordered pairs are just as real as other numbers and there is nothing "imaginary" about them.

Two complex numbers $a + bi$ and $a - bi$ differing only in the signs of their imaginary parts are called *conjugate*. If one of them is denoted by a single letter, say A, the conjugate number is denoted by A_0, or, as is sometimes written, $\overline{A}$. The product of two conjugate numbers

$$A = a + bi, \qquad A_0 = a - bi$$

is a real number

$$AA_0 = (a+bi)(a-bi) = a^2 + b^2$$

called the *norm* of A. The positive square root of the norm of A is called the *absolute value* or *modulus* of A and is denoted either by the sign $|A|$ or by the sign mod A, and the use of one or the other notation depends on considerations of convenience in writing or in printing. Thus,

$$|a+bi| = \sqrt{a^2+b^2},$$

or

$$\text{mod } (a + bi) = \sqrt{a^2 + b^2}.$$

If C is the sum of the complex numbers A and B

$$C = A + B,$$

then,

$$C_0 = A_0 + B_0;$$

that is, the sum of the conjugates of two complex numbers is equal to the conjugate of their sum. Similarly, if C is the product of complex numbers A and B

$$C = AB,$$

then,

$$C_0 = A_0 B_0;$$

that is, the product of the conjugates of two complex numbers is equal to the conjugate of their product. Both these propositions are verified directly by comparing the sum or the product of two complex numbers with the sum or the product of their conjugates, and from this it follows that the conjugate of the difference or quotient of two complex numbers is equal, respectively, to the difference or quotient of their conjugates, which, by means of the adopted signs, is expressed as follows:

$$(A - B)_0 = A_0 - B_0, \qquad \left(\frac{A}{B}\right)_0 = \frac{A_0}{B_0}.$$

By repeated application of these rules the following general and important conclusion can be derived: If in applying rational operations in finite number to the complex numbers $A, B, C, \ldots$ a complex number X is obtained, then in performing the same operations on the conjugates $A_0, B_0, C_0, \ldots$ the result will be X_0, the conjugate of X.

Real numbers coincide with their conjugates, and, conversely, a number that is equal to its conjugate is real. In fact, the equality

$$a + bi = a - bi$$

requires

$$b = -b \qquad \text{or} \qquad b = 0.$$

Problems

Find the absolute values of the following numbers:

1. i.

2. $-\dfrac{1}{2} + i\dfrac{\sqrt{3}}{2}$.

3. $3 + 4i$.

4. $\dfrac{1 - i}{\sqrt{2}}$.

5. $x^2 - 1 + 2ix$ where x is real.

6. $2x - 1 + (2x^2 - 2x)i$ where x is real.

7. $x^3 - x^2 - x + 1 + (2x^2 - 2x)i$ where x is real.

8. What is the real part of the number $(1 - x)/(1 + x)$ if $x = \cos \phi + i \sin \phi$?

9. Find a complex number ϵ such that $|\epsilon| = 1$ and $R(\epsilon^2) = 0$.

10. What are the complex numbers equal to (a) the square of their conjugates and (b) the cube of their conjugates?

★8. Theorem. *The absolute value of a product is equal to the product of the absolute values of its factors or, using the adopted signs,*

$$|ABC \cdots L| = |A| \cdot |B| \cdot |C| \cdots |L|.$$

PROOF.   Consider first the product of two factors

$$X = AB.$$

The norm of X is

$$XX_0 = (AB) \cdot (AB)_0,$$

but

$$(AB)_0 = A_0 B_0$$

and so

$$XX_0 = (AB)(A_0 B_0)$$

whence, making use of the associative and commutative laws for multiplication,

$$XX_0 = (AA_0) \cdot (BB_0)$$

and, taking positive square roots of both sides,

$$\sqrt{XX_0} = \sqrt{AA_0} \cdot \sqrt{BB_0}.$$

But

$$\sqrt{XX_0} = |X|, \qquad \sqrt{AA_0} = |A|, \qquad \sqrt{BB_0} = |B|$$

and, consequently,

$$|X| = |A| \cdot |B|$$

or

$$|AB| = |A| \cdot |B|.$$

Now considering the product of three factors

$$X = ABC,$$

let us set

$$Y = AB$$

so that

$$X = YC.$$

By what has been already proved

$$|Y| = |A| \cdot |B|, \qquad |X| = |Y| \cdot |C|;$$

hence,

$$|X| = |A| \cdot |B| \cdot |C|$$

or

$$|ABC| = |A| \cdot |B| \cdot |C|.$$

In the same way the theorem can be extended to any number of factors.

An important consequence can be drawn from this theorem: *If a product of complex numbers is 0, then at least one of the factors is 0.* From the assumption

$$ABC \cdot \cdot \cdot L = 0$$

it follows that

$$|A| \cdot |B| \cdot |C| \cdot \cdot \cdot |L| = |0| = 0$$

and since the factors on the left-hand side are real, one of them must be 0, say

$$|A| = 0.$$

But, writing $A = a + bi$, we have

$$|A| = \sqrt{a^2 + b^2} = 0,$$

that is, $a^2 + b^2 = 0$, which, with real a and b, is possible only if $a = 0$, $b = 0$ or $A = 0 + 0i = 0$. The same conclusion can be drawn from the fact that the quotient is uniquely determined when the divisor is not 0. Let the student develop the proof along this line.★

Problems

1. Prove that $\left|\dfrac{A}{B}\right| = \dfrac{|A|}{|B|}$.

2. What is the absolute value of $(4 + 3i)(1 + i)/(7 - i)$?

3. What is the absolute value of $(1 + xi)/(1 - xi)$ if x is real? And what can be said of the absolute value of the same number if $x = \alpha + i\beta$ is a complex number with $\beta > 0$?

4. Show that

$$|t - \tfrac{3}{4}i| = \tfrac{1}{4}$$

if

$$t = \frac{i - \tau}{1 + 2i\tau}$$

and τ is real.

5. If

$$\tau' = \frac{-3\tau + 2}{4\tau - 3}$$

and τ is such a complex number that

$$|\tau - \tfrac{3}{4}| = \tfrac{1}{4},$$

show that

$$|\tau' + \tfrac{3}{4}| = \tfrac{1}{4}.$$

★9. Inequality for the Absolute Value of the Sum. The absolute value of the sum of complex numbers does not depend simply on the absolute values of these numbers; thus, there is no such precise theorem for the sum as that in Sec. 8. Instead, we have the following less precise proposition, which is nevertheless of great importance and usefulness.

THEOREM. *The absolute value of the sum is not greater than the sum of the absolute values of its terms, or*

$$|A + B + \cdots + L| \leq |A| + |B| + \cdots + |L|,$$

and the equality sign holds only if either all numbers A, B, . . . , L are equal to 0 or, in case one of them, say A, is not equal to 0, all the ratios

$$\frac{B}{A}, \cdots, \frac{L}{A}$$

are nonnegative real numbers.

PROOF. We start with the following observation: If $A = a + bi$, then

$$a = R(A) \leq |A|$$

and the equality sign holds only if $b = 0$ and $a \geq 0$. In fact,

$$|A| = \sqrt{a^2 + b^2}$$

and certainly

$$a < \sqrt{a^2 + b^2}$$

if $b \neq 0$. If, on the other hand, $b = 0$ and $a < 0$,

$$|A| = \sqrt{a^2} = -a, \qquad a < -a.$$

Finally, if $b = 0$ and $a \geq 0$,

$$a = \sqrt{a^2} = |A|.$$

Consider now the sum of two complex numbers A and B. By the definition of the absolute value

$$|A + B|^2 = (A + B)(A + B)_0 = (A + B)(A_0 + B_0)$$

or

$$|A + B|^2 = AA_0 + BB_0 + (AB_0 + A_0B) = |A|^2 + |B|^2 + (AB_0 + A_0B).$$

Now the conjugate of AB_0 is A_0B, and the sum $AB_0 + A_0B$ of two conjugate numbers is double the real part of AB_0 or

$$AB_0 + A_0B = 2R(AB_0).$$

By the previous remark

$$R(AB_0) \leq |AB_0| = |A| \cdot |B_0| = |A| \cdot |B|$$

since conjugate numbers have the same absolute value. Hence,

$$|A + B|^2 \leqq |A|^2 + |B|^2 + 2|A| \cdot |B| = (|A| + |B|)^2.$$

But the numbers $|A + B|$ and $|A| + |B|$ are nonnegative and consequently the preceding inequality implies:

$$|A + B| \leqq |A| + |B|.$$

The equality sign holds here if and only if

$$R(AB_0) = R(A_0B) = |A_0B|,$$

and this is true only if A_0B is a real nonnegative number. Assuming $A \neq 0$, the product AA_0 is a positive number and

$$\frac{A_0B}{AA_0} = \frac{B}{A}$$

is a real nonnegative number. Conversely, if this equation holds, then $A_0B = (B/A)(AA_0)$ will be real and nonegative.

Consider now the sum of three numbers

$$A + B + C = (A + B) + C.$$

By what has been proved

$$|(A + B) + C| \leqq |A + B| + |C|$$
$$|A + B| \leqq |A| + |B|.$$

Hence,

$$|A + B + C| \leqq |A| + |B| + |C|.$$

The equality sign will hold here if and only if simultaneously

$$|A + B| = |A| + |B|,$$
$$|(A + B) + C| = |A + B| + |C|.$$

Supposing $A \neq 0$, the first of these equalities is valid only in case the ratio B/A is real and nonnegative. If such is the case, the number

$$A + B = A\left(1 + \frac{B}{A}\right)$$

is not 0, and

$$1 + \frac{B}{A}$$

is a positive number. The second equality requires that the ratio

$$\frac{C}{A + B} = \frac{C}{A}\left(1 + \frac{B}{A}\right)^{-1}$$

be real and nonnegative, which in turn is equivalent to the requirement that C/A be real and nonnegative. It is clear that, if the reasoning is continued in the same way, the theorem will be proved for sums of more than three terms.★

Problems

1. Show that
$$|A - B| \geqq |A| - |B| \quad \text{and also} \quad |A - B| \geqq |B| - |A|.$$
HINT: Write $A = B + (A - B)$.

2. If z is a complex number with $|z| \leqq 2$, what is the maximum of
$$|1 + z + z^2 + z^3|$$
and for what z is this maximum reached?

*3. If x and y are any two complex numbers, show that
$$|x + y|^2 + |x - y|^2 = 2|x|^2 + 2|y|^2.$$

*4. Show that the equality
$$|z_2 - z_1|^2 = |z_2 - z_0|^2 + |z_1 - z_0|^2$$
implies
$$z_2 - z_0 = i\lambda(z_1 - z_0)$$
where λ is real and vice versa.

10. Square Root of a Complex Number. Finding the square root of a complex number A is equivalent to finding the solution X of the quadratic equation
$$X^2 = A.$$

Let $A = a + bi$ and $X = x + iy$. Then, the real numbers x and y must be such that
$$(x + iy)^2 = a + bi.$$
Now
$$(x + iy)^2 = x^2 - y^2 + 2xyi;$$
consequently, real numbers x and y must satisfy the system of equations
$$x^2 - y^2 = a, \qquad 2xy = b. \tag{1}$$
The identity
$$(x^2 + y^2)^2 = (x^2 - y^2)^2 + 4x^2y^2$$
combined with equations (1) gives
$$x^2 + y^2 = \sqrt{a^2 + b^2},$$
the root being positive, whence and from the first equation of the system (1) it follows that
$$x^2 = \frac{\sqrt{a^2 + b^2} + a}{2}, \qquad y^2 = \frac{\sqrt{a^2 + b^2} - a}{2}. \tag{2}$$

Those equations are necessary consequences of the system (1), but they may have solutions that do not satisfy it. To segregate solutions of (2) satisfying (1), the equation

$$2xy = b$$

must be taken into account. In case $b \neq 0$ the equation determines the sign of y corresponding to a chosen sign of x; that is, x and y must be of the same sign when $b > 0$ and of opposite signs when $b < 0$. Accordingly, solutions of the equation

$$X^2 = a + bi$$

are

$$X = \pm \left(\sqrt{\frac{\sqrt{a^2 + b^2} + a}{2}} + i\sqrt{\frac{\sqrt{a^2 + b^2} - a}{2}} \right)$$

in case $b > 0$, and

$$X = \pm \left(\sqrt{\frac{\sqrt{a^2 + b^2} + a}{2}} - i\sqrt{\frac{\sqrt{a^2 + b^2} - a}{2}} \right)$$

in case $b < 0$. It remains to examine the case $b = 0$. Since

$$\sqrt{a^2} = a \quad \text{or} \quad -a$$

according as $a > 0$ or $a < 0$, it follows that

$$x = \pm \sqrt{a}, \qquad y = 0$$

if $a > 0$, and then the equation

$$X^2 = a$$

has two real roots

$$X = \pm \sqrt{a}.$$

If $a < 0$,

$$x = 0, \qquad y = \pm \sqrt{-a}$$

and in this case the same equation has two pure imaginary roots

$$X = \pm i\sqrt{-a}.$$

Finally, when $a = b = 0$, there is only one trivial solution $X = 0$.

This discussion shows that, once we introduce complex numbers, every complex number has square roots. The general quadratic equation

$$AX^2 + BX + C = 0$$

with arbitrary complex coefficients can be solved by means of the known formula

$$X = \frac{-B \pm \sqrt{B^2 - 4AC}}{2A}$$

the deduction of which is based on the fundamental laws of operations and on the existence of square roots.

NOTE: When A is a positive real number, the symbol $\sqrt{A}$, according to custom, always means the positive square root, and with this convention the law of multiplication of roots:

$$\sqrt{A} \cdot \sqrt{B} = \sqrt{AB}$$

is valid as long as A and B are positive. When, however, A is a negative real or imaginary number, there is no way to attribute to the symbol $\sqrt{A}$, by means of a simple convention, such a meaning that the law of multiplication of roots will always hold. In case of a negative real or imaginary A it is always necessary to specify which of the two roots the symbol $\sqrt{A}$ denotes by an additional condition, as, for instance, that the root should have positive real or imaginary parts. Thus, $\sqrt{-4}$ may mean either $2i$ or $-2i$, but if it is specified that this root must have a positive imaginary part, then the symbol $\sqrt{-4}$ will stand for $2i$. Notice that the relation of magnitude expressed by the words greater and smaller is defined only for real numbers and cannot be extended to complex numbers with the preservation of all the properties of this relation.

Example 1.   To solve the equation

$$X^2 = -i.$$

In this case $a = 0$, $b = -1$, $\sqrt{a^2 + b^2} = 1$, and, b being negative,

$$X = \pm \left(\sqrt{\frac{1}{2}} - i\sqrt{\frac{1}{2}} \right) = \pm \frac{1-i}{\sqrt{2}}.$$

Example 2.   To solve the equation

$$X^2 = -5 + 12i.$$

In this case

$$a = -5, \qquad b = 12, \qquad \sqrt{a^2 + b^2} = \sqrt{169} = 13$$

$$\frac{\sqrt{169} - 5}{2} = 4, \qquad \frac{\sqrt{169} + 5}{2} = 9$$

and, b being positive,

$$X = \pm (2 + 3i).$$

Example 3.   Solve the quadratic equation

$$iX^2 - (2 + 2i)X + 2 - i = 0.$$

Applying the formula we have

$$X = \frac{2 + 2i \pm \sqrt{-4}}{2i}$$

and, taking $\sqrt{-4} = 2i$, the roots are found to be

$$-i,\; 2 - i.$$

Problems

Find the square roots of the numbers:

1. i.

2. $\omega = -\frac{1}{2} + i\frac{\sqrt{3}}{2}$.

3. $\dfrac{1}{2} + i\dfrac{\sqrt{3}}{2}.$

4. $- 3 - 4i.$

5. $- 13 - 84i.$

6. $- 1 + i\sqrt{24}.$

7. $x^2 - 1 + 2xi$, x being real.

Solve the quadratic equations:

8. $x^2 + x + 1 = 0.$

9. $2x^2 - 3x + 2 = 0.$

10. $x^2 - (2 + 3i)x - 1 + 3i = 0.$

11. $(2 - 2i)x^2 - (11 + 9i)x - 16 + 6i = 0.$

Solve the equations:

12. $x^4 = 1.$

13. $x^4 = - 1.$

14. $x^4 + 4 = 0.$

15. $x^4 = 119 - 120i.$

16. $x^3 - 1 = 0.$ Notice that $x^3 - 1 = (x - 1)(x^2 + x + 1).$

17. $x^3 = i.$

18. $x^6 - 1 = 0.$

19. $x^6 + 1 = 0.$

20. $x^3 = 1 + i.$ Set $x = a + bi$; then $a^3 - 3ab^2 = 1$, $3a^2b - b^3 = 1$, and, be-sides, $a^2 + b^2 = \sqrt[3]{2}.$

21. Prove the following proposition: If a, b, a', b' are rational numbers but $\sqrt{b}$ is not rational and

$$a + \sqrt{b} = a' + \sqrt{b'},$$

then $a' = a$, $b' = b.$

22. Find all quadratic equations $x^2 + px + q = 0$ with rational coefficients but without rational roots if (a) one root is a square of another, (b) one root is a cube of another.

*23. If $a \neq 0$, $b \neq 0$ are two complex numbers such that the ratio b/a is not a negative real number, the square root $\sqrt{ab}$ can be chosen so that the ratio of the "geometric mean" of a and b

$$b_1 = \sqrt{ab}$$

to the "arithmetic mean"

$$a_1 = \frac{a + b}{2}$$

has a positive real part. Show that then also $R(b_1/a) > 0.$

*24. Under the same conditions show that

$$|b_1 - a_1| < \tfrac{1}{2}|b - a|.$$

11. Geometric Representation of Complex Numbers.

Relations between complex numbers and their handling are often made intuitive by means of a simple geometric representation. Having chosen two perpendicular lines OX, OY as coordinate axes, we attribute to OX a certain direction indicated in the figure by an arrow and then choose a direction on the axis OY so that after rotating OX through a right angle counterclockwise its direction coincides with that of OY. With reference to the chosen coordinate system

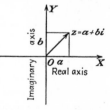

each point of the plane has definite coordinates, say a and b, and with this point the complex number $a + bi$ is associated. Vice versa, to any complex number we let correspond a point whose coordinates are, respectively, the real and imaginary parts of that complex number. In this manner between complex numbers and points of the plane there exists a one-to-one correspondence by virtue of which complex numbers are represented by points. Complex numbers of the form $a + 0i$—real numbers—are represented by points of the axis OX, which for this reason is called the *real axis*. Complex numbers of the form $0 + bi$ or pure imaginary numbers are represented by points of the axis OY, called the *imaginary axis*. It is customary to denote the point representing the complex number z by the same letter and to call it simply the point z. Thus, we can speak of the point O (origin), the point 1, the point i, the point $3 - 2i$, etc. Point O together with z determines a directed line segment $\overrightarrow{Oz}$ or vector $\overrightarrow{Oz}$, extending from the origin O to the end point z. Conversely, the end point of any vector $\overrightarrow{Oz}$ determines some complex number. Thus, we have another geometric representation of complex numbers by means of vectors with the common origin at O. Projections of the vector representing $z = a + bi$ on axis OX and OY are, respectively, a and b, and the length of the vector $\overrightarrow{Oz}$, or the distance from O to z, by the Pythagorean theorem is $\sqrt{a^2 + b^2}$ and thus gives an intuitive meaning of the absolute value or the modulus of z.

Problems

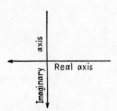

1. If the direction of the real axis is chosen as shown, locate points representing the complex numbers: (a) -1, (b) i, (c) $1 - i$, (d) $1 + 2i$, (e) $-3 - 2i$, (f) $-\frac{1}{2} + 2i$.

2. If $R(z) = \frac{1}{2}$, what can be said about the locus of the points z?

3. Work Prob. 2 if the complex numbers z satisfy instead the condition $-\frac{1}{2} < R(z) \leqq \frac{1}{2}$.

4. What is the relative position of points representing conjugate complex numbers?

5. What is the relative position of points representing the complex numbers $a + bi$ and $b + ai$?

6. What is the geometric meaning of (a) $|z| = 1$, (b) $|z| < 1$, (c) $|z| > 1$?

7. Where are the points representing z if $-\frac{1}{2} \leqq R(z) < \frac{1}{2}$ and $|z| \geqq 1$?

12. Angle between Directed Lines. The figure on page 19 represents two directed lines l and l' passing through a point S with their directions indicated by arrows. By an angle between l and l' measured from l to l', which we shall denote by (ll'), we understand an angle through which l

must be turned about S in order to coincide with l' in position and direction, this angle being considered as positive or negative according as l is rotated counterclockwise or in a positive sense, or clockwise or in a negative sense. From this point of view the angle (ll') is not uniquely determined but has infinitely many values, the connection between which can be found as follows: Let ϕ be the smallest positive angle through which l is to be turned to coincide with l' in position and direction. If rotation is continued in the positive sense, another coincidence occurs when the angle of rotation is $\phi + 2\pi$ (measuring angles in radian measure), still another when this angle is $\phi + 4\pi$, and in general $\phi + 2k\pi$ where k is any positive integer. Again, after rotating l in the negative sense through the angle whose absolute magnitude is $2\pi - \phi$ the lines l and l' coincide in position and direction; and the same happens after a rotation, in the negative sense, of magnitude $2k\pi - \phi$,

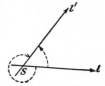

where k is any positive integer. Taking all these angles negatively, we may say that l' forms with l the negative angles $\phi - 2\pi, \phi - 4\pi, \phi - 6\pi,$ . . . . Thus, the general expression for the angle (ll') is $\phi + 2k\pi$ where $k = 0, \pm 1, \pm 2, . . .$ is an arbitrary integer and ϕ is the smallest positive angle between l and l'. It is easy to see that ϕ may mean any angle between l and l' and still all possible values of this angle will be of the same form. Angles that differ by multiples of 2π are said to be congruent modulo 2π (an expression borrowed from the theory of numbers), and the sign $\equiv$ is used to denote the congruence. In this sense we have an almost evident congruence

$$(ll') \equiv - (l'l).$$

Again, if three directed lines l, l', l'' pass through the same point, it is easy to verify that

$$(ll') + (l'l'') + (l''l) \equiv 0$$

whence, by virtue of the congruence $(l''l) \equiv - (ll'')$, it follows that

$$(ll'') \equiv (ll') + (l'l'').$$

Despite the multiplicity of the angle (ll') the trigonometric functions of this angle

$$\sin (ll'), \qquad \cos (ll')$$

have completely determined values owing to the fact that $\sin x$ and $\cos x$ are periodic functions with the period 2π.

Problems

1. The sides of an equilateral triangle are given directions as shown. What are the numerically smallest values of the angles (a) (ll'), (b) (ll''), (c) $(l'l'')$?

2. A square $ABCD$ is inscribed in a circle and a point P is taken on the arc BC. What are the angles (a) (l_1l_3), (b) (l_4l_2), (c) (l_1l_2), formed by the pairs of directed lines l_1, l_2, l_3, l_4 shown in the figure?

3. Three directed lines l, l', l'' go through the same point. If $(l'l) = 230°$, $(l''l) = -100°$, find the numerically smallest value of the angle $(l''l')$.

4. If five directed lines l_1, l_2, l_3, l_4, l_5 go through the same point and $(l_2l_1) = 30°$, $(l_3l_1) = -200°$, $(l_4l_2) = 300°$, $(l_5l_4) = -90°$, what is the numerically smallest value of the angle (l_3l_5)?

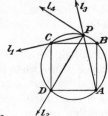

13. Trigonometric Form of a Complex Number. Returning to the geometric representation of complex numbers explained in Sec. 11, let the angle between the real axis and the vector $\overrightarrow{Oz}$ be θ. This angle is called the *argument* or *amplitude* of z. It is defined only for $z \neq 0$ and has infinitely many values differing from each other by multiples of 2π. If $z = a + bi$, then a is the projection of $\overrightarrow{Oz}$ on the real axis. Denoting by r the modulus of z,

$$r = \sqrt{a^2 + b^2},$$

from the definition of $\cos \theta$ it follows that

$$a = r \cos \theta.$$

Since the angle between the real and imaginary axes is $\pi/2$, the angle between $\overrightarrow{Oz}$ and the imaginary axis will be $(\pi/2) - \theta$, and the projection of $\overrightarrow{Oz}$ on it will be

$$b = r \cos\left(\frac{\pi}{2} - \theta\right) = r \sin \theta.$$

Consequently, $z = a + bi$ can be written in the form

$$z = r \left(\cos \theta + i \sin \theta\right),$$

which is the so-called *trigonometric form* of a complex number. It is immaterial which of the possible values of θ is taken; in practice, however, it is convenient to choose the numerically smallest value of the argument. To present a given complex number $a + bi$ in trigonometric form find first the modulus r from the formula

$$r = \sqrt{a^2 + b^2}$$

and then find the numerically smallest angle θ such that

$$\cos \theta = \frac{a}{r}, \qquad \sin \theta = \frac{b}{r}.$$

In general, it will be necessary to use trigonometric tables for this purpose, and then it is always more advantageous to determine the angle by its tangent. Accordingly, in case $b/a > 0$ determine the acute angle ω by its tangent

$$\tan \omega = \frac{b}{a}$$

and take $\theta = \omega$ if $a > 0$ and $\theta = \omega - \pi$ if $a < 0$. In case $b/a < 0$, the acute angle ω is determined by

$$\tan \omega = \frac{-b}{a}$$

and $\theta = -\omega$ if $a > 0$, $\theta = \pi - \omega$ if $a < 0$.

Problems

Present in trigonometric form the following complex numbers:

1. -4.

2. i.

3. $-6i$.

4. $-1 + i$.

5. $\dfrac{1}{2} - i\dfrac{\sqrt{3}}{2}$.

6. $-\dfrac{1}{2} + i\dfrac{\sqrt{3}}{2}$.

7. $\sqrt{3} - i$.

8. $1 - \sqrt{3} - i(1 + \sqrt{3})$.

9. $-4 - 3i$.

10. $-2 + i$.

11. $1 + \cos \alpha + i \sin \alpha$.

12. $\cos \alpha + \cos \beta + i(\sin \alpha + \sin \beta)$.

14. Multiplication and Division of Complex Numbers in Trigonometric Form. De Moivre's Formula. Rules for multiplication and division are particularly simple when complex numbers are taken in trigonometric form. Let

$$A = r(\cos \theta + i \sin \theta), \qquad B = r'(\cos \theta' + i \sin \theta').$$

On multiplying and rearranging factors on the right-hand side, we have

$$AB = rr'(\cos \theta + i \sin \theta)(\cos \theta' + i \sin \theta').$$

But

$$(\cos \theta + i \sin \theta)(\cos \theta' + i \sin \theta')$$
$$= \cos \theta \cos \theta' - \sin \theta \sin \theta' + i(\sin \theta \cos \theta' + \sin \theta' \cos \theta)$$

and, on the other hand,

$$\cos \theta \cos \theta' - \sin \theta \sin \theta' = \cos (\theta + \theta')$$
$$\sin \theta \cos \theta' + \sin \theta' \cos \theta = \sin (\theta + \theta');$$

hence,

$$AB = rr'[\cos (\theta' + \theta) + i \sin (\theta' + \theta)],$$

which means that *the modulus of the product is the product of the moduli of factors and the argument is the sum of their arguments.* By repeated application this rule extends to any number of factors. The product of n factors

$$\cos \theta_1 + i \sin \theta_1, \qquad \cos \theta_2 + i \sin \theta_2, \qquad \ldots, \qquad \cos \theta_n + \sin \theta_n$$

whose moduli are each 1 is

$$(\cos \theta_1 + i \sin \theta_1)(\cos \theta_2 + i \sin \theta_2) \cdots (\cos \theta_n + i \sin \theta_n)$$
$$= \cos (\theta_1 + \theta_2 + \cdots + \theta_n) + i \sin (\theta_1 + \theta_2 + \cdots + \theta_n).$$

In particular, when $\theta_1 = \theta_2 = \cdots = \theta_n = \theta$, this formula gives immediately an important identity

$$(\cos \theta + i \sin \theta)^n = \cos n\theta + i \sin n\theta$$

known as de Moivre's formula. Here, of course, n means a positive integer. Noticing that

$$(\cos \theta + i \sin \theta)^{-1} = \frac{1}{\cos \theta + i \sin \theta} = \frac{\cos \theta - i \sin \theta}{\cos^2 \theta + \sin^2 \theta}$$
$$= \cos \theta - i \sin \theta = \cos (- \theta) + i \sin (- \theta),$$

and raising each side of the equation to the power n, we have

$$(\cos \theta + i \sin \theta)^{-n} = \cos (- n\theta) + i \sin (- n\theta).$$

Thus, de Moivre's formula holds also for negative integer exponents.
As to the quotient of two complex numbers,

$$A = r(\cos \theta + i \sin \theta), \qquad B = r'(\cos \theta' + i \sin \theta'),$$

it can be presented thus:

$$\frac{A}{B} = \frac{r(\cos \theta + i \sin \theta)}{r'(\cos \theta' + i \sin \theta')}$$
$$= \frac{r}{r'} (\cos \theta + i \sin \theta)(\cos \theta' + i \sin \theta')^{-1}.$$

But

$$(\cos \theta' + i \sin \theta')^{-1} = \cos (- \theta') + i \sin (- \theta'),$$

and so by the previously stated rule for multiplication

$$\frac{A}{B} = \frac{r}{r'}[\cos (\theta - \theta') + i \sin (\theta - \theta')].$$

Hence, *the modulus of the quotient is equal to the quotient of the moduli, and the argument to the difference of arguments of the dividend and divisor.*

Problems

Find the general expression for the following expressions (n being an integer):

1. $(\sqrt{3} + i)^n$

2. $[1 + \sqrt{3} + i(1 - \sqrt{3})]^n$

3. $\left(\dfrac{1 + \sin\phi + i\cos\phi}{1 + \sin\phi - i\cos\phi}\right)^n$

4. $[\sin\theta - \sin\phi + i(\cos\theta - \cos\phi)]^n$

5. Setting

$$(1 + x)^n = p_0 + p_1 x + p_2 x^2 + \cdots,$$

where

$$p_0 = 1, \qquad p_1 = \frac{n}{1}, \qquad p_2 = \frac{n(n-1)}{1 \cdot 2}, \qquad \cdots$$

are binomial coefficients, and taking $x = i$, prove that

$$p_0 - p_2 + p_4 - \cdots = 2^{\frac{1}{2}n} \cos\frac{n\pi}{4},$$

$$p_1 - p_3 + p_5 - \cdots = 2^{\frac{1}{2}n} \sin\frac{n\pi}{4}.$$

*6. Taking in the same expansion $x = 1, \omega, \omega^2$ where

$$\omega = \frac{-1 + i\sqrt{3}}{2},$$

find the sums

(a) $p_0 + p_3 + p_6 + \cdots,$
(b) $p_1 + p_4 + p_7 + \cdots,$
(c) $p_2 + p_5 + p_8 + \cdots.$

Notice that $1 + \omega^n + \omega^{2n} = 0$ if n is not a multiple of 3, but it is equal to 3 if n is a multiple of 3.

*7. Taking $z = \cos\theta + i\sin\theta$ in the identity

$$1 + z + z^2 + \cdots + z^{n-1} = \frac{1 - z^n}{1 - z},$$

show that

$$1 + 2\cos\theta + 2\cos 2\theta + \cdots + 2\cos(n-1)\theta = \frac{\sin\left(n - \frac{1}{2}\right)\theta}{\sin\frac{1}{2}\theta}$$

$$\sin\theta + \sin 2\theta + \cdots + \sin(n-1)\theta = \frac{\cos\frac{1}{2}\theta - \cos\left(n - \frac{1}{2}\right)\theta}{2\sin\frac{1}{2}\theta}.$$

*8. Using an analogous method, show that

$$\cos\theta + \cos 3\theta + \cdots + \cos(2n-1)\theta = \frac{\sin 2n\theta}{2\sin\theta},$$

$$\sin\theta + \sin 3\theta + \cdots + \sin(2n-1)\theta = \frac{1 - \cos 2n\theta}{2\sin\theta}.$$

*9. By means of de Moivre's formula express (a) $\cos 3\phi$ through $\cos\phi$ and $\sin 3\phi$ through $\sin\phi$; (b) $\cos 5\phi$ through $\cos\phi$ and $\sin 5\phi$ through $\sin\phi$; (c) $\cos 4\phi$ through $\cos\phi$ and $\sin 4\phi/\sin\phi$ through $\sin\phi$.

*10. Express (a) $\sin^5\phi$ linearly through $\sin\phi$, $\sin 3\phi$, $\sin 5\phi$; (b) $\sin^4\phi$ linearly through $\cos 2\phi$, $\cos 4\phi$.

15. Trigonometric Solution of Binomial Equations. By means of de Moivre's formula it is possible to represent all roots of the *binomial* equation

$$X^n = A,$$

where $A \neq 0$ is any complex number, in trigonometric form. Let

$$A = r(\cos \theta + i \sin \theta)$$

and take the unknown number X also in trigonometric form:

$$X = R(\cos \Theta + i \sin \Theta).$$

Then,

$$X^n = R^n(\cos n\Theta + i \sin n\Theta),$$

and this must be equal to

$$A = r(\cos \theta + i \sin \theta).$$

Since equal complex numbers have equal moduli, we must have

$$R^n = r,$$

whence R is determined without ambiguity as the positive nth root of r:

$$R = \sqrt[n]{r}.$$

Again arguments of equal complex numbers may differ only by multiples of 2π so that

$$n\Theta = \theta + 2k\pi$$

where k is an integer. Hence, the expression for the roots X is

$$X = \sqrt[n]{r}\left(\cos \frac{\theta + 2k\pi}{n} + i \sin \frac{\theta + 2k\pi}{n}\right).$$

Here k can be any integer, but the number of distinct roots will be only n. To obtain them all it suffices to take in this formula $k = 0, 1, 2, \ldots, n - 1$. For if k is any integer, dividing it by n and calling the remainder l, we have

$$k = nq + l$$

where $0 \leq l < n$, so that l will be one of the numbers $0, 1, 2, \ldots, n - 1$. But

$$\frac{\theta + 2k\pi}{n} = \frac{\theta + 2l\pi}{n} + 2\pi q;$$

hence,

$$\cos \frac{\theta + 2k\pi}{n} = \cos \frac{\theta + 2l\pi}{n},$$

$$\sin \frac{\theta + 2k\pi}{n} = \sin \frac{\theta + 2l\pi}{n},$$

which proves the statement. On the other hand, the n roots obtained by taking $k = 0, 1, 2, \ldots, n - 1$ are distinct. For suppose that for two such values of k, say k' and k'', we find equal roots; then,

$$\cos \frac{\theta + 2k'\pi}{n} = \cos \frac{\theta + 2k''\pi}{n}, \qquad \sin \frac{\theta + 2k'\pi}{n} = \sin \frac{\theta + 2k''\pi}{n},$$

and this is possible only if

$$\frac{\theta + 2k''\pi}{n} = \frac{\theta + 2k'\pi}{n} + 2\pi q$$

where q is an integer, or

$$k'' - k' = nq.$$

But $k'' - k'$ is numerically less than n and cannot be divisible by n unless it is equal to 0; then, however, k'' and k' would not be two different numbers as was supposed.

Thus, all roots of the binomial equation

$$X^n = r(\cos \theta + i \sin \theta)$$

are given by the formula

$$X = \sqrt[n]{r}\left(\cos \frac{\theta + 2k\pi}{n} + i \sin \frac{\theta + 2k\pi}{n}\right),$$

taking in it $k = 0, 1, 2, \ldots, n - 1$.

Example 1. Solve the equation

$$x^4 = -4.$$

Since

$$-4 = 4(\cos \pi + i \sin \pi),$$

the formula for the roots is

$$x = \sqrt[4]{4}\left(\cos \frac{\pi + 2k\pi}{4} + i \sin \frac{\pi + 2k\pi}{4}\right).$$

Taking in it $k = 0, 1, 2, 3$, we find the four roots to be

$$\sqrt{2}\left(\cos \frac{\pi}{4} + i \sin \frac{\pi}{4}\right) = 1 + i,$$

$$\sqrt{2}\left(\cos \frac{3\pi}{4} + i \sin \frac{3\pi}{4}\right) = -1 + i,$$

$$\sqrt{2}\left(\cos \frac{5\pi}{4} + i \sin \frac{5\pi}{4}\right) = -1 - i,$$

$$\sqrt{2}\left(\cos \frac{7\pi}{4} + i \sin \frac{7\pi}{4}\right) = 1 - i.$$

Example 2. Solve the equation

$$x^3 = -8i.$$

In trigonometric form

$$-8i = 8\left[\cos\left(-\frac{\pi}{2}\right) + i \sin\left(-\frac{\pi}{2}\right)\right];$$

hence,

$$x = 2\left[\cos\frac{(4k-1)\pi}{6} + i\sin\frac{(4k-1)\pi}{6}\right].$$

Taking here $k = 0, 1, 2$, we have the following roots:

$$2\left(\cos\frac{\pi}{6} - i\sin\frac{\pi}{6}\right) = \sqrt{3} - i,$$

$$2\left(\cos\frac{\pi}{2} + i\sin\frac{\pi}{2}\right) = 2i,$$

$$2\left(\cos\frac{7\pi}{6} + i\sin\frac{7\pi}{6}\right) = -\sqrt{3} - i.$$

Problems

Present in trigonometric form the roots of the following equations:

1. $x^4 = -16i$.   2. $x^4 = 1 + i$.

3. $x^3 = -2i$.   4. $x^3 = 1 - i$.

5. $x^4 = \omega$, $\omega = -\dfrac{1}{2} + i\dfrac{\sqrt{3}}{2}$.   6. $x^3 = \omega$.

7. $x^6 = -4$.   8. $x^6 = 1 + \sqrt{3} + (1 - \sqrt{3})i$.

9. Solving Prob. 4 algebraically, find expressions for $\cos 15°$, $\sin 15°$.

10. Do the same, solving algebraically and trigonometrically the equation

$$x^4 = \frac{1}{2} + i\frac{\sqrt{3}}{2}.$$

11. By solving the equation $x^5 = i$ both algebraically and trigonometrically show that

$$\cos 18° = \frac{\sqrt{5 + 2\sqrt{5}}}{\sqrt[5]{176 + 80\sqrt{5}}}, \qquad \sin 18° = \frac{1}{\sqrt[5]{176 + 80\sqrt{5}}}.$$

On the other hand (see Sec. 16),

$$\sin 18° = \frac{\sqrt{5} - 1}{4}, \qquad \cos 18° = \frac{\sqrt{10 + 2\sqrt{5}}}{4}.$$

How can these expressions be reconciled?

16. Roots of Unity.
The particular binomial equation

$$x^n = 1,$$

defining the so-called roots of unity of degree n, is of special interest. Since in this case $r = 1$ and $\theta = 0$, all the nth roots of unity are obtained from the formula

$$\cos\frac{2k\pi}{n} + i\sin\frac{2k\pi}{n}$$

by taking in it $k = 0, 1, 2, \ldots, n - 1$. For $k = 0$ we have an evident root $x = 1$, and the other $n - 1$ roots, by de Moivre's formula, are powers

$$\omega^k, \quad k = 1, 2, \ldots, n - 1$$

of the root

$$\omega = \cos\frac{2\pi}{n} + i\sin\frac{2\pi}{n}.$$

Since

$$x^n - 1 = (x - 1)(x^{n-1} + x^{n-2} + \cdots + 1),$$

which is verified by direct multiplication, ω, ω^2, . . . , ω^{n-1} are roots of the equation

$$x^{n-1} + x^{n-2} + \cdots + x + 1 = 0.$$

For some particular values of n this equation can be easily solved algebraically, and by comparison of the algebraic and trigonometric solutions in such cases algebraic expressions for $\cos(2\pi/n)$ and $\sin(2\pi/n)$ can be found as will be seen from the following examples:

Example 1. The cube root of unity,

$$\omega = \cos\frac{2\pi}{3} + i\sin\frac{2\pi}{3},$$

satisfies the equation

$$x^2 + x + 1 = 0.$$

Roots of this equation found algebraically are

$$-\frac{1}{2} + i\frac{\sqrt{3}}{2}, \qquad -\frac{1}{2} - i\frac{\sqrt{3}}{2}.$$

Since $\cos(2\pi/3)$ is negative and $\sin(2\pi/3)$ is positive, it follows that

$$\cos\frac{2\pi}{3} + i\sin\frac{2\pi}{3} = -\frac{1}{2} + i\frac{\sqrt{3}}{2}$$

whence

$$\cos\frac{2\pi}{3} = -\frac{1}{2}, \qquad \sin\frac{2\pi}{3} = \frac{\sqrt{3}}{2},$$

as it is known from trigonometry.

Example 2. The fifth root of unity,

$$\omega = \cos\frac{2\pi}{5} + i\sin\frac{2\pi}{5},$$

satisfies the equation

$$x^4 + x^3 + x^2 + x + 1 = 0.$$

This equation belongs to the class of *reciprocal* equations of the type

$$ax^4 + bx^3 + cx^2 + bx + a = 0,$$

and any such equation can be solved in the following manner: Divided through by x^2, the proposed equation takes the form

$$x^2 + \frac{1}{x^2} + x + \frac{1}{x} + 1 = 0.$$

Now take

$$x + \frac{1}{x} = y.$$

Then,

$$x^2 + \frac{1}{x^2} = y^2 - 2,$$

and y has to be found from the quadratic equation

$$y^2 + y - 1 = 0,$$

whose roots are

$$y_1 = \frac{-1 + \sqrt{5}}{2}, \qquad y_2 = \frac{-1 - \sqrt{5}}{2}.$$

It remains to determine x by solving the two equations

$$x + \frac{1}{x} = y_1, \qquad x + \frac{1}{x} = y_2$$

which are equivalent to

$$x^2 - y_1 x + 1 = 0, \qquad x^2 - y_2 x + 1 = 0.$$

The four roots found by solving these are

$$\frac{-1 + \sqrt{5}}{4} + i \frac{\sqrt{10 + 2\sqrt{5}}}{4},$$

$$\frac{-1 + \sqrt{5}}{4} - i \frac{\sqrt{10 + 2\sqrt{5}}}{4},$$

$$\frac{-1 - \sqrt{5}}{4} + i \frac{\sqrt{10 - 2\sqrt{5}}}{4},$$

$$\frac{-1 - \sqrt{5}}{4} - i \frac{\sqrt{10 - 2\sqrt{5}}}{4}.$$

Now $\cos (2\pi/5) = \cos 72° $ and $\sin (2\pi/5) = \sin 72°$ are both positive, so that necessarily

$$\omega = \cos 72° + i \sin 72° = \frac{-1 + \sqrt{5}}{4} + i \frac{\sqrt{10 + 2\sqrt{5}}}{4},$$

whence

$$\cos 72° = \frac{-1 + \sqrt{5}}{4}, \qquad \sin 72° = \frac{\sqrt{10 + 2\sqrt{5}}}{4}.$$

The other roots in the order they are written are ω^4, ω^2, ω^3.

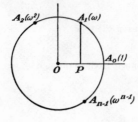

The division of the circumference of a circle into equal parts or the construction of regular polygons is intimately connected with the roots of unity. In fact, if $A_0, A_1, \ldots, A_{n-1}$ are vertices of a regular polygon of n sides inscribed in a circle of radius 1 and OA_0 is chosen for the real axis, then the angles that OA_1, $OA_2, OA_3, \ldots$ form with it are $2\pi/n$, $4\pi/n$,

$6\pi/n,\ \ldots$, and consequently the vertices $A_0,\ A_1,\ A_2,\ \ldots$ are represented by the complex numbers

$$1,\ \omega,\ \omega^2,\ \ldots$$

where $\omega = \cos\ (2\pi/n) + i\ \sin\ (2\pi/n)$. To construct the polygon it suffices to construct the abscissa

$$OP = \cos\frac{2\pi}{n}$$

or the real part of ω. The construction with the ruler and compass will be possible if it turns out that the algebraic expression of ω is composed of only square roots. Thus, the problem of constructing regular polygons requires the algebraic solution of the equation $x^n = 1$ and the investigation of conditions under which its roots are expressible through quadratic radicals. This constitutes an important chapter of algebra called *cyclotomy*. The reader is referred to the problems for the most convenient construction of regular polygons with 3, 4, 5, 6, 10 sides.

Problems

1. Show that the 24th root of unity, $\cos 15° + i \sin 15°$, satisfies the equation

$$x^8 - x^4 + 1 = 0,$$

and find the expression for the roots of this equation both in trigonometric and algebraic form. Notice that

$$x^{24} - 1 = (x^{12} - 1)(x^4 + 1)(x^8 - x^4 + 1).$$

2. Show that

$$\epsilon_1 = 2\cos\frac{2\pi}{7}, \qquad \epsilon_2 = 2\cos\frac{4\pi}{7}, \qquad \epsilon_3 = 2\cos\frac{6\pi}{7}$$

are roots of the cubic equation $y^3 + y^2 - 2y - 1 = 0$.

HINT: The seventh roots of unity satisfy the equation

$$x^6 + x^5 + x^4 + x^3 + x^2 + x + 1 = 0.$$

Divide it by x^3 and set $x + (1/x) = y$.

3. Show that

$$\epsilon_1 = 2\cos\frac{2\pi}{9}, \qquad \epsilon_2 = 2\cos\frac{4\pi}{9}, \qquad \epsilon_3 = 2\cos\frac{8\pi}{9}$$

are roots of the cubic equation $y^3 - 3y + 1 = 0$.

HINT: The ninth roots of unity that are not cube roots of unity satisfy the equation $x^6 + x^3 + 1 = 0$.

4. Setting

$$\eta = \cos 24° + i \sin 24°,$$

$$\omega = -\frac{1}{2} + i\frac{\sqrt{3}}{2},$$

$$\epsilon = -\frac{1 + \sqrt{5}}{4} + i\frac{\sqrt{10 + 2\sqrt{5}}}{4},$$

show that $\eta = \epsilon^2\omega^{-1}$ and express $\cos 24°$ and $\sin 24°$ in algebraic form.

5. To divide a circumference into 3, 4, 6, 8 equal parts the following construction can be used: For the sake of conciseness a circle with the center C and radius R will be denoted by $C(R)$. Points B, C, D, E are intersections of the circle $O(OA)$ with the circles $A(OA)$, $B(OA)$, $C(OA)$, $D(OA)$. Point X is the intersection of the circles $A(AC)$ and $D(AC)$. Point Y is obtained as the intersection of $X(OA)$ and $O(OA)$. Show that AC, OX, AB, AY are the sides of regular polygons with 3, 4, 6, 8 sides inscribed in $O(OA)$.

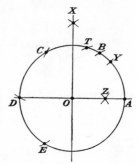

6. Keeping the notations of the preceding problem, describe arcs $C(OX)$ and $E(OX)$ that intersect at Z. Describe $Z(OA)$ intersecting $O(OA)$ at T. Show that AT and OZ are, respectively, sides of the regular pentagon and decagon inscribed in $O(OA)$.

7. Devise a construction of the regular polygon with 15 sides.

17. Geometric Meaning of Operations on Complex Numbers. The geometric representation of complex numbers explained in Sec. 11 opens the way to applications of complex numbers to geometry. It is clear that a certain geometric construction will result as a counterpart to any operation performed on complex numbers. Here we shall confine ourselves only to the examination of constructions that correspond to addition, subtraction, multiplication, and division of complex numbers together with a few additional facts that may be useful in the solution of the appended problems. Complex numbers are represented either

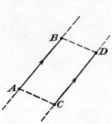

by points or by vectors all having their origin at O. In what follows it will be necessary to consider vectors with arbitrarily placed origins and to explain in this connection the notion of *equipollence* or equality of vectors. Two vectors $\overrightarrow{AB}$ and $\overrightarrow{CD}$ with origins at A and C, respectively, are called equipollent if they lie on the same or parallel lines and have the same direction and the same length. In the figure $\overrightarrow{AB}$ and $\overrightarrow{CD}$ are equipollent vectors. On joining origins and end points of equipollent vectors, in general a parallelogram is obtained. The addition of several vectors, say of three vectors

$$a = \overrightarrow{AB}, \qquad b = \overrightarrow{CD}, \qquad c = \overrightarrow{EF},$$

is performed in the following way: At the end point B of a, place the vector $\overrightarrow{BG}$ with the origin B and equipollent to b; taking G as the origin, construct the vector $\overrightarrow{GH}$ equipollent to c. Then, the vector $\overrightarrow{AH}$ or any equipollent vector is, by definition, the sum of vectors $a + b + c$. From the figure it is clear that the projection of the sum of vectors on any directed line l is equal to the sum of projection of these vectors, each projection being taken positively or negatively according as the direction of the corresponding line segment (like $A'B'$, $G'H'$, etc.) is the same as that of l or opposite to it. It will be easy now to describe the construction corresponding to the addition

of complex numbers. The complex numbers z_1 and z_2, being represented by the points z_1 and z_2, add according to the rule just ex-

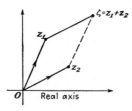

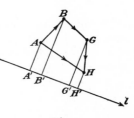

plained for adding vector $\overrightarrow{Oz_2}$ to $\overrightarrow{Oz_1}$; the resulting vector $\overrightarrow{O\zeta}$ will be their sum, and the point ζ will represent the complex number $z_1 + z_2$. In fact, if

$$z_1 = a_1 + b_1 i, \qquad z_2 = a_2 + b_2 i,$$

the projections of $\overrightarrow{Oz_1}$ and $\overrightarrow{Oz_2}$ on the real and imaginary axes will be, respectively,

$$a_1, a_2; \quad b_1, b_2;$$

hence, the projections of $\overrightarrow{O\zeta}$ are

$$a_1 + a_2 \qquad \text{and} \qquad b_1 + b_2,$$

and, consequently,

$$\zeta = a_1 + a_2 + i(b_1 + b_2) = z_1 + z_2.$$

Notice that the figure $Oz_1\zeta z_2$ in general is a parallelogram unless the points O, z_1, z_2 are collinear. In the triangle $Oz_1\zeta$ the side $O\zeta$ is less than the sum of the two other sides, which leads immediately to the inequality

$$|z_1 + z_2| < |z_1| + |z_2|$$

provided the points O, z_1, z_2 are not collinear. The same inequality holds even when O, z_1, z_2 are collinear but z_1 and z_2 are on opposite sides of O; but if they are on the same side, then

$$|z_1 + z_2| = |z_1| + |z_2|.$$

Now, if z_1 and z_2 are on the same side of O, the arguments of the complex numbers z_1 and z_2 are equal and their quotient is a positive real number; conversely, in such a case the arguments of z_1 and z_2 are equal, O, z_1, z_2

are collinear, and z_1 and z_2 are on the same side of O. Thus, by means of geometric representation we prove again and in an intuitive manner the proposition established algebraically in Sec. 9. The geometric construc-

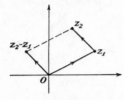

tion for the sum of two complex numbers leads immediately to the corresponding construction of the difference $z_2 - z_1$. This difference is represented by the fourth vertex of the parallelogram, three consecutive vertices of which are O, z_1, z_2. Clearly the vector representing the difference $z_2 - z_1$ is equipollent to the vector $\overrightarrow{z_1 z_2}$.

The rule for the multiplication of complex numbers in trigonometric form provides a simple construction for the product of two complex numbers z_1 and z_2. Before explaining this construction it is necessary to explain what is meant by a *sense* attached to a triangle ABC whose vertices are taken in the order indicated. Going from A to B, from B to C, and from C back to A, the interior of the triangle is situated either to the left or to the right; in the former case we say that it has a positive sense and in the latter that it has a negative sense. Thus, the

triangle ABC represented in the figure has a positive sense, but the same triangle if its vertices are taken in the order ACB will have a negative sense. On joining the point z_1 to O and 1 a triangle $O1z_1$ is formed. Now taking Oz_2 as the side corresponding to $O1$, construct another triangle $Oz_2\zeta$ directly similar to $O1z_1$, that is, having the same sense and equal angles at corresponding vertices. If ϕ_1 and ϕ_2 are the angles between the real axis and the vectors $\overrightarrow{Oz_1}$ and $\overrightarrow{Oz_2}$, it follows from the construction that the angle between the real axis and $\overrightarrow{O\zeta}$ is $\phi_1 + \phi_2$. The argument of ζ is thus $\phi_1 + \phi_2$. Moreover, denoting by ρ, r_1, r_2

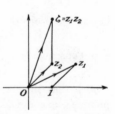

the distances from O to ζ, z_1, z_2, respectively, it follows from the similitude of the triangles $O1z_1$ and $Oz_2\zeta$ that

$$\frac{\rho}{r_1} = \frac{r_2}{1}$$

whence $\rho = r_1 r_2$, the modulus of ζ. Thus, ζ actually represents the product $z_1 z_2$. A similar construction can be devised for representing the quotient z_1/z_2.

Let the complex numbers z_1, z_2, ζ be represented by three collinear points. Then, as can be seen from the figure, the points O, $\zeta - z_1$,

$z_2 - z_1$ are also collinear, and hence the arguments of $\zeta - z_1$ and $z_2 - z_1$ are either equal or differ by π. Consequently,

$$\zeta - z_1 = \lambda(z_2 - z_1)$$

or

$$\zeta = (1 - \lambda)z_1 + \lambda z_2$$

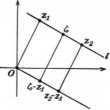

where λ is a real number; and it is evident that the converse is also true, that is, if λ is real, points ζ, z_1, z_2 are collinear. The number λ in the preceding formula has a simple meaning. Denoting by r the distance between z_1 and z_2 and by ρ the segment $z_1\zeta$, taken positively or negatively according as the direction of $\overrightarrow{z_1\zeta}$ coincides with the direction $\overrightarrow{z_1z_2}$ or is opposite to it, obviously λ is equal to the ratio r/ρ. In particular, if $\lambda = \frac{1}{2}$, the point ζ is the mid-point of the segment z_1z_2, and it is represented by the complex number

$$\zeta = \frac{z_1 + z_2}{2}.$$

The vector corresponding to $i(z_2 - z_1)$ is perpendicular to the line l joining z_1 and z_2, whence it is easy to see that the complex numbers

$$\zeta = a + i\lambda(z_2 - z_1),$$

where λ is real, represent points on the line through an arbitrary point a and perpendicular to l.

Problems

1. To construct a triangle XYZ given the mid-points P, Q, R of the sides XY, YZ, ZX.

HINT: Let x, y, z be complex numbers representing the unknown vertices X, Y, Z and p, q, r complex numbers representing P, Q, R. Then, $x + y = 2p$, $y + z = 2q$, $z + x = 2r$. For simplicity of construction the origin can be placed, for example, at P.

2. To construct a quadrangle $XYZT$ given the mid-points P, Q, R, S of the sides XY, YZ, ZT, TX. The problem is possible only if P, Q, R, S fulfill a certain condition. What is the geometric meaning of this condition? This condition being fulfilled, the problem is indeterminate.

3. Five given points are mid-points of the sides of a hexagon; how can one locate the sixth in order that there shall exist hexagons with mid-points of their sides thus located?

4. Points P, Q, R are given such that they divide the sides of a triangle XYZ in the ratios

$$\frac{XP}{PY} = \frac{1}{1}, \qquad \frac{YQ}{QZ} = \frac{2}{1}, \qquad \frac{ZR}{RX} = \frac{1}{2}.$$

Construct the triangle. Discuss the condition for the existence of a true triangle.

5. The complex numbers $z = a + (b - a)t$, where a and b are given complex numbers and t a variable real number, represent points of the line through a and b. Show that the two lines

$$z = a + (b - a)t, \qquad z = c + (d - c)t'$$

are parallel if

$$I\left(\frac{b - a}{d - c}\right) = 0$$

and perpendicular if

$$R\left(\frac{b - a}{d - c}\right) = 0.$$

6. How can one find the point of intersection of the lines of Prob. 5 if they are not parallel? The value of t for the point of intersection is given by

$$tI\left(\frac{b - a}{d - c}\right) + I\left(\frac{a - c}{d - c}\right) = 0.$$

7. If z_1, z_2, z_3 represent the vertices of a triangle, show that the medians pass through the same point and find the complex number corresponding to this point.

8. The complex numbers z_1, z_2, z_3 represent the vertices of a triangle $z_1z_2z_3$ with positive sense. If a, b, c are lengths of the sides z_1z_2, z_2z_3, z_1z_3 and A, B, C the opposite angles, show that

$$\frac{z_1 - z_2}{z_3 - z_2} = \frac{a}{b}\,(\cos C + i \sin C), \qquad \frac{z_3 - z_1}{z_3 - z_2} = \frac{c}{b}\,(\cos A - i \sin A).$$

Also, using the identity

$$(z_1 - z_2) + (z_2 - z_3) + (z_3 - z_1) = 0,$$

show that

$$\frac{\sin A}{a} = \frac{\sin B}{b} = \frac{\sin C}{c},$$

and

$$b = a \cos C + c \cos A, \text{ etc.}$$

*9. The geometric mean $\sqrt{z_1z_2}$ of the complex numbers z_1 and z_2 can be constructed as follows: Draw the bisector l of the angle between $\overrightarrow{Oz_1}$ and $\overrightarrow{Oz_2}$ and through O the line l' perpendicular to l. Take z_3 symmetrically to z_2 with respect to l', and through z_1, z_2, z_3 draw a circle that intersects l at the points ζ and $-\zeta$. These two points represent two possible values of the geometric mean.

*10. On the base $AB = 2a$ construct a triangle ABX knowing the product of the sides $AX \cdot BX = m^2$ and the difference of the angles $\angle ABX - \angle BAX = \delta$.

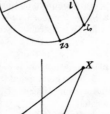

HINT: Let the base AB extended be the real axis, its midpoint O the origin, and X the complex number representing the unknown vertex X. The conditions of the problem lead to the equation

$$X^2 = a^2 - m^2\,(\cos \delta - i \sin \delta)$$

or

$$X^2 = \left[a + m\left(\cos \frac{\delta}{2} - i \sin \frac{\delta}{2}\right)\right]\left[a - m\left(\cos \frac{\delta}{2} - i \sin \frac{\delta}{2}\right)\right]$$

The construction of X follows from Prob. 9.

CHAPTER II

POLYNOMIALS IN ONE VARIABLE

1. Integral Rational Functions or Polynomials. An expression of the form

$$a_0 x^n + a_1 x^{n-1} + \cdots + a_n$$

in which $a_0, a_1, \ldots, a_n$ are given numbers (real or imaginary) and with x as the variable is called an *integral rational function of x* or a *polynomial in x*. The constants $a_0, a_1, \ldots, a_n$ are called *coefficients*, and the single monomials

$$a_0 x^n, a_1 x^{n-1}, \ldots, a_n$$

are called the *terms* of the polynomial. If $a_0 \neq 0$, the polynomial is of *degree n* and $a_0 x^n$ is the *leading* term. Terms with coefficients equal to 0 are usually omitted while, on the other hand, before the leading term it is permissible to add as many terms with zero coefficients as we wish, and all polynomials thus obtained are considered as identical. Though strictly speaking a polynomial must involve the variable x, yet, for the sake of convenience, it is customary to consider constants, different from 0, as polynomials of degree 0. A polynomial all of whose coefficients are equal to 0 is called an *identically vanishing* polynomial and is replaced by 0. No degree is attributed to identically vanishing polynomials. Two polynomials are called *equal* if they are identical term for term; that is, the equality

$$a_0 x^n + a_1 x^{n-1} + \cdots + a_n = b_0 x^n + b_1 x^{n-1} + \cdots + b_n$$

implies

$$a_0 = b_0, \qquad a_1 = b_1, \qquad \ldots, \qquad a_n = b_n.$$

Often to denote polynomials it is convenient to use functional signs $f(x), g(x), \phi(x)$, etc., and even to omit x, writing simply f, g, etc., if by so doing no misunderstanding can arise. The result of the substitution of a number a for x into a polynomial $f(x)$ is a number called the value of this polynomial for $x = a$ and is denoted by $f(a)$. Thus, for the polynomials

$$f(x) = 3x^3 - x + 2, \qquad g(x) = 4x^4 - x^3 + 2x - 1,$$
$$h(x) = \sqrt{2}x^2 - (3 + \sqrt{2})x + 4,$$

35

we have

$$f(-1) = 0, \qquad g(i) = 3 + 3i, \qquad h(1) = 1.$$

2. Multiplication of Polynomials. The addition, subtraction, and multiplication of polynomials are sufficiently well known from elementary courses of algebra. In regard to multiplication, only one remark of a practical nature may be added. If we want, for example, to multiply the polynomials

$$x^2 - x + 1 \qquad \text{and} \qquad x^2 + x + 1,$$

the usual arrangement of calculations is as follows:

$$\frac{(x^2 - x + 1) \times (x^2 + x + 1)}{x^4 - x^3 + x^2}$$
$$x^3 - x^2 + x$$
$$\frac{x^2 - x + 1}{x^4 \quad + \quad x^2 \quad + \quad 1}$$

This procedure, especially when the polynomials to be multiplied have many terms, entails a good deal of useless work in writing powers of x. This can be avoided by using the method of *detached coefficients*. In this method we write only sequences of coefficients of the polynomials we wish to multiply, beginning with the leading ones and without omitting zero coefficients. Then, the coefficients of one polynomial are multiplied in order by the first, second, third, etc., coefficients of the second polynomial, and the resulting lines of numbers are placed one below another in such a way that each line is shifted one place to the right with respect to the preceding line. Adding the numbers standing in the same column, we obtain in order the coefficients of the product, and finally we restore the missing powers of x. For instance, the example given above can be worked out as follows:

1	-1	1	$\times$	1	1	1
1	-1	1				
	1	-1	1			
		1	-1	1		
1	0	1	0	1		

so that the product is

$$x^4 + 0x^3 + x^2 + 0x + 1 = x^4 + x^2 + 1.$$

As another example let us multiply

$$x^5 + x^3 - 2x^2 + 3 \qquad \text{by} \qquad 2x^4 - 3x^3 + 4x^2 - 1.$$

In this example the calculation is arranged as follows:

1	0	1	-2	0	3	$\times$	2	-3	4	0	-1
2	0	2	-4	0	6						
	-3	0	-3	6	0	-9					
		4	0	4	-8	0	12				
			-1	0	-1	2	0	-3			
2	-3	6	-7	9	-2	-10	14	0	-3		

so that the product is

$$2x^9 - 3x^8 + 6x^7 - 7x^6 + 9x^5 - 2x^4 - 10x^3 + 14x^2 - 3.$$

One line consisting of zeros obviously could be omitted.

One further remark of theoretical importance should be added. If two nonidentically vanishing polynomials $f(x)$ and $g(x)$ have the leading terms a_0x^n and b_0x^m, the leading term of the product will be

$$a_0b_0x^{n+m}$$

and the coefficient differs from 0; hence, $f(x)g(x)$ is a nonidentically vanishing polynomial. Consequently, if

$$f(x)g(x) = 0,$$

one of the factors must be an identically vanishing polynomial.

Problems

By the method of detached coefficients multiply

1. $x^4 + x^3 + x^2 + x + 1$ by $x^4 - x^3 + x^2 - x + 1$.
2. $2x^4 - 3x^3 + x - 1$ by $x^3 + 3x^2 - 1$.
3. $x^4 + 4x^3 - 5x^2 - 2$ by $x^4 - 4x^3 - 5x^2 - 2$.
4. $x^5 - 3x^4 + x^3 - x + 1$ by $3x^3 + 7x^2 - x + 1$.

3. Division of Polynomials. The division of polynomials requires more explanation. Let

$$f(x) = a_0x^n + a_1x^{n-1} + \cdots + a_n,$$
$$g(x) = b_0x^m + b_1x^{m-1} + \cdots + b_m$$

be two polynomials of degrees n and m, respectively, so that $a_0 \neq 0$, $b_0 \neq 0$, and assume that $n \geq m$. By choosing properly a constant c_0 we can obtain the polynomial

$$f(x) - c_0x^{n-m}g(x) = f_1(x),$$

which, if not vanishing identically, will have degree $n_1 < n$; for this it suffices to take

$$c_0 = \frac{a_0}{b_0}.$$

As long as $n_1 \geqq m$, a constant c_1 can be found so that

$$f_1(x) - c_1 x^{n_1-m} g(x) = f_2(x)$$

which, if not vanishing identically, will have degree $n_2 < n_1$. If $n_2 \geqq m$, the same process can be repeated. Now the degrees of the "partial remainders" $f_1(x)$, $f_2(x)$, . . . form a decreasing sequence so that there will be some first partial remainder $f_{k+1}(x)$ that either vanishes identically or is of degree $n_{k+1} < m$. By eliminating $f_1(x)$, $f_2(x)$, . . . , $f_k(x)$ from the identities

$$f(x) - c_0 x^{n-m} g(x) = f_1(x),$$
$$f_1(x) - c_1 x^{n_1-m} g(x) = f_2(x),$$
$$\cdots\cdots\cdots\cdots\cdots\cdots$$
$$f_k(x) - c_k x^{n_k-m} g(x) = f_{k+1}(x),$$

and setting for brevity

$$c_0 x^{n-m} + c_1 x^{n_1-m} + \cdots + c_k x^{n_k-m} = q(x), \qquad f_{k+1}(x) = r(x),$$

we obtain the identity

$$f(x) = g(x)q(x) + r(x),$$

in which $r(x)$ has a degree $< m$ or vanishes identically. The polynomials $q(x)$ and $r(x)$ are called the quotient and the remainder in the division of $f(x)$ by $g(x)$ and are found by the above described process, which is essentially the same as the one taught in elementary courses of algebra.

In practice again it is profitable to avoid writing the powers of x, using instead the method of "detached coefficients." For example, let us divide

$$x^8 + x^7 + 3x^4 - 1 \qquad \text{by} \qquad x^4 - 3x^3 + 4x + 1.$$

In writing the detached coefficients one must not forget the coefficients 0 of the missing terms. The operation itself is arranged as follows:

		Dividend								Divisor			
1	1	0	0	3	0	0	0	−1	1	−3	0	4	1
1	−3	0	4	1					1	4	12	32	82 quotient
	4	0	−4	2	0								
	4	−12	0	16	4								
		12	−4	−14	−4	0							
		12	−36	0	48	12							
			32	−14	−52	−12	0						
			32	−96	0	128	32						
				82	−52	−140	−32	−1					
				82	−246	0	328	82					
					194	−140	−360	−83 remainder					

Thus, the quotient and the remainder are

$$x^4 + 4x^3 + 12x^2 + 32x + 82, \text{ quotient}$$
$$194x^3 - 140x^2 - 360x - 83, \text{ remainder}$$

and identically

$$x^8 + x^7 + 3x^4 - 1 = (x^4 - 3x^3 + 4x + 1)(x^4 + 4x^3 + 12x^2 + 32x + 82)$$
$$+ 194x^3 - 140x^2 - 360x - 83.$$

If the remainder in the division of $f(x)$ by $g(x)$ is 0, that is, if

$$f(x) = g(x)q(x),$$

where $q(x)$ is a polynomial, it is said that $f(x)$ is divisible by $g(x)$ or that $g(x)$ is a divisor of $f(x)$. Clearly no polynomial that is not identically vanishing can be divisible by another of higher degree. From this it can be inferred that in an identity of the form

$$f(x) = g(x)q_1(x) + r_1(x),$$

where $q_1(x)$ and $r_1(x)$ are polynomials and $r_1(x)$ is either 0 or has lower degree than $g(x)$, $q_1(x)$ and $r_1(x)$ coincide with the quotient and the remainder obtained by division. In fact, if

$$f(x) = g(x)q_1(x) + r_1(x) = g(x)q(x) + r(x),$$

then,

$$g(x)[q_1(x) - q(x)] = r(x) - r_1(x),$$

which shows that $r(x) - r_1(x)$ is divisible by $g(x)$. It is impossible that $r(x) - r_1(x)$ does not vanish identically; for in this case its degree is less than that of $g(x)$ and it cannot be divisible by $g(x)$. Hence, $r_1(x) = r(x)$ and also $q_1(x) = q(x)$.

The following simple remark will be needed later: If two polynomials f and f_1 are divisible by g, then for arbitrary polynomials l and l_1 the polynomial

$$lf + l_1 f_1$$

will be divisible by g. In fact, by hypothesis

$$f = gq, \qquad f_1 = gq_1$$

where q and q_1 are polynomials; hence,

$$lf + l_1 f_1 = g(lq + l_1 q_1)$$

is divisible by g.

Problems

By the method of detached coefficients divide

1. $x^7 + 3x^6 - 2x^3 + 3x^2 - x + 1$ by $x^4 - x + 1$.

2. $2x^7 - 3x^6 + x^5 - 3x^4 + 5x^3 - 4x^2 + 2x - 1$ by $2x^3 - 3x^2 + x - 1$.

3. $x^5 - 3x^2 + 6x - 1$ by $x^2 + x + 1$.

4. $x^{10} + x^5 + 1$ by $x^2 + x + 1$.
5. $(x + 1)^7 - x^7 - 1$ by $(x^2 + x + 1)^2$.

4. The Remainder Theorem. The remainder in the division of a polynomial by a binomial $x - c$, where c is an arbitrary number, can be found without actually performing the division by means of the following theorem, important despite its simplicity:

The Remainder Theorem. The remainder obtained in dividing $f(x)$ by $x - c$ is the value of the polynomial $f(x)$ for $x = c$, that is, $f(c)$.

PROOF. Since the divisor is of the first degree, the remainder will be a constant r. Calling the quotient $q(x)$, we have the identity

$$f(x) = (x - c)q(x) + r.$$

On substituting the number c in place of x into this identity we must get equal numbers. Now, since r is a constant, it is not affected by this substitution and the value of the right-hand member for $x = c$ will be

$$(c - c)q(c) + r = r,$$

whereas the value of the left-hand member is $f(c)$; hence,

$$r = f(c),$$

which means also that identically in x

$$f(x) = (x - c)q(x) + f(c).$$

It follows from this theorem that $f(x)$ is divisible by $x - c$ if and only if $f(c) = 0$.

Example 1. To show that $f(x) = x^3 + x^2 - 5x + 3$ is divisible by $x + 3$. In this case $c = -3$, and thus we have to calculate

$$f(-3) = -27 + 9 + 15 + 3 = 0;$$

hence, $f(x)$ is divisible by $x + 3$.

Example 2. To show that $x^n - c^n$ is divisible by $x - c$. This is true since $c^n - c^n = 0$; the quotient found by ordinary division is

$$x^{n-1} + cx^{n-2} + c^2x^{n-3} + \cdots + c^{n-1}.$$

Example 3. Under what conditions is $x^n + c^n$ divisible by $x + c$? In this case $x = -c$ must be substituted into $x^n + c^n$; the result of the substitution is

$$(-c)^n + c^n = c^n + c^n = 2c^n \text{ if } n \text{ is even},$$
$$(-c)^n + c^n = -c^n + c^n = 0 \text{ if } n \text{ is odd}.$$

Hence, $x^n + c^n$ is divisible by $x + c$ (for $c \neq 0$) only if n is odd, and for an even n the remainder after the division is $2c^n$.

Problems

Without actual division show that

1. $x^4 + 3x^3 + 3x^2 + 3x + 2$ is divisible by $x + 2$.

2. $x^5 - 3x^4 + x^2 - 2x - 3$ is divisible by $x - 3$.

3. If a and b are different and $f(x)$ is separately divisible by $x - a$ and $x - b$, show that $f(x)$ is divisible by $(x - a)(x - b)$.

Without actual division show that

4. $2x^4 - 7x^3 - 2x^2 + 13x + 6$ is divisible by $x^2 - 5x + 6$.

5. $2x^6 + 2x^5 + x^4 + 2x^3 + x^2 + 2$ is divisible by $x^2 + 1$.

6. $x^6 + 4x^5 + 3x^4 + 2x^3 + x + 1$ is divisible by $x^2 + x + 1$.

7. Show that $(x + 1)^n - x^n - 1$ is divisible by $x^2 + x + 1$ only if n is an odd number nondivisible by 3.

5. Synthetic Division. The quotient in the division by $x - c$ can be found by a very convenient process known as *synthetic division*. Into the identity of Sec. 4

$$f(x) = (x - c)q(x) + r$$

let us substitute

$$q(x) = b_0 x^{n-1} + b_1 x^{n-2} + \cdots + b_{n-1}$$

where $b_0, b_1, \ldots, b_{n-1}$ are coefficients to be determined. Performing the multiplication, we have

$$(x - c)q(x) = b_0 x^n + (b_1 - cb_0)x^{n-1} + (b_2 - cb_1)x^{n-2} + \cdots$$
$$+ (b_{n-1} - cb_{n-2})x - cb_{n-1},$$

and

$$(x - c)q(x) + r = b_0 x^n + (b_1 - cb_0)x^{n-1} + \cdots$$
$$+ (b_{n-1} - cb_{n-2})x + r - cb_{n-1}.$$

Since this polynomial must be identical to

$$a_0 x^n + a_1 x^{n-1} + \cdots + a_n,$$

to determine $b_0, b_1, \ldots, b_{n-1}$ and r we equate coefficients of like powers of x, getting the set of equations

$$b_0 = a_0, \qquad b_1 - cb_0 = a_1, \qquad b_2 - cb_1 = a_2, \qquad \ldots,$$
$$b_{n-1} - cb_{n-2} = a_{n-1}, \qquad r - cb_{n-1} = a_n,$$

from which it follows that $b_0, b_1, \ldots, b_{n-1}, r$ are found one after another as follows:

$$b_0 = a_0, \qquad b_1 = a_1 + cb_0, \qquad b_2 = a_2 + cb_1, \qquad \ldots,$$
$$b_{n-1} = a_{n-1} + cb_{n-2}, \qquad r = a_n + cb_{n-1}.$$

The calculation is of a recursive nature and in practice can be arranged more conveniently thus:

$c)$

a_0	a_1	a_2	$\cdots$	a_{n-1}	a_n	
	$b_0 c$	$b_1 c$	$\cdots$	$b_{n-2} c$	$b_{n-1} c$	
$a_0 = b_0$	b_1	b_2	$\cdots$	b_{n-1}	r	remainder

$$\underbrace{\qquad\qquad\qquad\qquad\qquad\qquad\qquad}_{\text{coefficients of the quotient}}$$

Here in the first line all the coefficients of $f(x)$ are written without omission starting with a_0. The third line begins with $b_0 = a_0$, which is multiplied by c, the product placed in the second line and added to a_1; the sum b_1 is placed in the third line. Again, b_1 is multiplied by c, the product placed in the second line and added to a_2; the sum is placed in the third line and the same process is repeated until in the last column and in the third line the remainder r is found. The independent expressions for $b_0, b_1, \ldots, b_{n-1}, r$ obtained by successive substitutions are

$$b_0 = a_0, \qquad b_1 = a_0 c + a_1, \qquad b_2 = a_0 c^2 + a_1 c + a_2, \qquad \ldots ,$$
$$b_{n-1} = a_0 c^{n-1} + a_1 c^{n-2} + \cdots + a_{n-1}$$

and

$$r = a_0 c^n + a_1 c^{n-1} + \cdots + a_n = f(c),$$

which gives the second proof of the remainder theorem. Considering the sequence of polynomials

$$f_0 = a_0, \qquad f_1 = x f_0 + a_1, \qquad f_2 = x f_1 + a_2, \qquad \ldots , \qquad f_n(x) = x f_{n-1}(x) + a_n,$$

it is clear that

$$f_i(x) = a_0 x^i + a_1 x^{i-1} + \cdots + a_i.$$

Hence,

$$b_i = f_i(c), \qquad i = 0, 1, 2, \ldots , i - 1$$

and, moreover,

$$f_i(x) = (x - c)[f_0(c) x^{i-1} + f_1(c) x^{i-2} + \cdots + f_{i-1}(c)] + f_i(c).$$

The above described process of finding the quotient and the remainder when dividing $f(x)$ by $x - c$ is known as *synthetic division*. Since the remainder is $f(c)$, the synthetic division provides a convenient means for calculating the value of a polynomial for a given value of the variable.

Example 1. To find the quotient and the remainder when dividing
$$3x^6 - 7x^5 + 5x^4 - x^2 - 6x - 8 \text{ by } x + 2.$$

The necessary calculations are arranged as follows:

$-2)$	3	-7	5	0	-1	-6	-8
		-6	26	-62	124	-246	504
	3	-13	31	-62	123	-252	496

Hence, the quotient is

$$3x^5 - 13x^4 + 31x^3 - 62x^2 + 123x - 252$$

and the remainder is

$$r = 496.$$

Example 2. To divide

$$5x^5 - 7x^3 + 6x^2 - 2x + 4 \text{ by } x - 1.$$

In this case the process of synthetic division is reduced to additions and can be arranged in two lines as shown:

$$
\begin{array}{r|rrrrrr}
1) & 5 & 0 & -7 & 6 & -2 & 4 \\
 & 5 & 5 & -2 & 4 & 2 & 6 \\
\end{array}
$$

so that the quotient is

$$5x^4 + 5x^3 - 2x^2 + 4x + 2$$

and the remainder is

$$r = 6.$$

Similar simplification occurs when $c = -1$ in which case the whole process consists of subtractions.

Problems

By synthetic division find the quotient and the remainder in the division of

1. $2x^4 - 6x^3 + 7x^2 - 5x + 1$ by $x + 2$.
2. $-x^4 + 7x^3 - 4x^2$ by $x - 3$.
3. $6x^3 - 10x^2 + 5x + 3$ by $x - 1.2$.
4. $10x^3 - 2x^2 + 3x - 1$ by $2x - 3$.
5. $x^4 + x^3 - x^2 + 1$ by $3x + 2$.
6. $5x^6 - 6x^4 + 1$ by $x + 1$.
7. $(n - 1)x^n - nx^{n-1} + 1$ by $(x - 1)^2$.
8. Compute $f(0.75)$ if $f(x) = -3x^3 + 6x^2 - x + 1$.
9. Compute $f(-0.3)$ if $f(x) = -2x^4 + 6x^3 - x^2 + 2$.

6. Horner's Process. Since by the binomial theorem any power

$$x^m = [c + (x - c)]^m = c^m + mc^{m-1}(x - c) + \frac{m(m-1)}{1 \cdot 2} c^{m-2}(x - c)^2 + \cdots$$

can be expanded in powers of $x - c$, c being an arbitrary number, any polynomial can be similarly expanded. Let

$$f(x) = A_0 + A_1(x - c) + A_2(x - c)^2 + \cdots + A_n(x - c)^n.$$

The coefficients in this expansion can be determined very conveniently by repeated applications of synthetic division. In fact, writing

$$f(x) = A_0 + (x - c)f_1(x), \quad f_1(x) = A_1 + A_2(x - c) + \cdots + A_n(x - c)^{n-1}$$
$$f_1(x) = A_1 + (x - c)f_2(x), \quad f_2(x) = A_2 + \cdots + A_n(x - c)^{n-2}$$
$$\cdots \cdots \cdots \cdots \cdots \cdots \cdots \cdots \cdots \cdots \cdots \cdots \cdots$$

it is clear that A_0 is obtained as the remainder in dividing $f(x)$ by $x - c$; A_1 is the remainder obtained in dividing the first quotient $f_1(x)$ by $x - c$; A_2 is the remainder obtained in dividing the second quotient $f_2(x)$ by

$x - c$, etc. The arrangement of this process known as Horner's process is best understood from examples.

Example 1. Expand in powers of $x - 1$

$$f(x) = 4x^5 - 6x^4 + 3x^3 + x^2 - x - 1.$$

In this case Horner's process is simplified and reduced to additions, thus.

$$
\begin{array}{r|rrrrrr}
1) & 4 & -6 & 3 & 1 & -1 & -1 \\
 & 4 & -2 & 1 & 2 & 1 & \underline{0} \\
 & 4 & 2 & 3 & 5 & \underline{6} \\
 & 4 & 6 & 9 & \underline{14} \\
 & 4 & 10 & \underline{19} \\
 & 4 & \underline{14} \\
 & \underline{4}
\end{array}
$$

The underlined numbers are read downward and represent the requested coefficients $A_0, A_1, A_2, \ldots$. Thus,

$$f(x) = 0 + 6(x - 1) + 14(x - 1)^2 + 19(x - 1)^3 + 14(x - 1)^4 + 4(x - 1)^5.$$

Example 2. Expand

$$f(x) = x^4 - 6x^2 + 1$$

in powers of $x + 2$. The arrangement of the Horner's process in this example is as follows:

$$
\begin{array}{r|rrrrr}
-2) & 1 & 0 & -6 & 0 & 1 \\
 & & -2 & 4 & 4 & -8 \\
\hline
 & 1 & -2 & -2 & 4 & \underline{-7} \\
 & & -2 & 8 & -12 \\
\hline
 & 1 & -4 & 6 & \underline{-8} \\
 & & -2 & 12 \\
\hline
 & 1 & -6 & \underline{18} \\
 & & -2 \\
\hline
 & 1 & \underline{-8} \\
 & 1
\end{array}
$$

Hence,

$$x^4 - 6x^2 + 1 = -7 - 8(x + 2) + 18(x + 2)^2 - 8(x + 2)^3 + (x + 2)^4$$

is the requested expansion.

Problems

Expand

1. $x^5 - 1$ in powers of $x - 1$.

2. $x^5 - 6x^3 + x^2 - 1$ in powers of $x + 1$.

3. $-4x^6 + 2x^5 - x + 1$ in powers of $x + 2$.

4. $3x^4 + 6x^3 + x^2 - 1$ in powers of $x - 0.3$.

7. Taylor's Formula. The coefficients in the expansion of a polynomial in powers of $x - c$ depend in a simple manner on the values of this polynomial and its derivatives at $x = c$. While the notion of a derivative for functions in general is intimately connected with the idea of a limit and therefore properly belongs to differential calculus, in the special case of polynomials derivatives can be introduced algebraically without any reference to limits. The derivative $f'(x)$ of a polynomial $f(x)$ may be defined as the coefficient of the first power of h in the expansion of $f(x + h)$ in ascending powers of an auxiliary letter h. From this definition and the binomial expansion

$$(x - c + h)^n = (x - c)^n + n(x - c)^{n-1}h + \cdots$$

it follows at once that the derivative of $(x - c)^n$ is $n(x - c)^{n-1}$, and in particular nx^{n-1} is the derivative of x^n. Furthermore, it is clear that on multiplying a polynomial by a constant its derivative is multiplied by that constant and the derivative of the sum of polynomials is equal to the sum of their derivatives. From this it is easy to conclude that the derivative of

$$f(x) = a_0x^n + a_1x^{n-1} + \cdots + a_{n-1}x + a_n$$

is

$$f'(x) = na_0x^{n-1} + (n - 1)a_1x^{n-2} + \cdots + a_{n-1}.$$

Since $f'(x)$ is a polynomial, we can consider its derivative, which is called the second derivative, $f''(x)$, of $f(x)$. Similarly, the derivative of the second derivative is the third derivative, $f'''(x)$, of $f(x)$, etc. Now take the expansion

$$f(x) = A_0 + A_1(x - c) + A_2(x - c)^2 + \cdots + A_n(x - c)^n$$

and form successive derivatives of both members

$$f'(x) = A_1 + 2A_2(x - c) + 3A_3(x - c)^2 + \cdots + nA_n(x - c)^{n-1},$$
$$f''(x) = 2A_2 + 3 \cdot 2A_3(x - c) + \cdots + n(n - 1)A_n(x - c)^{n-2},$$
$$f'''(x) = 3 \cdot 2A_3 + 4 \cdot 3 \cdot 2A_4(x - c) + \ldots,$$
$$\cdots \cdots \cdots \cdots \cdots \cdots \cdots \cdots \cdots \cdots \cdots \cdots$$

Taking here $x = c$, we find

$$A_0 = f(c), \qquad A_1 = f'(c), \qquad A_2 = \frac{f''(c)}{1 \cdot 2}, \qquad A_3 = \frac{f'''(c)}{1 \cdot 2 \cdot 3}$$

and in general

$$A_i = \frac{f^{(i)}(c)}{1 \cdot 2 \cdot 3 \cdots i}.$$

Thus, the expansion of $f(x)$ in powers of $x - c$ takes the form

$$f(x) = f(c) + \frac{f'(c)}{1}(x - c) + \frac{f''(c)}{1 \cdot 2}(x - c)^2 + \cdots$$
$$+ \frac{f^{(n)}(c)}{1 \cdot 2 \cdot 3 \cdots n}(x - c)^n$$

and presented in this form it is known as Taylor's formula. Horner's process, which serves to calculate the coefficients in this expansion, provides a convenient means for computing values of a polynomial and its derivatives for a given value of the variable.

Problems

Calculate the values of the following polynomials and their derivatives for the value of x indicated:

1. $-x^4 + 6x^3 + x - 1$ for $x = 1$.
2. $2x^5 - 7x^3 - 10x^2 + 2x - 6$ for $x = -1$.
3. $4x^3 - 7x^2 + 5x + 3$ for $x = -2$.
4. $\frac{1}{4}x^4 - \frac{1}{2}x^3 + \frac{1}{6}x^2 - \frac{1}{3}x + 1$ for $x = -\frac{1}{2}$.

8. Highest Common Divisor of Two Polynomials.

Two polynomials may be divisible by the same third polynomial, which is then called their common divisor. For instance, the polynomials

$$x^4 + 4x^3 + 4x^2 - x - 2 = (x + 1)(x + 2)(x^2 + x - 1),$$
$$x^6 + 2x^5 + x^3 + 3x^2 + 3x + 2 = (x + 1)(x + 2)(x^4 - x^3 + x^2 + 1)$$

have the common divisors

$$x + 1, \qquad x + 2, \qquad (x + 1)(x + 2) = x^2 + 3x + 2.$$

Of all the divisors common to two polynomials, special interest is attached to the common divisor of highest degree. This expression is called the *highest common divisor*. We shall see presently that the highest common divisor is essentially unique and that it can be found by a series of regular operations. Let two given polynomials be f and f_1. Dividing f by f_1, let q_1 be the quotient and f_2 the remainder so that

$$f = f_1 q_1 + f_2.$$

If f_2 is not an identically vanishing polynomial, we may further divide f_1 by f_2, obtaining the quotient q_2 and the remainder f_3 so that

$$f_1 = f_2 q_2 + f_3.$$

Again, if f_3 does not vanish identically, the division of f_2 by f_3 leads to another identity

$$f_2 = f_3 q_3 + f_4,$$

etc. Since the degrees of the polynomials $f_1, f_2, f_3, \ldots$ diminish and operations can be continued as long as the last remainder obtained is not an identically vanishing polynomial, we must come to some remainder f_r that divides the preceding remainder so that we shall have r identities

$$f = f_1 q_1 + f_2,$$
$$f_1 = f_2 q_2 + f_3,$$
$$\cdots \cdots \cdots$$
$$f_{r-2} = f_{r-1} q_{r-1} + f_r,$$
$$f_{r-1} = f_r q_r.$$

From these identities it can be inferred first that f_r is a common divisor of f and f_1, and, second, that any common divisor of these polynomials divides f_r. To prove the first point we observe that f_r divides f_{r-1}, hence, it divides also

$$f_{r-2} = f_{r-1} q_{r-1} + f_r.$$

Again, since f_r divides both f_{r-1} and f_{r-2}, it will divide f_{r-3}, and, continuing in the same manner, we may conclude finally that f_r divides both f_1 and f. To prove the second point, suppose that d divides both f and f_1. Then, as seen from the identity

$$f_2 = f - f_1 q_1,$$

d will divide both f_1 and f_2. Again, the identity

$$f_3 = f_1 - f_2 q_2$$

shows that d divides both f_2 and f_3. Continuing the same reasoning, we conclude that d divides f_{r-1} and f_r. Since every common divisor of f and f_1, as has been just proved, divides f_r, no one of them can have degree higher than the degree of f_r. Hence, f_r is a common divisor of the highest degree; and, if d is any other common divisor of the same degree, it divides f_r and the quotient is a constant. Thus, there are infinitely many common divisors of f and f_1 of the highest degree, but all of them are of the form

$$cf_r$$

where c is an arbitrary constant. In questions of divisibility, polynomials differing only by a constant factor may be considered as not essentially different. In this sense there is an essentially unique highest common divisor of two polynomials for which can be taken either f_r, as given by the above process, or cf_r, with the constant c so chosen as to obtain the simplest result. It may happen that f_r itself is a constant; in this case polynomials do not have common divisors in a proper sense and are called polynomials without common divisor, or relatively prime polynomials. The process of successive divisions by means of which the highest common divisor is found is similar to Euclid's algorithm, which in arithmetic is used to find the greatest common divisor of two integers. Hence, it is also called the Euclid algorithm applied to polynomials.

Example 1. To find the highest common divisor of

$$f = x^6 + 2x^5 + x^3 + 3x^2 + 3x + 2$$

and

$$f_1 = x^4 + 4x^3 + 4x^2 - x - 2.$$

The first step in the Euclid algorithm is to divide f by f_1. This division is performed with detached coefficients as follows:

1	2	0	1	3	3	2	1	4	4	-1	-2
1	4	4	-1	-2			1	-2	4		
	-2	-4	2	5	3						
	-2	-8	-8	2	4						
		4	10	3	-1	2					
		4	16	16	-4	-8					
			-6	-13	3	10					

The first remainder is

$$f_2 = -6x^3 - 13x^2 + 3x + 10.$$

Now we have to divide f_1 by f_2. This division will introduce fractional coefficients and, to avoid this inconvenience, we may multiply f_1 by 6; thereby f_3 will be multiplied by a constant factor, but this is of no importance for our purposes. The next operation is therefore the following:

6	24	24	-6	-12	-6	-13	3	10
6	13	-3	-10		-1	-11		
	11	27	4	-12				

Now to avoid fractions again we multiply the numbers of the last line by 6; this changes the final remainder so that instead of f_3, as given by the regular procedure, we shall have f_3 multiplied by a constant. The operation continues thus:

66	162	24	-72
66	143	-33	-110
	19	57	38

Here all coefficients have the factor 19; suppressing it, we can take for f_3

$$f_3 = x^2 + 3x + 2.$$

Notice that in the line where the coefficients of the quotient are written the numbers no longer represent these coefficients. But this has no importance, since we are not interested in quotients but only in remainders and then only save for constant factors. Now we have to divide f_2 by f_3. This division

-6	-13	3	10	1	3	2
-6	-18	-12		-6	5	
	5	15	10			
	5	15	10			
		0				

succeeds without a remainder. Hence, the operations stop here and the requested highest common divisor can be taken as

$$x^2 + 3x + 2.$$

Example 2. Find the highest common divisor of

$$f = x^5 - x^4 - 2x^3 + 2x^2 + x - 1$$

and

$$f_1 = 5x^4 - 4x^3 - 6x^2 + 4x + 1.$$

Here, at the beginning, all coefficients of the first polynomial are multiplied by 5; then, the procedure continues as follows:

	5	− 5	− 10	10	5	− 5	5	− 4	− 6	4	1
	5	− 4	− 6	4	1		1	− 1			
Multiply by 5:		− 1	− 4	6	4	− 5					
		− 5	− 20	30	20	− 25					
		− 5	4	6	− 4	− 1					
Suppress factor − 24:			− 24	24	24	− 24					
			1	− 1	− 1	1					

We may take

$$f_2 = x^3 - x^2 - x + 1.$$

Next,

	5	− 4	− 6	4	1	1	− 1	− 1	1
	5	− 5	− 5	5		5	1		
		1	− 1	− 1	1				
		1	− 1	− 1	1				
			0						

and since there is no remainder in this division,

$$f_2 = x^3 - x^2 - x + 1$$

is the requested highest common divisor.

NOTE: If all the identities in Euclid's algorithm applied to f and f_1 are multiplied by an arbitrary polynomial g, it is clear that

$$gf_2, \ gf_3, \ \ldots, \ gf_r$$

will be the successive remainders of which the last, gf_r, divides the preceding, gf_{r-1}. Hence, the conclusion: If the highest common divisor of f and f_1 is d, that of gf and gf_1 will be gd. In particular, if f and f_1 are relatively prime, g may be taken for the highest common divisor of gf and gf_1. In Chap. XII reference will be made to this remark.

Problems

Find the highest common divisor of the following polynomials:

1. $f = 2x^4 + 2x^3 - 3x^2 - 2x + 1,$ $f_1 = x^3 + 2x^2 + 2x + 1.$
2. $f = x^4 - 6x^2 - 8x - 3,$ $f_1 = x^3 - 3x - 2.$
3. $f = 2x^5 + 4x^4 + x^3 - x^2 + x + 1,$ $f_1 = 6x^5 - 2x^4 + x^3 + 2x^2 - x + 1.$
4. $f = 2x^6 + 3x^5 + x^4 + 7x^3 + 4x^2 + 4x + 5,$
                             $f_1 = x^4 - x^3 - x - 1.$
5. $f = 10x^6 - 9x^5 - 12x^4 + 2x^2 - x - 1,$ $f_1 = 4x^5 + x^4 - 7x^3 - 8x^2 - x + 1.$

CHAPTER III

ALGEBRAIC EQUATIONS AND THEIR ROOTS

1. Algebraic Equations. Let $f(x)$ be a polynomial, with real or complex coefficients, or degree ≥ 1. On equating it to zero we get an equation

$$f(x) = 0,$$

which is called an *algebraic* equation. In this equation x stands for an unknown *number* that *satisfies* it, that is, when substituted into $f(x)$ gives 0 as a result. Any number satisfying the proposed equation is called its *root*. The problem of solving an equation consists in finding all its roots. Roots of an equation $f(x) = 0$ are often called the roots of the polynomial $f(x)$. If the degree of this polynomial is n, the corresponding equation is said to be of degree n. Corresponding to $n = 1$, 2, 3, 4, etc., we have equations of the form

$$a_0 x + a_1 = 0,$$
$$a_0 x^2 + a_1 x + a_2 = 0,$$
$$a_0 x^3 + a_1 x^2 + a_2 x + a_3 = 0,$$
$$a_0 x^4 + a_1 x^3 + a_2 x^2 + a_3 x + a_4 = 0,$$
$$\text{etc.}$$

of degrees 1, 2, 3, 4, etc., or linear, quadratic, cubic, biquadratic, etc., equations. The leading coefficient a_0 is supposed to be different from 0, but no condition is imposed on the other coefficients.

If c is a root of the equation

$$f(x) = 0,$$

it follows from the remainder theorem that $f(x)$ is divisible by $x - c$ so that

$$f(x) = (x - c)f_1(x),$$

$f_1(x)$ being a polynomial of degree $n - 1$; and, conversely, in case $f(x)$ has the factor $x - c$, the number c is a root of this polynomial. If c_1 is another root different from c, so that $f(c_1) = 0$, then, substituting c_1 into the preceding identity, we have

$$(c_1 - c)f_1(c_1) = 0,$$

whence, since $c_1 - c \neq 0$, it follows that $f_1(c_1) = 0$, that is, $f_1(x)$ is divisible by $x - c_1$, or

$$f_1(x) = (x - c_1)f_2(x),$$

$f_2(x)$ being a polynomial of degree $n - 2$. Consequently,

$$f(x) = (x - c)(x - c_1)f_2(x),$$

which shows that, having two distinct roots c and c_1, the polynomial $f(x)$ is divisible by $(x - c)(x - c_1)$. Continuing in the same manner, we may conclude that $f(x)$ will be divisible by

$$(x - c)(x - c_1) \cdots (x - c_{m-1})$$

if the equation $f(x) = 0$ has m distinct roots $c, c_1, c_2, \ldots, c_{m-1}$.

Two important results may be derived from the last conclusion: First, it shows that an equation of degree n cannot have more than n distinct roots. For suppose that $c, c_1, \ldots, c_{n-1}$ are n distinct roots of the equation $f(x) = 0$; then, $f(x)$ is of degree n and is divisible by $(x - c)(x - c_1) \cdots (x - c_{n-1})$, which is also a polynomial of degree n. Therefore, the quotient is necessarily a constant that is equal to the leading coefficient a_0 of $f(x)$, so that

$$f(x) = a_0(x - c)(x - c_1) \cdots (x - c_{n-1}),$$

and the product on the right-hand side cannot vanish for other values of x except $c, c_1, \ldots, c_{n-1}$. Second, if a certain number $m < n$ of distinct roots $c, c_1, \ldots, c_{m-1}$ is known, the remaining roots will be found by solving the *depressed* equation

$$f_m(x) = \frac{f(x)}{(x - c)(x - c_1) \cdots (x - c_{m-1})} = 0$$

of degree $n - m$. In forming the depressed equation it is profitable to divide by the single binomials $x - c, x - c_1, \ldots, x - c_{m-1}$ in succession, using synthetic division.

Example. Solve the biquadratic equation

$$x^4 - 5x^2 - 10x - 6 = 0$$

having the two roots -1 and 3. The calculations to obtain the depressed equation are arranged as follows:

$$
\begin{array}{r|rrrrr}
-1) & 1 & 0 & -5 & -10 & -6 \\
3) & 1 & -1 & -4 & -6 & 0 \\
& & 3 & 6 & 6 & \\
\hline
& 1 & 2 & 2 & 0 &
\end{array}
$$

The depressed equation is

$$x^2 + 2x + 2 = 0$$

and has two roots

$$-1 + i, \quad -1 - i.$$

Thus, in addition to the two real roots -1 and 3, the proposed equation has two imaginary roots $-1 + i$ and $-1 - i$ and, these four roots being distinct, there are no other roots. At the same time we have the factorization

$$x^4 - 5x^2 - 10x - 6 = (x + 1)(x - 3)(x + 1 + i)(x + 1 - i)$$

of the given polynomial into linear factors, two of which have imaginary coefficients. However, on performing the multiplication we have

$$(x + 1 + i)(x + 1 - i) = (x + 1)^2 + 1 = x^2 + 2x + 2,$$

and so the same polynomial is factorized now into linear and quadratic factors but with real coefficients:

$$x^4 - 5x^2 - 10x - 6 = (x + 1)(x - 3)(x^2 + 2x + 2).$$

Although an equation of degree n cannot have more than n distinct roots, sometimes the number of distinct roots can be less than the degree of equation. Thus, in the equations

$$x^2 + 2x + 1 = 0,$$
$$x^3 - x^2 - x + 1 = 0,$$
$$x^4 + x^3 = 0,$$

the first has only one root -1, the second only two roots 1 and -1, and the third two 0 and -1. Notice, however, that the corresponding polynomials

$$x^2 + 2x + 1 = (x + 1)(x + 1),$$
$$x^3 - x^2 - x + 1 = (x - 1)(x - 1)(x + 1),$$
$$x^4 + x^3 = x \, x \, x \, (x + 1),$$

are factorized into 2, 3, and 4 linear factors, respectively, some of them occurring repeatedly.

Problems

1. Write a cubic equation with the roots $0, 1, 2$.

2. Write a cubic equation with the roots $1, 1 + i, 1 - i$.

3. Write a biquadratic equation with the roots $i, -i, 1 + i, 1 - i$.

4. Solve $20x^3 - 30x^2 + 12x - 1 = 0$ given that $\frac{1}{2}$ is a root.

5. One root of the cubic $x^3 - (2a + 1)x^2 + a(a + 2)x - a(a + 1) = 0$ is $a + 1$. Find the other roots.

6. Solve $2x^4 - x^3 - 17x^2 + 15x + 9 = 0$ if $1 + \sqrt{2}$ and $1 - \sqrt{2}$ are roots.

7. Find the polynomial of the lowest degree that vanishes for $x = -1, 0, 1$ and takes the value 1 for $x = 2$.

8. Find the polynomial of the lowest degree that vanishes for $x = 0, 2 + i, 2 - i$ and takes the values 1 and -1 for $x = -1$ and $x = 1$.

9. Solve $x^3 - 2(1 + i)x^2 - (1 - 2i)x + 2(1 + 2i) = 0$ given one root $1 + 2i$.

10. Solve $x^4 - (1 + 2i)x^3 + (-4 + i)x^2 + (3 + 6i)x + 3 - 3i = 0$ given two roots i and $\sqrt{3}$.

2. Identity Theorem.

From the fact that the number of distinct roots of an equation cannot exceed its degree, the following important theorem can easily be deduced:

Identity Theorem. If two polynomials $f(x)$ and $f_1(x)$, both of degree not *exceeding n, have equal values for more than n distinct values of x, then they are identical.*

PROOF. Suppose that c_1, c_2, . . . , c_m are the $m > n$ distinct numbers for which $f(x) = f_1(x)$, that is, for which

$$f(c_1) = f_1(c_1), \qquad f(c_2) = f_1(c_2), \qquad \ldots, \qquad f(c_m) = f_1(c_m).$$

If

$$F(x) = f(x) - f_1(x)$$

is not an identically vanishing polynomial, then its degree does not exceed n. However, the equation

$$F(x) = 0$$

has m roots c_1, c_2, . . . , c_m, all distinct by hypothesis. Since $m > n$, the number of distinct roots of this equation exceeds its degree, which is impossible. Hence, $F(x)$ vanishes identically and the polynomials $f(x)$ and $f_1(x)$ are identical term for term.

The theorem just proved has useful applications. Here we shall show only one application of it.

Example. By the binomial theorem, for any positive integer exponent n the following expansion holds:

$$(1 + x)^n = 1 + \frac{n}{1}x + \frac{n(n - 1)}{1 \cdot 2}x^2 + \frac{n(n - 1)(n - 2)}{1 \cdot 2 \cdot 3}x^3 + \cdots.$$

The coefficients

$$1, \frac{n}{1}, \frac{n(n - 1)}{1 \cdot 2}, \frac{n(n - 1)(n - 2)}{1 \cdot 2 \cdot 3}, \ldots$$

are the binomial coefficients. Introducing the usual notation

$$\binom{t}{r} = \frac{t(t - 1) \cdots (t - r + 1)}{1 \cdot 2 \cdots r}, \text{ for } r \geqq 1,$$

the binomial expansion can be written thus:

$$(1 + x)^n = 1 + \binom{n}{1}x + \binom{n}{2}x^2 + \cdots + \binom{n}{n}x^n.$$

Write a similar expansion for another integer exponent m:

$$(1 + x)^m = 1 + \binom{m}{1}x + \binom{m}{2}x^2 + \cdots + \binom{m}{m}x^m,$$

and multiply them together. The coefficient of any given power x^k must be the same in both members. Multiplying the right-hand sides, this coefficient is found to be

$$\binom{m}{k} + \binom{m}{k - 1}\binom{n}{1} + \binom{m}{k - 2}\binom{n}{2} + \cdots + \binom{n}{k}$$

and it must be the same as the coefficient of x^k in the expansion of the product of the left-hand sides, namely,

$$(1 + x)^n(1 + x)^m = (1 + x)^{m+n}.$$

This coefficient is easily seen to be

$$\binom{m+n}{k}.$$

Thus, we have a numerical identity

$$\binom{m}{k} + \binom{m}{k-1}\binom{n}{1} + \binom{m}{k-2}\binom{n}{2} + \cdots + \binom{n}{k} = \binom{m+n}{k} \qquad (1)$$

for arbitrary positive integers m, n, and k. Replace now the integers m and n by a variable x; the left- and right-hand sides of (1) will become polynomials

$$f(x) = \binom{x}{k} + \binom{x}{k-1}\binom{x}{1} + \cdots + \binom{x}{k},$$

$$f_1(x) = \binom{2x}{k},$$

of degree k. For all integral values $x = 1, 2, 3, \ldots$, by what has been proved, these polynomials take equal values; hence, they are identical and therefore the identity

$$\binom{x}{k} + \binom{x}{k-1}\binom{x}{1} + \cdots + \binom{x}{k} = \binom{2x}{k}$$

will hold for all values of x and not only for positive integers. In particular, taking here $x = k$ and noticing that in general

$$\binom{k}{k-i} = \binom{k}{i},$$

we find an interesting expression for the sum of squares of the binomial coefficients:

$$1 + \binom{k}{1}^2 + \left(\frac{k(k-1)}{1\cdot 2}\right)^2 + \left(\frac{k(k-1)(k-2)}{1\cdot 2\cdot 3}\right)^2 + \cdots = \frac{(k+1)(k+2)\cdots 2k}{1\cdot 2\cdots k}.$$

Here, of course, k is a positive integer.

By similar but slightly more complicated reasoning, it can be proved that for arbitrary x and y

$$\binom{x}{k} + \binom{x}{k-1}\binom{y}{1} + \binom{x}{k-2}\binom{y}{2} + \cdots + \binom{y}{k} = \binom{x+y}{k},$$

which generalizes (1). The direct proof of this identity, without resorting to the identity theorem, would not be easy.

Problems

Using the identities of this section, show that

1. $1 + \dfrac{k}{1}\dfrac{1}{2k-1} + \dfrac{k(k-1)}{1\cdot 2}\dfrac{1\cdot 3}{(2k-1)(2k-3)}$

$$+ \frac{k(k-1)(k-2)}{1\cdot 2\cdot 3}\frac{1\cdot 3\cdot 5}{(2k-1)(2k-3)(2k-5)} + \cdots$$

$$= \frac{2\cdot 4\cdot 6\cdots 2k}{1\cdot 3\cdot 5\cdots (2k-1)}$$

for any positive integer k. Take $x = -\frac{1}{2}$.

2. $1 + \dfrac{2k}{2k - 3} \cdot \dfrac{1}{2} + \dfrac{2k(2k - 2)}{(2k - 3)(2k - 5)} \dfrac{1}{2 \cdot 4}$

$$+ \frac{2k(2k - 2)(2k - 4)}{(2k - 3)(2k - 5)(2k - 7)} \frac{1 \cdot 3}{2 \cdot 4 \cdot 6} + \cdots = 2$$

for any positive integer $k > 1$. Take $x = \frac{1}{2}$.

3. Taking $x = -2$, prove that

$$1^2 + 2^2 + \cdots + k^2 = \frac{k(k + 1)(2k + 1)}{6}.$$

3. The Fundamental Theorem of Algebra.

It cannot be sufficiently emphasized that not every mathematical problem requiring something to be found is solvable. For instance, if it is required to find a real number whose square is -1, obviously such a requirement is impossible to satisfy. Therefore, when an algebraic equation is given and we seek to find its roots, even admitting complex numbers for roots, it is not evident a priori that the problem is solvable. In this case, however, all doubts are removed by a theorem which, on account of its importance, is called the fundamental theorem of algebra.

Fundamental Theorem. Every algebraic equation with arbitrarily given complex coefficients has always at least one real or imaginary root.

A great many proofs of this theorem are known, but none is sufficiently simple to be given at this place. Therefore, we shall take it as a basis for further development but without proof. The proof will be found in Appendix I.

Let $f(x)$ be a polynomial with complex coefficients, of degree n, and with the leading coefficient a_0. By the fundamental theorem there is a real or imaginary number α_1 such that $f(\alpha_1) = 0$. Then, $f(x)$ is divisible by $x - \alpha_1$, and we can set

$$f(x) = (x - \alpha_1)f_1(x),$$

$f_1(x)$ being a polynomial of degree $n - 1$ whose leading coefficient is a_0. By the same theorem the equation

$$f_1(x) = 0$$

has a real or imaginary root α_2, and, consequently,

$$f_1(x) = (x - \alpha_2)f_2(x)$$

where $f_2(x)$ is of degree $n - 2$ and with the leading coefficient a_0. If $n > 2$, the same reasoning can be repeated until we come to some polynomial of the first degree, $f_{n-1}(x)$, with the root α_n and the leading coefficient a_0, so that

$$f_{n-1}(x) = a_0(x - \alpha_n).$$

From the chain of identities

$$f(x) = (x - \alpha_1)f_1(x), \qquad f_1(x) = (x - \alpha_2)f_2(x), \qquad \cdots \, ,$$

$$f_{n-1}(x) = a_0(x - \alpha_n),$$

by successive substitutions, we find

$$f(x) = a_0(x - \alpha_1)(x - \alpha_2) \cdots (x - \alpha_n).$$

This means that every polynomial of degree n can be factorized into n linear factors, the constant a_0 not being counted among them. These factors need not be distinct. Suppose that among the numbers α_1, $\alpha_2, \ldots, \alpha_n$ we have

$$\alpha \text{ numbers equal to } a,$$
$$\beta \text{ numbers equal to } b,$$
$$\cdots \cdots \cdots \cdots \cdots$$
$$\lambda \text{ numbers equal to } l.$$

Then, combining the equal factors, we see that

$$f(x) = a_0(x - a)^\alpha (x - b)^\beta \cdots (x - l)^\lambda.$$

The numbers $a, b, \ldots, l$ represent all the distinct roots of the equation $f(x) = 0$. Their number may be less than n—the degree of the equation—whereas the total number of linear factors is n. To restore the correspondence between the number of roots and the number of linear factors the notion of a multiple root is introduced. A root a, corresponding to which the linear factor $x - a$ occurs α times, is said to be a root of *multiplicity* α and is counted as α roots equal to a. In case $\alpha = 1$ the root a is called a *simple* root; in case $\alpha = 2, 3, 4, \ldots$ it is called a *double, triple, quadruple*, etc., root. If each root is counted according to its multiplicity, the proposition that an equation of degree n always has n equal or unequal roots is universally valid. If a is a root of multiplicity α, then, in the factorization of $f(x)$ the factor $x - a$ occurs α times. Hence, $f(x)$ is divisible by $(x - a)^\alpha$, but not divisible by $(x - a)^{\alpha+1}$. For the quotient

$$\phi(x) = \frac{f(x)}{(x - a)^\alpha} = a_0(x - b)^\beta \cdots (x - l)^\lambda$$

does not vanish for $x = a$, because all the differences $a - b, \cdots, a - l$ are different from zero, and hence $\phi(x)$ is not divisible by $x - a$, which would be the case were $f(x)$ divisible by $(x - a)^{\alpha+1}$. The condition for a root a to be of multiplicity α can be expressed in a different way. Expanding $f(x)$ in powers of $x - a$, by Taylor's formula,

$$f(x) = f(a) + \frac{f'(a)}{1}(x - a) + \frac{f''(a)}{1 \cdot 2}(x - a)^2 + \cdots$$
$$+ \frac{f^{(n)}(a)}{1 \cdot 2 \cdot 3 \cdots n}(x - a)^n,$$

it is clear that the divisibility of $f(x)$ by $(x - a)^\alpha$ requires the fulfillment of the following conditions:

$$f(a) = 0, \qquad f'(a) = 0, \qquad \ldots, \qquad f^{(\alpha-1)}(a) = 0,$$

and once these conditions are satisfied,

$$f(x) = \frac{f^{(\alpha)}(a)}{1 \cdot 2 \cdots \alpha}(x - a)^\alpha + \cdots + \frac{f^{(n)}(a)}{1 \cdot 2 \cdots n}(x - a)^n.$$

Hence, if $f(x)$ is not divisible by $(x - a)^{\alpha+1}$, it must happen that $f^{(\alpha)}(a)$ is not zero. Consequently, the conditions for a root a to be of multiplicity α are that

$$f(a) = 0, \qquad f'(a) = 0, \qquad \ldots, \qquad f^{(\alpha-1)}(a) = 0,$$

but

$$f^{(\alpha)}(a) \neq 0.$$

Thus, if a is a simple root,

$$f(a) = 0 \qquad \text{but} \qquad f'(a) \neq 0;$$

if a is a double root,

$$f(a) = f'(a) = 0 \qquad \text{but} \qquad f''(a) \neq 0;$$

if a is a triple root,

$$f(a) = f'(a) = f''(a) = 0 \qquad \text{but} \qquad f'''(a) \neq 0;$$

etc.

Example 1. The equation

$$f(x) = x^n - nx + n - 1 = 0, \qquad n > 1$$

is satisfied by $x = 1$. What is the multiplicity of this root? We have

$$f'(x) = nx^{n-1} - n,$$
$$f''(x) = n(n - 1)x^{n-2}.$$

Hence,

$$f(1) = 0, \qquad f'(1) = 0, \qquad f''(1) \neq 0$$

and 1 is therefore a double root. The polynomial $f(x)$ is divisible by $(x - 1)^2$ but not by $(x - 1)^3$.

Example 2. Can any other root of the same equation be a multiple root? Suppose that some root α, differing from 1, is a multiple root. Then,

$$f(\alpha) = \alpha^n - n\alpha + n - 1 = 0, \qquad f'(\alpha) = n(\alpha^{n-1} - 1) = 0.$$

From the second condition

$$\alpha^{n-1} = 1, \qquad \alpha^n = \alpha.$$

and, substituting α for α^n in the first equation, we find

$$\alpha^n - n\alpha + n - 1 = (1 - n)(\alpha - 1) = 0.$$

But this is impossible, since $\alpha - 1 \neq 0$ and $n > 1$.

Problems

1. Write a polynomial of the lowest degree that for $x = 0$ takes the value 1 and has the following roots: 1 and -1 as simple roots, 2 as a double root, -3 as a triple root.

2. Write a polynomial of the seventh degree with 0 and 1 as double roots, -1 as triple root, if for $x = 2$ the polynomial takes the value -1.

Factorize into linear factors the following polynomials:

3. $x^3 - 1.$ **4.** $x^4 - 1.$

5. $x^6 - 1.$ **6.** $x^4 + 1.$

7. $x^3 - i.$ **8.** $x^3 - \dfrac{1 + i}{\sqrt{2}}.$

9. $x^4 + x^2 + 1.$ **10.** $x^4 - x^2 + 1.$

11. $x^8 - x^4 + 1.$ **12.** $(1 + xi)^4 + (1 - xi)^4.$

13. $x^4 + 3x^3 + x^2 - 3x - 2.$ **14.** $2x^5 - 3x^4 - 2x^3 + 4x^2 - 1.$

15. $(x + 1)^7 - x^7 - 1.$

*16. Writing $\omega = \cos (2\pi/n) + i \sin (2\pi/n)$, show that

$$x^{n-1} + x^{n-2} + \cdots + x + 1 = (x - \omega)(x - \omega^2) \cdots (x - \omega^{n-1}).$$

*17. Show that

$$\sin \frac{\pi}{n} \sin \frac{2\pi}{n} \cdots \sin \frac{(n-1)\pi}{n} = \frac{n}{2^{n-1}}.$$

HINT: Set $x = 1$ in the identity of Prob. 16 and take the absolute value of both members.

*18. Show that the roots of the polynomial

$$1 - \frac{x}{1} + \frac{x(x-1)}{1 \cdot 2} - \frac{x(x-1)(x-2)}{1 \cdot 2 \cdot 3} + \cdots + (-1)^n \frac{x(x-1) \cdots (x - n + 1)}{1 \cdot 2 \cdot 3 \cdots n}$$

are $1, 2, 3, \ldots, n$, and factorize it.

*19. If $f(0) \neq 0$ and $\alpha_1, \alpha_2, \ldots, \alpha^n$ are the roots of $f(x) = 0$, show that

$$f(x) = f(0) \left(1 - \frac{x}{\alpha_1}\right)\left(1 - \frac{x}{\alpha_2}\right) \cdots \left(1 - \frac{x}{\alpha_n}\right).$$

*20. Find the roots of the equation

$$(1 + xi)^n + (1 - xi)^n = 0,$$

and write the factorization of

$$\frac{(1 + xi)^n + (1 - xi)^n}{2}.$$

*21. Show that the only polynomial of degree $n - 1$ that vanishes for $x = x_2,$ $x_3, \ldots, x_n$ and takes the value 1 for $x = x_1$ is

$$\frac{g(x)}{(x - x_1)g'(x_1)}$$

where

$$g(x) = (x - x_1)(x - x_2) \cdots (x - x_n).$$

*22. Let $x_1, x_2, \ldots, x_n$ be n distinct numbers and

$$g(x) = (x - x_1)(x - x_2) \cdots (x - x_n).$$

Show that the following polynomial is of degree not exceeding $n - 1$:

$$f(x) = \frac{g(x)}{(x - x_1)g'(x_1)}\, y_1 + \frac{g(x)}{(x - x_2)g'(x_2)}\, y_2 + \cdots + \frac{g(x)}{(x - x_n)g'(x_n)}\, y_n$$

and for $x = x_1, x_2, \ldots, x_n$ takes the values $y_1, y_2, \ldots, y_n$. Show also that such a polynomial is unique. This formula—Lagrange's interpolation formula—solves the interpolation problem: To find a polynomial of the lowest degree that for n given values of x: $x_1, x_2, \ldots, x_n$ takes the prescribed values $y_1, y_2, \ldots, y_n$.

4. Imaginary Roots of Equations with Real Coefficients.

All previously established results hold for equations with arbitrary complex coefficients. Concerning the imaginary roots of equations with real coefficients we have the following theorem:

THEOREM. *If an equation with real coefficients has an imaginary root $a + bi$ of multiplicity α, it has also the conjugate root $a - bi$ of the same multiplicity, or, imaginary roots occur in conjugate parts.*

PROOF. Let

$$f(x) = a_0 x^n + a_1 x^{n-1} + \cdots + a_n = 0$$

be an equation with real coefficients and having an imaginary root $a + bi$ of multiplicity α. Then,

$$f(a + bi) = 0, \qquad f'(a + bi) = 0, \qquad \ldots,$$
$$f^{(\alpha-1)}(a + bi) = 0, \qquad f^{(\alpha)}(a + bi) \neq 0.$$

The first equality means

$$a_0(a + bi)^n + a_1(a + bi)^{n-1} + \cdots + a_n = 0.$$

When each number in the left-hand member is replaced by its conjugate, the result will be a number conjugate to 0, that is, 0 again (Chap. I, Sec. 7). On the other hand, $a_0, a_1, \ldots, a_n$ as real numbers coincide with their own conjugates; hence, the conjugate of the above equation is

$$a_0(a - bi)^n + a_1(a - bi)^{n-1} + \cdots + a_n = 0,$$

or

$$f(a - bi) = 0.$$

Similarly, it is proved that

$$f'(a - bi) = 0, \qquad \ldots, \qquad f^{(\alpha-1)}(a - bi) = 0,$$

and it remains to show that

$$f^{(\alpha)}(a - bi) \neq 0.$$

By hypothesis

$$f^{(\alpha)}(a + bi) = A + Bi$$

and A, B are not both equal to 0. The same reasoning as before shows that

$$f^{(\alpha)}(a - bi) = A - Bi,$$

and this number is not equal to 0.

From the theorem just proved it follows that imaginary roots of *real* equations (that is, equations with real coefficients) occur always in pairs of conjugate roots, and so their number is even. If the number of imaginary roots is $2s$ and that of the real roots is r,

$$r + 2s = n,$$

n being the degree of the equation. In case n is odd, r must be odd and therefore at least 1, which means that a real equation of an odd degree has at least one real root. In case of an even degree it can very well happen that all roots are imaginary.

To each linear factor

$$x - (a + bi) = x - a - bi$$

corresponding to an imaginary root $a + bi$, there is a companion factor

$$x - (a - bi) = x - a + bi$$

corresponding to the conjugate root $a - bi$, and their product

$$(x - a - bi)(x - a + bi) = (x - a)^2 + b^2 = x^2 - 2ax + a^2 + b^2$$

is a quadratic factor with real coefficients. Hence, it is possible to draw the conclusion that any real polynomial can be factorized into real linear and quadratic factors.

Example.　The roots of the equation

$$x^4 + 1 = 0$$

are

$$\frac{1 + i}{\sqrt{2}}, \quad \frac{1 - i}{\sqrt{2}}, \quad \frac{-1 - i}{\sqrt{2}}, \quad \frac{-1 + i}{\sqrt{2}}.$$

Since

$$\left(x - \frac{1}{\sqrt{2}} - \frac{i}{\sqrt{2}}\right)\left(x - \frac{1}{\sqrt{2}} + \frac{i}{\sqrt{2}}\right) = x^2 - x\sqrt{2} + 1,$$

$$\left(x + \frac{1}{\sqrt{2}} - \frac{i}{\sqrt{2}}\right)\left(x + \frac{1}{\sqrt{2}} + \frac{i}{\sqrt{2}}\right) = x^2 + x\sqrt{2} + 1,$$

the factorization of $x^4 + 1$ into real quadratic factors is

$$x^4 + 1 = (x^2 - x\sqrt{2} + 1)(x^2 + x\sqrt{2} + 1).$$

Problems

Factorize into real linear and quadratic factors:

1. $x^4 + 4$.

2. $x^4 + x^2 + 1$.

3. $x^4 - x^2 + 1$.

4. $x^6 - 1$.

5. $x^6 + 1$.

6. $x^4 + x^3 + x^2 + x + 1$.

Solve:

7. $x^4 - 2x^3 + 6x^2 + 22x + 13 = 0$ having the root $2 + 3i$.

8. $x^5 - 3x^4 + 4x^3 - 4x + 4$ having the root $1 + i$.

9. $x^6 - 3x^5 + 4x^4 - 6x^3 + 5x^2 - 3x + 2 = 0$ having the root i.

10. $x^7 + 2x^5 - x^4 + x^3 - 2x^2 - 1 = 0$ having the root i.

*11. $x^6 - x^5 - 8x^4 + 2x^3 + 21x^2 - 9x - 54 = 0$ having the root $\sqrt{2} + i$.

*12. Points representing the roots of the equation $3x^3 + 4x^2 + 8x + 24 = 0$ are on a circle with the center 0. Solve it.

*13. If p, q, r are real numbers and the roots of the equation $x^3 + px^2 + qx + r = 0$ have equal moduli, show that

$$p^3 r - q^3 = 0 \qquad \text{and} \qquad (p^2 - q)^2 \leqq 4q^2.$$

*14. The equation $2x^4 + x^3 - 2x - 8 = 0$ has four distinct roots of equal moduli. Solve it.

*15. The equation $6x^4 - x^3 + 10x^2 - x + 6 = 0$ has four distinct roots of equal moduli. Solve it.

5. Relations between Roots and Coefficients. Between the roots and coefficients of an equation there are relations that are important to know. In order to discover them let us consider first the expansion of the product

$$(x + b_1)(x + b_2) \cdots (x + b_n)$$

in descending powers of x, beginning with the particular cases $n = 2, 3, 4$. By direct multiplication it is found that

$$(x+b_1)(x+b_2) = x^2 + (b_1+b_2)x + b_1 b_2,$$
$$(x+b_1)(x+b_2)(x+b_3) = x^3 + (b_1+b_2+b_3)x^2 + (b_1 b_2 + b_1 b_3 + b_2 b_3)x + b_1 b_2 b_3,$$
$$(x+b_1)(x+b_2)(x+b_3)(x+b_4) = x^4 + (b_1+b_2+b_3+b_4)x^3$$
$$+ (b_1 b_2 + b_1 b_3 + b_1 b_4 + b_2 b_3 + b_2 b_4 + b_3 b_4)x^2$$
$$+ (b_1 b_2 b_3 + b_1 b_2 b_4 + b_1 b_3 b_4 + b_2 b_3 b_4)x + b_1 b_2 b_3 b_4.$$

On examining these results we observe that

1. When $n = 2$, the leading term is x^2, the coefficient of x is the sum of quantities b_1, b_2, and the term independent of x is their product.

2. When $n = 3$, the leading term is x^3, the coefficient of x^2 is the sum of quantities b_1, b_2, b_3, the coefficient of x is the sum of products of these quantities taken two at a time, and the term independent of x is their product.

3. When $n = 4$, the leading term is x^4, the coefficient of x^3 is the sum of quantities b_1, b_2, b_3, b_4, the coefficient of x^2 is the sum of products of the same taken two at a time, the coefficient of x is the sum of products of the same taken three at a time, and the term independent of x is their product.

It appears that the general law for any number n of factors is the following: Let

s_1 be the sum of quantities $b_1, b_2, \ldots, b_n$.

s_2 be the sum of products of these quantities taken two at a time.

. . . .

s_i be the sum of products of these quantities taken i at a time.

. . . .

s_n be the product of all of them.

Then,

$$P = (x + b_1)(x + b_2) \cdots (x + b_n) = x^n + s_1 x^{n-1} + s_2 x^{n-2} + \cdots$$
$$+ s_i x^{n-i} + \cdots + s_n.$$

To prove that this law is general we use the method of proof by induction. Assuming the law to hold for n factors, we shall prove that it holds for $n + 1$ factors. Once this is done, the validity of the law will be established in general. For being true for 2, 3, 4 factors, as we have seen, it will hold for 5 factors, then again for 6 factors, etc. To start the proof, multiply the assumed expression of P by $x + b_{n+1}$, getting

$$P(x+b_{n+1}) = x^{n+1} + s_1 \begin{vmatrix} x^n + s_2 \\ +b_{n+1} \end{vmatrix} \begin{vmatrix} x^{n-1} + \cdots + s_i \\ +b_{n+1}s_1 \end{vmatrix} \begin{vmatrix} x^{n+1-i} + \cdots + s_n b_{n+1}. \\ +b_{n+1}s_{i-1} \end{vmatrix}$$

Now

$$s_1 + b_{n+1}$$

is the sum of all $n + 1$ quantities $b_1, b_2, \ldots, b_{n+1}$, and $s_n b_{n+1}$ is their product as is evident from the definition of s_n. For $1 < i < n + 1$

$$s_i + b_{n+1}s_{i-1}$$

is the sum of products of quantities $b_1, b_2, \ldots, b_{n+1}$ taken i at a time. In fact, in this sum we can consider first the terms not containing b_{n+1}; their sum will be clearly s_i. The terms containing b_{n+1} are the products of b_{n+1} by the products of $i - 1$ quantities taken among $b_1, b_2, \ldots, b_n$; hence, the sum of all such terms is $b_{n+1}s_{i-1}$. Consequently, the coefficient of x^{n+1-i};

$$s_i + b_{n+1}s_{i-1},$$

is the sum of all products that can be formed taking i factors from among $b_1, b_2, \ldots, b_{n+1}$ as the law requires. Thus, this law retains its validity in passing from n factors to $n + 1$ factors, which establishes the proof by induction.

It will be convenient to introduce short and expressive notations to designate the sums previously denoted by $s_1, s_2, \ldots$. Using the sign of summation Σ, we shall denote them thus:

$$s_1 = \Sigma b_1, \qquad s_2 = \Sigma b_1 b_2, \qquad \ldots, \qquad s_i = \Sigma b_1 b_2 \cdots b_i.$$

For instance,

$$\Sigma b_1 b_2 \cdots b_i$$

means the sum that consists of all terms resulting from the *typical term* $b_1 b_2 \cdots b_i$ by replacing the indices $1, 2, \ldots, i$ by all selections of i numbers taken from $1, 2, \ldots, n$. Since there are

$$C_n^i = \frac{n(n-1) \cdots (n-i+1)}{1 \cdot 2 \cdots i},$$

such selections or combinations, the sum

$$\Sigma b_1 b_2 \cdots b_i$$

consists of that many terms.

Considering now a polynomial

$$f(x) = a_0 x^n + a_1 x^{n-1} + \cdots + a_n$$

with roots $\alpha_1, \alpha_2, \ldots, \alpha_n$ (equal or unequal), we have

$$f(x) = a_0 (x - \alpha_1)(x - \alpha_2) \cdots (x - \alpha_n).$$

On the other hand, replacing in the expressions $s_1, s_2, \ldots, s_n$ the letters $b_1, b_2, \ldots, b_n$ by $-\alpha_1, -\alpha_2, \ldots, -\alpha_n$, we have

$$(x - \alpha_1)(x - \alpha_2) \cdots (x - \alpha_n) = x^n - x^{n-1} \Sigma \alpha_1 + x^{n-2} \Sigma \alpha_1 \alpha_2$$
$$- \cdots + (-1)^n \alpha_1 \alpha_2 \cdots \alpha_n.$$

Consequently,

$$- a_0 \Sigma \alpha_1 = a_1,$$
$$+ a_0 \Sigma \alpha_1 \alpha_2 = a_2,$$
$$- a_0 \Sigma \alpha_1 \alpha_2 \alpha_3 = a_3,$$
$$\cdots \cdots \cdots \cdots \cdots$$
$$(-1)^n a_0 \alpha_1 \alpha_2 \cdots \alpha_n = a_n,$$

whence finally can be deduced the desired relations between the roots and coefficients of an equation:

$$\Sigma \alpha_1 = - \frac{a_1}{a_0},$$

$$\Sigma \alpha_1 \alpha_2 = + \frac{a_2}{a_0},$$

$$\cdots \cdots \cdots \cdots$$

$$\Sigma \alpha_1 \alpha_2 \cdots \alpha_i = (-1)^i \frac{a_i}{a_0},$$

$$\cdots \cdots \cdots \cdots \cdots$$

$$\alpha_1 \alpha_2 \cdots \alpha_n = (-1)^n \frac{a_n}{a_0}.$$

These relations can be used to solve problems that are of the following type:

Example 1. Solve the equation

$$3x^3 - 16x^2 + 23x - 6 = 0$$

if the product of two roots is 1. Let the roots be a, b, c; then,

$$a + b + c = 16\tfrac{2}{3},$$
$$ab + ac + bc = 23\tfrac{2}{3},$$
$$abc = 6\tfrac{2}{3} = 2.$$

In addition to these general relations we have to take into consideration the specific condition that the product of two roots is 1. Letters a and b may denote these roots so that

$$ab = 1.$$

Then, the third root c is found immediately from the third relation to be equal to **2**, and for a and b we have three equations

$$a + b = 16\tfrac{2}{3},$$
$$a + b = 16\tfrac{2}{3},$$
$$ab = 1,$$

two of which are identical; a and b will be roots of the quadratic equation

$$t^2 - 16\tfrac{2}{3}t + 1 = 0,$$

which has the roots 3 and $\tfrac{1}{3}$. Thus, the roots of the proposed equation are

$$2, 3, \tfrac{1}{3}.$$

Example 2. Find the sum of squares of roots of the equation

$$2x^4 - 8x^3 + 6x^2 - 3 = 0.$$

If the roots are a, b, c, d, we have

$$a + b + c + d = \tfrac{8}{2} = 4,$$
$$\Sigma ab = ab + ac + ad + bc + bd + cd = \tfrac{6}{2} = 3.$$

On the other hand,

$$(a + b + c + d)^2 = a^2 + b^2 + c^2 + d^2 + 2\Sigma ab = 4^2,$$

whence

$$a^2 + b^2 + c^2 + d^2 = 16 - 6 = 10.$$

Problems

Solve the cubic equations whose roots are a, b, c:

1. $x^3 + 2x^2 + 3x + 2 = 0$ if $a = b + c$.
2. $2x^3 - x^2 - 18x + 9 = 0$ if $a + b = 0$.
3. $3x^3 + 2x^2 - 19x + 6 = 0$ if $a + b = -1$.
4. $2x^3 - x^2 - 5x - 2 = 0$ if $ab = -1$.
5. $x^3 - 7x^2 - 42x + 216 = 0$ if $c^2 = ab$.
6. $x^3 + 9x^2 + 6x - 56 = 0$ if $b = -2a$.

7. $9x^3 - 36x^2 + 44x - 16 = 0$ if the roots form an arithmetic progression $\alpha - \beta$, α, $\alpha + \beta$.

8. $3x^3 - 26x^2 + 52x - 24 = 0$ if the roots form a geometric progression $\alpha\beta^{-1}$, α, $\alpha\beta$.

9. $2x^3 - 6x^2 + 3x + k = 0$. Determine k and solve the equation if $a = 2b + 2c$.

10. $x^3 - 2x^2 + kx + 46 = 0$. Determine k and solve the equation if the roots are in an arithmetic progression.

11. What relation exists between p, q, r if the roots of $x^3 + px^2 + qx + r = 0$ are in a geometric progression?

12. What is the relation between p and q if the equation $x^3 + px + q = 0$ has a multiple root?

13. Show that $(2q - p^2)^3 r = (pq - 4r)^3$ if the roots of $x^3 + px^2 + qx + r = 0$ satisfy the condition $c^2 = -ab$.

Solve the biquadratic equations whose roots are a, b, c, d:

14. $x^4 - 2x^3 + 2x^2 - x - 2 = 0$ if $a + b = 1$.

15. $2x^4 - 3x^3 - 9x^2 + 15x - 5 = 0$ if $a = -b$.

16. $x^4 - 7x^3 + 18x^2 - 22x + 12 = 0$ if $ab = 6$.

17. $x^4 + x^3 - 2x^2 + 3x - 1 = 0$ if $ab = -1$.

18. $2x^4 + 13x^3 + 25x^2 + 15x + 9 = 0$ if $a = b$.

19. $9x^4 + 9x^3 + 2x^2 - 14x + 4 = 0$ if $a = 2b$.

20. $4x^4 - 4x^3 - 21x^2 + 11x + 10 = 0$ if the roots are in an arithmetic progression. Represent the roots by $\alpha - 3\beta$, $\alpha - \beta$, $\alpha + \beta$, $\alpha + 3\beta$.

21. Determine k and solve the equation $2x^4 - 15x^3 + kx^2 - 30x + 8 = 0$ if its roots are in a geometric progression. The roots may be represented by $\alpha\beta^{-3}$, $\alpha\beta^{-1}$, $\alpha\beta$, $\alpha\beta^3$.

22. Find the sum of squares of roots for the equations:

$$(a) \qquad 2x^4 - 6x^3 + 5x^2 - 7x + 1 = 0;$$
$$(b) \qquad 3x^5 - 3x^3 + 2x^2 + x - 1 = 0.$$

23. For the same equations find the sum of reciprocals of roots; and also the sum of squares of these reciprocals.

If between the roots of $f(x) = 0$ there exists a relation such as $a = kb$ or $ab = k$, k being given, then $f(x)$ and $f_1(x) = f(kx)$ or $f_1(x) = x^n f(k/x)$ have common roots that can be determined by equating the highest common divisor of $f(x)$ and $f_1(x)$ to zero. Solve by this method:

24. Prob. 6.   **25.** Prob. 3.   **26.** Prob. 4.

27. Prob. 17.   **28.** Prob. 15.   **29.** Prob. 19.

6. Discovery of Multiple Roots.

Performing only rational operations, it is possible to discover whether an equation has multiple roots, to determine their multiplicities, and to reduce the finding of the roots themselves to the solution of equations with simple roots. Let a, b, $\ldots$, l be the distinct roots of an equation $f(x) = 0$ and let α, β, $\ldots$, λ be their respective multiplicities. Since a root of multiplicity k of $f(x) = 0$ is a root of multiplicity $k - 1$ (that is, no root in case $k = 1$) of the derived equation $f'(x) = 0$, it is clear that $f(x)$ and $f'(x)$ are both divisible by

$$(x - a)^{\alpha-1}(x - b)^{\beta-1} \cdots (x - l)^{\lambda-1}$$

and so will be their highest common divisor $D(x)$. We can, therefore, set

$$D(x) = (x - a)^{\alpha-1}(x - b)^{\beta-1} \cdots (x - l)^{\lambda-1}\phi(x).$$

If $\phi(x)$ is not a constant, the polynomial $\phi(x)$ will have some factor $x - m$ where m is some root of $f(x)$, say $m = a$. But then $f'(x)$ will be divisible by $(x - a)^{\alpha}$, which is impossible since a is a root of multiplicity $\alpha - 1$ for $f'(x)$. Therefore, $\phi(x)$ is a constant and so

$$(x - a)^{\alpha-1}(x - b)^{\beta-1} \cdots (x - l)^{\lambda-1}$$

is the highest common divisor of f and f'. This fact can be interpreted in a different way. Let X_1 be the product of all linear factors corresponding to simple roots, X_2 the product of all those that correspond to double roots, X_3 the product of all those that correspond to triple roots, etc., agreeing to set X_k equal to a constant if the equation has no roots of multiplicity k. Then,

$$X_1 X_2^2 X_3^3 \cdots$$

differs only by a constant factor from $f(x)$, and thus

$$D = X_2 X_3^2 X_4^3 \cdots$$

will be the highest common divisor of f and f'. Similarly,

$$D_1 = X_3 X_4^2 \cdots$$

will be the highest common divisor of D and its derivative D',

$$D_2 = X_4 \cdots$$

the highest common divisor of D_1 and D_1', etc. This sequence of highest common divisors

$$D, D_1, D_2, \cdots$$

of decreasing degrees ends with a term D_{m-1}, which is a constant. Then, it is clear that there are no roots of multiplicity higher than m. Again,

$$f_1 = \frac{f}{D} = X_1 X_2 \cdots X_m,$$

$$f_2 = \frac{D}{D_1} = X_2 \cdots X_m,$$

$$f_3 = \frac{D_1}{D_2} = X_3 \cdots X_m,$$

$$\cdots \cdots \cdots \cdots \cdots \cdots \cdots$$

$$f_m = \frac{D_{m-2}}{D_{m-1}} = X_m,$$

whence

$$\frac{f_1}{f_2} = X_1, \qquad \frac{f_2}{f_3} = X_2, \qquad \ldots, \qquad \frac{f_{m-1}}{f_m} = X_{m-1}, \qquad f_m = X_m.$$

The functions $X_1, X_2, \ldots, X_m$ found in this manner lead to the equations

$$X_1 = 0, \qquad X_2 = 0, \qquad \ldots, \qquad X_m = 0,$$

all of which have simple roots. These roots give at once the simple, double, triple, etc., roots of $f(x) = 0$. Naturally, if some X_k turns out to be a constant, this means that there are no roots of multiplicity k.

Example 1. To investigate for multiple roots:

$$f = x^5 - x^4 - 2x^3 + 2x^2 + x - 1 = 0.$$

The highest common divisor of f and f' was found in Example 2, page 49, to be

$$D = x^3 - x^2 - x + 1.$$

We seek now the highest common divisor, D_1, of D and

$$D' = 3x^2 - 2x - 1.$$

The corresponding operation is:

(Multiply by 3)	1	−1	−1	1	3	−2	−1	3	−2	−1	1 −1
	3	−3	−3	3	1	−1		3	−3		3 1
	3	−2	−1					1	−1		
(Multiply by 3)	−1	−2	3					1	−1		
	−3	−6	9						0		
	−3	2	1								
(Suppress factor − 8)	−8	8									
	1	−1									

Hence, $D_1 = x - 1$ and $D_2 = 1$, which shows that the proposed equation does not have roots of multiplicity higher than 3. To find X_1, X_2, X_3 we find the quotients

$$f_1 = \frac{f}{D} = x^2 - 1, \qquad f_2 = \frac{D}{D_1} = x^2 - 1, \qquad f_3 = \frac{D_1}{D_2} = x - 1,$$

and, hence,

$$X_1 = 1, \qquad X_2 = x + 1, \qquad X_3 = x - 1$$

so that the proposed equation has no simple roots, one double root − 1, and one triple root 1, and in fact

$$f = (x + 1)^2(x - 1)^3.$$

Example 2. To investigate for multiple roots:

$$f = x^5 - x^3 - 4x^2 - 3x - 2 = 0.$$

We begin by finding the highest common divisor D of f and f'. This operation is exhibited in full as follows:

(Multiply by 5)　　$\quad$ 1　0　−1　−4　−3　−2　$\Big|$　5　0　−3　−8　−3

$\qquad\qquad\qquad\qquad$ 5　0　−5　−20　−15　−10　$\Big|$　1

$\qquad\qquad\qquad\qquad$ 5　0　−3　−8　−3

(Suppress − 2)　$\qquad$ −2　−12　−12　−10

$\qquad\qquad\qquad\qquad\quad$ 1　6　6　5

$\qquad\qquad\qquad\qquad\quad$ $\cdots\cdots\cdots\cdots$

$\qquad\qquad\quad$ 5　0　−3　−8　−3　$\Big|$　1　6　6　5

$\qquad\qquad\quad$ 5　30　30　25　$\Big|$　5　10

(Suppress − 3)　$\quad$ −30　−33　−33　−3

$\qquad\qquad\qquad\quad$ 10　11　11　1

$\qquad\qquad\qquad\quad$ 10　60　60　50

(Suppress − 49)　$\quad$ −49　−49　−49

$\qquad\qquad\qquad\qquad$ 1　1　1

$\qquad\qquad\qquad\qquad$ $\cdots\cdots\cdots\cdots$

$\qquad\qquad\qquad\qquad$ 1　6　6　5　$\Big|$　1　1　1

$\qquad\qquad\qquad\qquad$ 1　1　1　$\Big|$　1　5

$\qquad\qquad\qquad\qquad\qquad$ 5　5　5

$\qquad\qquad\qquad\qquad\qquad$ 5　5　5

$\qquad\qquad\qquad\qquad\qquad\quad$ 0

Hence, $D = x^2 + x + 1$. The operation of finding D_1 is

(Multiply by 2)　$\quad$ 1　1　1　$\Big|$　2　1

$\qquad\qquad\qquad\quad$ 2　2　2　$\Big|$　1　1

$\qquad\qquad\qquad\quad$ 2　1

(Multiply by 2)　$\qquad$ 1　2

$\qquad\qquad\qquad\qquad$ 2　4

$\qquad\qquad\qquad\qquad$ 2　1

$\qquad\qquad\qquad\qquad\quad$ 3

Hence, D_1 = const. and we may take $D_1 = 1$, so that there are no roots of higher multiplicity than 2. Next,

$$f_1 = \frac{f}{D} = x^3 - x^2 - x - 2,$$

$$f_2 = \frac{D}{D_1} = x^2 + x + 1,$$

and

$$X_1 = \frac{f_1}{f_2} = x - 2, \qquad X_2 = f_2 = x^2 + x + 1.$$

Consequently, the proposed equation has one simple root 2, and two double roots

$$\omega = \frac{-1 + i\sqrt{3}}{2}, \qquad \omega^2 = \frac{-1 - i\sqrt{3}}{2};$$

and f is completely factorized as follows:

$$x^5 - x^4 - 2x^3 + 2x^2 + x - 1 = (x - 2)(x - \omega)^2(x - \omega^2)^2.$$

Problems

Solve the following equations each of which has multiple roots:

1. $x^3 - 7x^2 + 16x - 12 = 0.$ $\qquad\qquad$ **2.** $x^3 - 3x^2 - 9x + 27 = 0.$

3. $x^3 - x^2 - 8x + 12 = 0.$ $\qquad\qquad$ **4.** $x^3 - 5x^2 - 8x + 48 = 0.$

5. $x^3 - x - \dfrac{2}{3\sqrt{3}} = 0.$

6. $x^4 - \frac{1}{2}x + \frac{3}{16} = 0.$

7. $x^4 - 11x^2 + 18x - 8 = 0.$

8. $x^4 - 2x^3 - x^2 - 4x + 12 = 0.$

9. $x^4 - 4x^3 - 6x^2 + 36x - 27 = 0.$

10. $2x^4 - 12x^3 + 19x^2 - 6x + 9 = 0.$

11. $x^5 - x^4 - 2x^3 + 2x^2 + x - 1 = 0.$

12. $x^5 - 2x^4 - 6x^3 + 4x^2 + 13x + 6 = 0.$

13. $x^6 - 3x^5 + 6x^3 - 3x^2 - 3x + 2 = 0.$

14. $x^5 + 2x^4 - 8x^3 - 16x^2 + 16x + 32 = 0.$

15. $9x^5 + 96x^4 + 292x^3 + 48x^2 - 576x + 256 = 0.$

16. $x^7 - 3x^6 + 5x^5 - 7x^4 + 7x^3 - 5x^2 + 3x - 1 = 0.$

17. $x^8 + x^7 - 8x^6 - 6x^5 + 21x^4 + 9x^3 - 22x^2 - 4x + 8 = 0.$

CHAPTER IV

LIMITS OF ROOTS. RATIONAL ROOTS

1. Limits of Roots. An algebraic equation being given, it is often desirable to have an idea of how large its roots can be. Two cases must be considered. If the coefficients are real and we are only concerned with real roots, it may be of interest to find a number surpassing all the positive roots or a negative number smaller than all possible negative roots. Two such numbers, one positive and another negative, are called, respectively, an *upper limit* of the positive roots and a *lower limit* of the negative roots. The second case is that of an equation with complex coefficients when all its roots, real and imaginary, are taken into consideration. A positive number greater than the absolute values of all the roots may be called an upper limit of the roots. If this number is called r, then the circle with the center at the point O and radius r will contain inside all points representing the roots of the equation under consideration.

2. A Method to Find an Upper Limit of Positive Roots. Let

$$f(x) = a_0x^n + a_1x^{n-1} + \cdots + a_n = 0$$

be an equation with real coefficients of which the leading coefficient a_0 may and will be supposed positive. Of the various methods that may be used to find an upper limit of the positive roots we shall consider only one, which combines the advantages of giving comparatively low values of that limit together with ease in application.

When considering synthetic division (Chap. II, Sec. 5) we encounter polynomials

$$f_0 = a_0, \qquad f_1 = xf_0 + a_1, \qquad f_2 = xf_1 + a_2, \qquad \ldots, \qquad f_n = xf_{n-1} + a_n,$$

the last of which coincides with f. For any $i = 1, 2, \ldots, n$ we have identically

$$f_i(x) = (x - c)[f_0(c)x^{i-1} + f_1(c)x^{i-2} + \cdots + f_{i-1}(c)] + f_i(c). \quad (1)$$

The numbers

$$f_0(c), \qquad f_1(c), \qquad \ldots, \qquad f_n(c)$$

are those obtained in the process of synthetic division of the functions $f_i(x)$ by $x - c$. The method for finding an upper limit of the positive roots is based upon the following two properties of the polynomials $f_0, f_1, \ldots, f_n$: First, if for some positive number c the numbers

$$f_1(c), \qquad f_2(c), \qquad \ldots, \qquad f_{n-1}(c)$$

are nonnegative and $f_n(c) > 0$, then c can be taken as the requested upper limit of the positive roots. In fact, from the identity (1) for $i = n$ it follows then that $f_n(x) = f(x) > 0$ for $x \geqq c$, so that there cannot be any real root of the equation $f(x) = 0$ surpassing c or even equal to c. Second, if for some $c > 0$ the numbers

$$f_1(c), \qquad f_2(c), \qquad \ldots, \qquad f_k(c), \qquad k \leqq n$$

are nonnegative, then for $c' > c$

$$f_1(c'), \qquad f_2(c'), \qquad \ldots, \qquad f_k(c')$$

are positive. This again follows from the identity (1), in which we take $x = c'$; then,

$$f_i(c') = (c' - c)[f_0(c)c'^{i-1} + \cdots + f_{i-1}(c)] + f_i(c)$$

is a positive number for $i = 1, 2, \ldots, k$, since $c' > c$ and $f_0(c) = a_0 > 0$.

These two simple remarks suggest the following procedure for finding an upper limit of the positive roots: First, we start with some positive number c, preferably an integer, which makes $f_1(c)$ positive or zero. Such a number is easily found since $f_1(x)$ is of the first degree. If it turns out that all the numbers

$$f_2(c), \qquad f_3(c), \qquad \ldots, \qquad f_n(c)$$

are nonnegative, then in case $f_n(c) > 0$ we can take c for an upper limit. If it happens that $f(c) = 0$, then one root is found and the other positive roots will be less than c. But suppose that $f_{k+1}(c)$ is negative while the preceding numbers

$$f_1(c), \qquad f_2(c), \qquad \ldots, \qquad f_k(c)$$

are nonnegative. Then, the process can be repeated again, trying integers greater than c until one, say c_1, is found such that

$$f_{k+1}(c) \geqq 0.$$

At the same time all the numbers

$$f_1(c_1), \qquad f_2(c_1), \qquad \ldots, \qquad f_k(c_1)$$

will be positive. Now, if

$$f_1(c_1), \qquad f_2(c_1), \qquad \ldots, \qquad f_n(c_1)$$

are nonnegative and $f_n(c_1) > 0$, then c_1 can be taken for an upper limit. In the contrary case the process is repeated once more with a properly chosen larger integer c_2, etc. In this manner the requested limit can be found after comparatively few trials. When some coefficients are large negative numbers, it is advantageous to make preliminary trials, taking for c the values 10, 100, 1000, etc., and then reducing the limit found as much as is consistent with the method. If it is desirable to find a lower limit of the negative roots, we can first make the substitution $x = -y$ and then seek an upper limit of the positive roots of the transformed equation

$$a_0 y^n - a_1 y^{n-1} + a_2 y^{n-2} - \cdots + (-1)^n a_n = 0.$$

If this upper limit is c, then clearly $-c$ will be a lower limit of the negative roots of the original equation.

Example 1. To find an upper limit of the positive roots of the equation

$$2x^5 - 7x^4 - 5x^3 + 6x^2 + 3x - 10 = 0.$$

To make $f_1 = 2x - 7$ positive, we start with $c = 4$ and then apply the process of synthetic division as shown:

$$
\begin{array}{r|rrrrrr}
4) & 2 & -7 & -5 & 6 & 3 & -10 \\
   &   & 8 & 4 &   &   & \\
\hline
   & 2 & 1 & -1 & & & \\
\end{array}
$$

Since the third number is negative, we try 5:

$$
\begin{array}{r|rrrrrr}
5) & 2 & -7 & -5 & 6 & 3 & -10 \\
   &   & 10 & 15 & 50 & 280 & 1415 \\
\hline
   & 2 & 3 & 10 & 56 & 283 & 1405 \\
\end{array}
$$

Hence, 5 can be taken for an upper limit of the positive roots.

Example 2. To find an upper limit of the positive roots of the equation

$$x^5 - 7x^4 - 100x^3 - 1000x^2 + 10x - 50 = 0.$$

Since there are large negative coefficients we start the trials with the number 10:

$$
\begin{array}{r|rrrrrr}
10) & 1 & -7 & -100 & -1000 & 10 & -50 \\
    &   & 10 & 30   &   &   & \\
\hline
    & 1 & 3 & -70 & & & \\
\end{array}
$$

The presence of a negative number shows that we must try a larger number than 10, and so we try 20:

$$
\begin{array}{r|rrrrrr}
20) & 1 & -7 & -100 & -1000 & 10 & -50 \\
    &   & 20 & 260 &   &   & \\
\hline
    & 1 & 13 & 160 & & & \\
\end{array}
$$

and without continuing the process farther it is seen that the remaining numbers will be all positive. Thus, 20 is certainly an upper limit of the positive roots. If it is considered worth while to reduce this limit, we may examine the smaller numbers 19, 18, 17, etc. In this manner it is found that 17 can be taken for an upper limit, but not 16 if we are to satisfy the conditions imposed by the method.

Example 3. To find a lower limit of the negative roots of the equation

$$2x^6 + 20x^5 + 30x^3 + 50x + 1 = 0.$$

Set $x = -y$ and write the resulting equation for y:

$$2y^6 - 20y^5 - 30y^3 - 50y + 1 = 0.$$

After starting with $c = 10$ and trying next 11, we find

11)	2	-20	0	-30	0	-50	1
		22	22	242			
	2	2	22	212			

and conclude immediately that 11 can be taken for an upper limit of positive roots of the equation in y. Hence, -11 is a lower limit of the negative roots of the proposed equation.

Problems

Find limits of the roots for the equations:

1. $x^4 - 7x^3 + 10x^2 - 30 = 0.$
2. $x^4 - 8x^3 + 12x^2 + 16x - 50 = 0.$
3. $x^4 - 2x^3 - 3x^2 - 15x - 3 = 0.$
4. $3x^4 - 8x^3 - 9x^2 + 10x - 27 = 0.$
5. $-6x^4 + 20x^3 - 6x^2 - 5x + 10 = 0.$
6. $x^5 + 8x^4 - 14x^3 - 53x^2 + 56x - 18 = 0.$
7. $x^5 - 5x^4 - 13x^3 + 2x^2 + x - 70 = 0.$
8. $x^5 + x^4 + x^2 - 25x - 100 = 0.$
9. $x^6 - 5x^5 + x^4 + 12x^3 - 12x^2 + 1 = 0.$
10. $6x^5 + 27x^4 - 100x^3 - 200x - 50 = 0.$

3. Limit for Moduli of Roots.

Given an equation

$$f(x) = a_0x^n + a_1x^{n-1} + \cdots + a_n = 0$$

with arbitrary complex coefficients, the problem of finding an upper limit of the moduli of its roots can be made dependent on finding an upper limit of the positive roots of a certain auxiliary equation.

Let a and b be two complex numbers. Writing

$$a = (a + b) + (-b)$$

and applying the theorem concerning the absolute value of the sum (Chap. I, Sec. 7), we derive the inequality

$$|a| \leqq |a + b| + |-b| = |a + b| + |b|,$$

whence it follows that

$$|a + b| \geqq |a| - |b|.$$

Take in this inequality

$$a = a_0x^n, \qquad b = a_1x^{n-1} + \cdots + a_n,$$

x being an arbitrary complex number; then,

$$|f(x)| \geqq |a_0 x^n| - |a_1 x^{n-1} + \cdots + a_n|.$$

Now, if

$$|x| = r, \qquad |a_i| = A_i, \qquad i = 0, 1, 2, \ldots, n,$$

then,

$$|a_0 x^n| = A_0 r^n,$$

and

$$|a_1 x^{n-1} + \cdots + a_n| \leqq A_1 r^{n-1} + \cdots + A_n,$$

by the same theorem. Hence,

$$|f(x)| \geqq A_0 r^n - A_1 r^{n-1} - \cdots - A_n.$$

Let R be an upper limit of the positive roots of the auxiliary equation

$$A_0 x^n - A_1 x^{n-1} - \cdots - A_n = 0.$$

Then,

$$A_0 R^n - A_1 R^{n-1} - \cdots - A_n > 0,$$

and since

$$A_0 r^n - A_1 r^{n-1} - \cdots - A_n = r^n \left(A_0 - \frac{A_1}{r} - \cdots - \frac{A_n}{r^n} \right)$$

increases with increasing r, we shall have

$$A_0 r^n - A_1 r^{n-1} - \cdots - A_n > 0$$

for $r \geqq R$. Therefore,

$$|f(x)| > 0$$

as long as $|x| \geqq R$, and this means that the moduli of all roots of the proposed equation are less than R, and thus this number can be taken for the requested upper limit.

Example. To find an upper limit of the moduli of roots for the equation

$$2x^6 - 7x^5 - 10x^4 + 30x^3 - 60x^2 + 10x - 50 = 0.$$

The auxiliary equation in this case is

$$2x^6 - 7x^5 - 10x^4 - 30x^3 - 60x^2 - 10x - 50 = 0.$$

By the method of Sec. 2 it is found that $R = 6$ is an upper limit of its positive roots. Hence, all roots of the proposed equation have moduli less than 6.

Problems

Find limits of the moduli of roots for the equations:

1. $2x^4 - 7x^3 + 6x^2 - 5 = 0.$
2. $6x^5 - 10x^4 + 7x^3 + 8x - 10 = 0.$
3. $ix^4 + 4x^3 - (3 + 4i)x^2 + 4x - 1 - i = 0.$
4. $2x^5 - ix^3 + (5 + 5i)x^2 + (3 + 2i)x - 10 = 0.$

*5. If $a_0 \geq a_1 \geq a_2 \geq \cdots \geq a_n > 0$, show that no root of the equation

$$f(x) = a_0 x^n + a_1 x^{n-1} + \cdots + a_n = 0$$

has modulus greater than 1.

HINT: Notice that

$$|(1 - x)f(x)| \geq a_0|x|^n - [(a_0 - a_1)|x|^{n-1} + (a_1 - a_2)|x|^{n-2} + \cdots + a_n],$$

and that the right-hand side is positive if $|x| > 1$.

*6. If the coefficients of the equation

$$f(x) = a_0 x^n + a_1 x^{n-1} + \cdots + a_n = 0$$

are positive, and λ is the greatest of the numbers

$$\frac{a_1}{a_0}, \qquad \frac{a_2}{a_1}, \qquad \frac{a_3}{a_2}, \qquad \ldots, \qquad \frac{a_n}{a_{n-1}},$$

show that the moduli of the roots are not greater than λ.

HINT: Set $x = \lambda y$ and apply Prob. 5 to the equation in y.

4. Integral Roots.

In the rest of this chapter we shall consider equations with rational coefficients. Writing the coefficients of such an equation as fractions with a common denominator and multiplying the denominator times both sides of the equation, we replace the latter by an equivalent equation with integral coefficients. Let this equation be

$$f(x) = a_0 x^n + a_1 x^{n-1} + \cdots + a_n = 0.$$

It may have rational roots and the question is how to find such roots if they exist or to show their absence. We shall see that rational roots can be discovered once we can find integral roots, and thus it is necessary to explain first how integral roots can be found. Suppose that $x = c$ is an integral root so that

$$a_1 c^n + a_1 c^{n-1} + \cdots + a_{n-1}c + a_n = 0,$$

or

$$c(a_0 c^{n-1} + \cdots + a_{n-1}) = - a_n.$$

In the left-hand side both factors are integers, since c as well as a_0, $a_1, \ldots, a_{n-1}$ are integers; hence, a_n is divisible by c. Therefore, the integral roots, if there are any, are positive or negative divisors of the last term a_n. Thus, in principle the problem of finding integral roots is at once reduced to a finite number of trials: first, to exhibit all positive and negative divisors of a_n, and, then, to try each of them separately by direct substitution into the given equation. In this manner all integral roots will be found or it will be shown that there are no such roots. In practice, when the number of divisors to be tried is large, it is desirable to abbreviate trials by excluding divisors that cannot possibly be roots. To this end we determine first an upper limit of positive roots and a lower limit of negative roots and retain only those divisors that fall between these limits. Of the divisors retained, some may be excluded on the basis of the following remark: If a is any integer and c an integral root, then $f(a)$ is divisible by $c - a$. In fact, if c is a root, we have

$$f(x) = (x - c)f_1(x),$$

the coefficients of $f_1(x)$ being integers. Substituting here $x = a$, it follows that

$$f(a) = (a - c)f_1(a)$$

and, since $f_1(a)$ is an integer, $f(a)$ is divisible by $c - a$. Among the divisors to be tried there are always 1 and -1. Accordingly, we calculate $f(1)$ and $f(-1)$ by synthetic division; and if none of these numbers is zero, we exclude all divisors c such that $c - 1$ does not divide $f(1)$, or all divisors c such that $c + 1$ does not divide $f(-1)$. Then, picking some divisor d, we calculate $f(d)$ and, provided $f(d) \neq 0$, exclude further those divisors c for which $f(d)$ is not divisible by $c - d$. In this manner the number of divisors to be tried is reduced considerably. Once an integral root c is found, it is good to try it for multiplicity. Suppose it turns out to be of multiplicity α; then, other divisors will be tried for the depressed equation

$$F(x) = \frac{f(x)}{(x - c)^\alpha} = 0.$$

Example 1. Examine whether the equation

$$x^6 + 3x^5 - 36x^4 - 45x^3 + 93x^2 + 132x + 140 = 0$$

has integral roots or not. The first step is to find limits for the roots by the method of Sec. 2. It is found that the roots are less than 6 and greater than -8. Since

$$140 = 2^2 \cdot 5 \cdot 7,$$

positive divisors less than 6 are

$$1, \quad 2, \quad 4, \quad 5,$$

and negative divisors greater than -8 are

$$-1, \quad -2, \quad -4, \quad -5, \quad -7.$$

Next, 1 and -1 are tried by synthetic division:

1)	1	3	-36	-45	93	132	140
	1	4	-32	-77	16	148	$288 = f(1)$
-1)	1	3	-36	-45	93	132	140
	1	2	-38	-7	100	32	$108 = f(-1)$

Of the positive divisors, 4 must be excluded since $4 + 1 = 5$ does not divide $f(-1)$ $= 108$. Of the negative divisors, -4 must be excluded since $-4 - 1 = -5$ does not divide $f(1) = 288$. The following divisors remain to be tried:

$$2, \quad -2, \quad 5, \quad -5, \quad -7$$

We try first 2:

2)	1	3	-36	-45	93	132	140
		2	10	-52	-194	-202	-140
2)	1	5	-26	-97	-101	-70	$0 = f(2)$
		2	14	-24	-242	-686	
	1	7	-12	-121	-343	$-756 = f_1(2)$	

Hence, 2 is a simple root and the depressed equation is

$$f_1(x) = x^5 + 5x^4 - 26x^3 - 97x^2 - 101x - 70 = 0.$$

Next, we try -2:

$-2)$	1	5	-26	-97	-101	-70
		-2	-6	64	66	70
$-2)$	1	3	-32	-33	-35	$0 = f_1(-2)$
		-2	-2	68	-70	
	1	1	-34	35	$-105 = f_2(-2)$	

Thus, -2 is a simple root and the second depressed equation is

$$f_2(x) = x^4 + 3x^3 - 32x^2 - 33x - 35 = 0.$$

Next, we try 5:

5)	1	3	-32	-33	-35
		5	40	40	35
5)	1	8	8	7	$0 = f_2(5)$
		5	65	365	
	1	13	73	$372 = f_3(5)$	

Hence, 5 is a simple root and the third depressed equation is

$$f_3(x) = x^3 + 8x^2 + 8x + 7 = 0.$$

Since -5 does not divide 7, it is useless to try -5. So we try -7:

$-7)$	1	8	8	7
		-7	-7	-7
	1	1	1	$0 = f_3(-7)$

Consequently, -7 is another integral root and the fourth depressed equation

$$x^2 + x + 1 = 0$$

has imaginary roots

$$\omega = -\frac{1}{2} + i\frac{\sqrt{3}}{2}, \qquad \omega^2 = -\frac{1}{2} - i\frac{\sqrt{3}}{2}.$$

Thus, the roots of the proposed equation are

$$2, \quad -2, \quad 5, \quad -7, \quad \omega, \quad \omega^2.$$

Example 2. Examine whether the equation

$$x^5 + x^4 - 20x^3 - 44x^2 - 21x - 45 = 0$$

has integral roots or not. By the method of Sec. 2 it is found that the roots are less than 6 and greater than -5. Divisors of $45 = 3^2 \cdot 5$ contained between these limits are

$$1, \quad -1, \quad 3, \quad -3, \quad 5.$$

We try first 1 and -1:

1)	1	$1 \quad -20$	-44	-21	-45
	1	$2 \quad -18$	-62	-83	$-128 = f(1)$
$-1)$	1	$1 \quad -20$	-44	-21	-45
	1	$0 \quad -20$	-24	3	$-48 = f(-1)$

Since $3, -3, 5$, diminished by 1, divide -128, and augmented by 1, divide -48, none of the divisors $3, -3, 5$ is rejected. We try 3:

3)	1	$1 \quad -20$	-44	-21	-45
		$3 \quad 12$	-24	-204	-675
	1	$4 \quad -8$	-68	-225	$-720 = f(3)$

Thus, 3 is not a root but $-3-3 = -6$, $5-3 = 2$ are divisors of 720; hence, it is necessary to try -3 and 5. We try first -3:

$-3)$	1	$1 \quad -20$	-44	-21	-45
		$-3 \quad 6$	42	6	45
$-3)$	1	$-2 \quad -14$	-2	-15	$0 = f(-3)$
		$-3 \quad 15$	-3	15	
$-3)$	1	$-5 \quad 1$	-5	$0 = f'(-3)$	
		$-3 \quad 24$	-75		
	1	$-8 \quad 25$	-80		

Hence, -3 is a double root, and the depressed equation is

$$f_1(x) = x^3 - 5x^2 + x - 5 = 0.$$

Finally, try 5:

5)	1	-5	1	-5
		5	0	5
	1	0	1	$0 = f_1(5)$

and so 5 is a root and the second depressed equation is

$$x^2 + 1 = 0.$$

Thus, the proposed equation has the following roots: -3 double root, $5, i, -i$ simple roots.

Problems

Examine for integral roots:

1. $x^3 - 2x^2 - 25x + 50 = 0.$ **2.** $x^3 - 9x^2 + 22x - 24 = 0.$

3. $x^3 - 106x - 420 = 0.$ **4.** $x^4 - x^3 - 13x^2 + 16x - 48 = 0.$

5. $x^4 - x^3 - x^2 + 19x - 42 = 0.$

6. $x^4 + 8x^3 - 7x^2 - 49x + 56 = 0.$

7. $x^5 - 3x^4 - 9x^3 + 21x^2 - 10x + 24 = 0.$

8. $x^5 - 5x^4 + 2x^3 - 25x^2 + 21x + 270 = 0.$

9. $x^6 - 7x^5 + 11x^4 - 7x^3 + 14x^2 - 28x + 40 = 0.$

10. $x^6 + 3x^5 + 4x^4 + 3x^3 - 15x^2 - 16x + 20 = 0.$

11. Prove that, if both $f(0)$ and $f(1)$ are odd numbers, then the equation $f(x) = 0$ with integral coefficients cannot have integral roots.

12. The same holds if none of the three numbers $f(-1), f(0), f(1)$ is divisible by 3.

5. Rational Roots.

Rational roots of an equation

$$x^n + p_1 x^{n-1} + \cdots + p_n = 0,$$

with the leading coefficient 1 and the other coefficients integers, can be only integers. Let r/s be a rational root expressed in simplest terms so that r and s are relatively prime integers. On substituting this root into the equation and clearing of fractions we have

$$r^n + p_1 r^{n-1} s + \cdots + p_{n-1} r s^{n-1} + p_n s^n = 0,$$

or

$$r^n = s(- p_1 r^{n-1} - \cdots - p_n s^{n-1}).$$

Since

$$p_1 r^{n-1} + \cdots + p_n s^{n-1}$$

is an integer, r^n is divisible by s and this is possible only if $s = 1$, since the numbers r and s have no common divisors. Thus, the supposed rational root r/s is an integral root. Profiting by this remark, we see that the rational roots of an equation

$$a_0 x^n + a_1 x^{n-1} + \cdots + a_n = 0$$

can be found in the following manner: If x is a rational root,

$$y = a_0 x$$

will be a rational root of the equation

$$y^n + a_1 y^{n-1} + a_0 a_2 y^{n-2} + \cdots + a_0^{n-1} a_n = 0,$$

whose leading coefficient is 1 and the others are integers. Hence, y is an integral root and can be found (if there are integral roots) by the method of Sec. 4. Sometimes the work can be simplified by making the substitution

$$x = \frac{y}{k}$$

and choosing k as the smallest integer such that all the coefficients of the resulting equation

$$y^n + \frac{k a_1}{a_0} y^{n-1} + \frac{k^2 a_2}{a_0} y^{n-2} + \cdots + \frac{k^n a_n}{a_0} = 0$$

are integers. The choice $k = a_0$ is always possible, but sometimes a smaller value of k fulfills all the requirements.

Example 1. Find the rational roots of the equation

$$6x^4 - 7x^3 + 8x^2 - 7x + 2 = 0.$$

Setting

$$x = \frac{y}{6},$$

the transformed equation in y is

$$f(y) = y^4 - 7y^3 + 48y^2 - 252y + 432 = 0.$$

This equation has no negative roots, and all its positive roots are less than 7. Positive divisors of $432 = 2^4 \cdot 3^3$ below this limit are

$$1, \quad 2, \quad 3, \quad 4, \quad 6.$$

Since $f(1) = 222$ is not divisible by $6 - 1 = 5$, we need not try 6. On trying 2, 3, 4 it is found:

2)	1	-7	48	-252	432
		2	-10	76	-352
	1	-5	38	-176	$80 = f(2)$

3)	1	-7	48	-252	432
		3	-12	108	-432
	1	-4	36	-144	$0 = f(3)$

4)	1	-4	36	-144	
		4	0	144	
	1	0	36	$0 = f(4)$	

Hence, 3 and 4 are the only integral values of y, corresponding to which we have two rational roots of the proposed equation

$$\tfrac{3}{6} = \tfrac{1}{2}, \qquad \tfrac{4}{6} = \tfrac{2}{3}.$$

Example 2. Find the rational roots of the equation

$$25x^4 - 70x^3 - 126x^2 + 414x - 243 = 0.$$

After substituting

$$x = \frac{y}{k},$$

the equation in y is

$$y^4 - {}^{70}\!/_{25}\, ky^3 - {}^{126}\!/_{25}\, k^2y^2 + {}^{414}\!/_{25}\, k^3y - {}^{243}\!/_{25}\, k^4 = 0,$$

and it suffices to take $k = 5$ to make all its coefficients integral. With this choice of k the equation will be

$$f(y) = y^4 - 14y^3 - 126y^2 + 2070y - 6075 = 0.$$

All roots are contained between -15 and 21, and the divisors of $6075 = 3^5 \cdot 5^2$ between these limits are

$$1, \quad 3, \quad 9, \quad 5, \quad 15; \quad -1, \quad -3, \quad -9, \quad -5.$$

On calculating $f(1) = -4144$ and $f(-1) = -8256$ we reject -5, 9, and -9. Next, 3 and -3 are tried:

$$
\begin{array}{rrrrrr}
3) & 1 & -14 & -126 & 2070 & -6075 \\
& & 3 & -33 & -477 & 4779 \\
\hline
& 1 & -11 & -159 & 1593 & -1296 = f(3) \\
-3) & 1 & -14 & -126 & 2070 & -6075 \\
& & -3 & 51 & 225 & -6885 \\
\hline
& 1 & -17 & -75 & 2295 & -12960 = f(3)
\end{array}
$$

It remains to try 5 and 15:

$$
\begin{array}{rrrrrr}
5) & 1 & -14 & -126 & 2070 & -6075 \\
& & 5 & -45 & -855 & 6075 \\
15) & 1 & -9 & -171 & 1215 & 0 = f(5) \\
\hline
& & 15 & 90 & -1215 & \\
\hline
& 1 & 6 & -81 & 0 &
\end{array}
$$

Hence, $y = 5$ and $y = 15$ are roots of the auxiliary equation in y, and

$$ x = {}^{15}\!/_5 = 3, \qquad x = {}^5\!/_5 = 1 $$

are the only rational roots of the proposed equation. They could have been found with less computation if we had first sought the integral roots and after that passed to the examination of fractional roots.

Problems

Examine for rational roots the following equations:

1. $3x^3 - 26x^2 + 34x - 12 = 0.$ **2.** $2x^3 + 12x^2 + 13x + 15 = 0.$

3. $6x^3 - x^2 + x - 2 = 0.$ **4.** $10x^3 + 19x^2 - 30x + 9 = 0.$

5. $2x^3 - x^2 + 1 = 0.$ **6.** $x^3 - 3x + 1 = 0.$

7. $6x^4 - 11x^3 - x^2 - 4 = 0.$ **8.** $4x^4 - 11x^2 + 9x - 2 = 0.$

9. $2x^4 - 4x^3 + 3x^2 - 5x - 2 = 0.$

10. $6x^5 + x^4 - 14x^3 + 4x^2 + 5x - 2 = 0.$

11. $6x^5 + 11x^4 - x^3 + 5x - 6 = 0.$

12. $2x^6 + x^5 - 9x^4 - 6x^3 - 5x^2 - 7x + 6 = 0.$

*13. If a polynomial of degree ≤ 5 with rational coefficients has multiple roots, it has also a rational root, except in the case when the degree is 4 and the polynomial is a perfect square.

*14. How can one use this remark to discover the existence of multiple roots of equations whose degree does not exceed 5? Consider the examples:

(a) $x^5 - 2x^4 - 6x^3 + 4x^2 + 13x + 6 = 0;$

(b) $3x^5 - x^4 + 6x^3 - 2x^2 + 3x - 1 = 0.$

CHAPTER V

CUBIC AND BIQUADRATIC EQUATIONS

1. What is the "Solution" of an Equation? The main problem of algebra consists in the "solution" of algebraic equations, and it is important to understand clearly what is meant by this. The solution of an equation involves the determination of all its roots, whether real or imaginary, either exactly or approximately to any desired degree of approximation. Naturally, the difficulty of solving equations increases with their degree, if not for other reasons, because the higher the degree the more roots one has to compute. For equations of the first degree

$$ax + b = 0,$$

the solution is given by the formula

$$x = -\frac{b}{a},$$

which shows what arithmetical operation has to be performed on the arbitrary coefficients in order to present the root either exactly or approximately with any desired approximation. The solution of equations of the second degree,

$$ax^2 + bx + c = 0,$$

is given by the formula

$$\frac{-b \pm \sqrt{b^2 - 4ac}}{2a}$$

which shows clearly the nature of the operations to be performed on the arbitrary coefficients in order to obtain the value of roots to any desired approximation, if not exactly. On examining the formula we see that to calculate the roots of quadratic equations, besides rational operations, it is necessary to "extract" the square root of a given number. Now the extraction of a square root amounts again to the solution of a quadratic equation but of the following very special type:

$$x^2 = A,$$

and so the solution of a general quadratic equation by the foregoing formula is really a reduction of the original problem to another similar but simpler problem. To solve this simpler problem of "extracting" square root, either exactly or to any desired approximation, there are systematic methods not more difficult in applications than multiplication and division. This leads us to think of quadratic radicals $\sqrt{A}$ as something familiar and "known." Again, there are systematic methods for "extracting" cube roots of real numbers, that is, of solving special cubic equations of the form

$$x^3 = A.$$

This fact induces us to consider radicals $\sqrt[3]{A}$ as something familiar and "known." More generally, the root of equations of the type

$$x^n = A,$$

where A is a real number, can be readily found approximately by using, for instance, logarithmic tables. Even when A is a complex number, the values of its nth root $\sqrt[n]{A}$, as we have seen in Chap. I, can be computed approximately by means of tables of logarithms of numbers and trigonometric functions. Thus, we come to regard the radicals $\sqrt[n]{A}$ as familiar, readily computable, quantities. The solution of an equation that exhibits its roots by a combination of rational operations and root extractions is called an *algebraic solution* or *solution by radicals*. Thus, for instance, the equation

$$x^4 + x^3 + x^2 + x + 1 = 0$$

can be solved algebraically and its roots presented in radical form:

$$\frac{-1 + \sqrt{5}}{4} \pm i\frac{\sqrt{10 + 2\sqrt{5}}}{4}, \qquad \frac{-1 - \sqrt{5}}{4} \pm i\frac{\sqrt{10 - 2\sqrt{5}}}{4}.$$

A quadratic equation can be solved algebraically no matter what values are attributed to its coefficients. But what about cubic, biquadratic, and equations of higher degree? Can they be solved algebraically for arbitrary values of their coefficients? As far as cubic and biquadratic equations are concerned, it was shown by Italian algebraists (Scipio Ferro, Tartaglia, Cardan, and Ferrari) in the first half of the sixteenth century that they can be solved algebraically and their roots presented in radical form for arbitrary values of the coefficients. But all attempts during the following two centuries to find an algebraic solution of "general" (that is, without specializing coefficients in any way whatsoever) equations of degree higher than the fourth failed. The cause of this failure lies in the very nature of things and was not due to the oversight or the lack of inventiveness of those who occupied themselves with this problem. It was proved early in the nineteenth century first by Ruffini (whose proof, it is true, was not complete) and afterward by Abel that it is absolutely impossible to express, by means of a formula involving only rational operations and root extractons, the roots of an equation of degree higher than fourth, when the coefficients are left arbitrary. The proof of this impossibility belongs to higher parts of algebra and cannot be attempted in this course. As to the algebraic solution of cubic and biquadratic equations, the theory is comparatively simple and will be explained in this chapter and again resumed from a higher point of view in Chap. XII.

2. Cardan's Formulas. There is no loss of generality in taking the general cubic equation in the form

$$f(x) = x^3 + ax^2 + bx + c = 0$$

since division by the coefficient of x^3 does not modify the roots of the equation. By introduction of a new unknown this equation can be simplified, moreover, so that it will not contain the second power of the unknown. To this end we set

$$x = y + k$$

with k still arbitrary. By Taylor's formula

$$f(y + k) = f(k) + f'(k)y + \frac{f''(k)}{2} y^2 + \frac{f'''(k)}{6} y^3$$

and

$$f(k) = k^3 + ak^2 + bk + c, \qquad f'(k) = 3k^2 + 2ak + b,$$
$$\tfrac{1}{2}f''(k) = 3k + a, \qquad\qquad \tfrac{1}{6}f'''(k) = 1.$$

To get rid of the term involving y^2 it suffices to choose k so that

$$3k + a = 0 \qquad \text{or} \qquad k = -\frac{a}{3}.$$

Since then

$$f'\left(-\frac{a}{3}\right) = b - \frac{a^2}{3}, \qquad f\left(-\frac{a}{3}\right) = c - \frac{ba}{3} + \frac{2a^3}{27},$$

it follows that, after substituting

$$x = y - \frac{a}{3},$$

the proposed equation is transformed into

$$y^3 + py + q = 0 \tag{1}$$

where

$$p = b - \frac{a^2}{3}, \qquad q = c - \frac{ba}{3} + \frac{2a^3}{27}.$$

A cubic equation of the form (1) can be solved by means of the following device: We seek to satisfy it by setting

$$y = u + v,$$

thus introducing two unknowns u and v. On substituting this expression into (1) and arranging terms in a proper way, u and v have to satisfy the equation

$$u^3 + v^3 + (p + 3uv)(u + v) + q = 0 \qquad (2)$$

with two unknowns. This problem is indeterminate unless another relation between u and v is given. For this relation we take

$$3uv + p = 0$$

or

$$uv = -\frac{p}{3}.$$

Then, it follows from (2) that

$$u^3 + v^3 = -q,$$

so that the solution of the cubic (1) can be obtained by solving the system of two equations

$$u^3 + v^3 = -q, \qquad uv = -\frac{p}{3}. \qquad (3)$$

Taking the cube of the latter equation, we have

$$u^3 v^3 = -\frac{p^3}{27} \qquad (4)$$

and so, from equations (3) and (4), we know the sum and the product of the two unknown quantities u^3 and v^3. These quantities are the roots of the quadratic equation

$$t^2 + qt - \frac{p^3}{27} = 0.$$

Denoting them by A and B, we have then

$$A = -\frac{q}{2} + \sqrt{\frac{q^2}{4} + \frac{p^3}{27}},$$

$$B = -\frac{q}{2} - \sqrt{\frac{q^2}{4} + \frac{p^3}{27}},$$

where we are at liberty to choose the square root as we please. Now owing to the symmetry between the terms u^3 and v^3 in the system (3) we can set

$$u^3 = A, \qquad v^3 = B.$$

If some determined value of the cube root of A is denoted by $\sqrt[3]{A}$, the three possible values of u will be

$$u = \sqrt[3]{A}, \qquad u = \omega\sqrt[3]{A}, \qquad v = \omega^2\sqrt[3]{A}$$

where

$$\omega = \frac{-1 + i\sqrt{3}}{2}$$

is an imaginary cube root of unity. As to v it will have also three values:

$$v = \sqrt[3]{B}, \qquad v = \omega\sqrt[3]{B}, \qquad v = \omega^2\sqrt[3]{B}$$

but not every one of them can be associated with the three possible values of u, since u and v must satisfy the relation

$$uv = -\frac{p}{3}.$$

If $\sqrt[3]{B}$ stands for that cube root of B which satisfies the relation

$$\sqrt[3]{A} \cdot \sqrt[3]{B} = -\frac{p}{3},$$

then, the values of v that can be associated with

$$u = \sqrt[3]{A}, \qquad u = \omega\sqrt[3]{A}, \qquad u = \omega^2\sqrt[3]{A}$$

will be

$$v = \sqrt[3]{B}, \qquad v = \omega^2\sqrt[3]{B}, \qquad v = \omega\sqrt[3]{B}.$$

Hence, equation (1) will have the following roots:

$$y_1 = \sqrt[3]{A} + \sqrt[3]{B},$$
$$y_2 = \omega\sqrt[3]{A} + \omega^2\sqrt[3]{B},$$
$$y_3 = \omega^2\sqrt[3]{A} + \omega\sqrt[3]{B}.$$

These formulas are known as Cardan's formulas after the name of the Italian algebraist Cardan (1501–1576), who was the first to publish them. It must be remembered that $\sqrt[3]{A}$ can be taken arbitrarily among the three possible cube roots of A, but $\sqrt[3]{B}$ must be so chosen that

$$\sqrt[3]{A} \cdot \sqrt[3]{B} = -\frac{p}{3}.$$

3. Discussion of Solution. In discussing Cardan's formulas we shall suppose that p and q are real numbers. Then, the nature of the roots will be shown to depend on the function

$$\Delta = 4p^3 + 27q^2$$

Obviously, Δ will be positive, zero, or negative. Supposing first Δ to be positive, the square root

$$\sqrt{\frac{q^2}{4} + \frac{p^3}{27}} = \sqrt{\frac{\Delta}{108}}$$

will be real, and we shall take it positively. Then, A and B will be real, and by $\sqrt[3]{A}$ we shall mean the real cube root of A. Since p is real and

$$\sqrt[3]{A} \cdot \sqrt[3]{B} = -\frac{p}{3},$$

$\sqrt[3]{B}$ will be the real cube root of B. Hence, equation (1) has a real root

$$y_1 = \sqrt[3]{A} + \sqrt[3]{B},$$

but the other two roots,

$$y_2 = \omega\sqrt[3]{A} + \omega^2\sqrt[3]{B} = -\frac{\sqrt[3]{A} + \sqrt[3]{B}}{2} + i\sqrt{3}\frac{\sqrt[3]{A} - \sqrt[3]{B}}{2},$$

$$y_3 = \omega^2\sqrt[3]{A} + \omega\sqrt[3]{B} = -\frac{\sqrt[3]{A} + \sqrt[3]{B}}{2} - i\sqrt{3}\frac{\sqrt[3]{A} - \sqrt[3]{B}}{2},$$

will be imaginary conjugates since A and B are not equal and consequently

$$\sqrt[3]{A} - \sqrt[3]{B} \neq 0.$$

Example 1. Let the proposed cubic equation be

$$x^3 + x^2 - 2 = 0.$$

First, it must be transformed by the substitution

$$x = y - \tfrac{1}{3}.$$

The resulting equation for y (best found by synthetic division) is

$$y^3 - \tfrac{1}{3}y - \tfrac{52}{27} = 0,$$

so that

$$p = -\frac{1}{3}, \qquad q = -\frac{52}{27}, \qquad \Delta = \frac{52^2}{27} - \frac{4}{27} = 100.$$

Hence,

$$\sqrt{\frac{\Delta}{108}} = \frac{5}{\sqrt{27}}, \qquad A = \frac{26}{27} + \frac{5\sqrt{27}}{27}, \qquad B = \frac{26}{27} - \frac{5\sqrt{27}}{27},$$

$$\sqrt[3]{A} = \tfrac{1}{3}\sqrt[3]{26 + 15\sqrt{3}}, \qquad \sqrt[3]{B} = \tfrac{1}{3}\sqrt[3]{26 - 15\sqrt{3}},$$

and

$$y_1 = \frac{1}{3}\left(\sqrt[3]{26 + 15\sqrt{3}} + \sqrt[3]{26 - 15\sqrt{3}}\right),$$

$$y_2 = -\frac{1}{6}\left(\sqrt[3]{26 + 15\sqrt{3}} + \sqrt[3]{26 - 15\sqrt{3}}\right) + \frac{i\sqrt{3}}{6}\left(\sqrt[3]{26 + 15\sqrt{3}} - \sqrt[3]{26 - 15\sqrt{3}}\right),$$

$$y_3 = -\frac{1}{6}\left(\sqrt[3]{26 + 15\sqrt{3}} + \sqrt[3]{26 - 15\sqrt{3}}\right) - \frac{i\sqrt{3}}{6}\left(\sqrt[3]{26 + 15\sqrt{3}} - \sqrt[3]{26 - 15\sqrt{3}}\right).$$

Correspondingly, the roots of the proposed equation are

$$x_1 = \frac{1}{3}\left(\sqrt[3]{26 + 15\sqrt{3}} + \sqrt[3]{26 - 15\sqrt{3}} - 1\right),$$

$$x_2 = -\frac{1}{6}\left(\sqrt[3]{26 + 15\sqrt{3}} + \sqrt[3]{26 - 15\sqrt{3}} + 2\right) + \frac{i\sqrt{3}}{6}\left(\sqrt[3]{26 + 15\sqrt{3}} - \sqrt[3]{26 - 15\sqrt{3}}\right),$$

$$x_3 = -\frac{1}{6}\left(\sqrt[3]{26 + 15\sqrt{3}} + \sqrt[3]{26 - 15\sqrt{3}} + 2\right) - \frac{i\sqrt{3}}{6}\left(\sqrt[3]{26 + 15\sqrt{3}} - \sqrt[3]{26 - 15\sqrt{3}}\right).$$

The equation

$$x^3 + x^2 - 2 = 0$$

has, however, an integral root 1 and the remaining two roots,

$$-1 \pm i,$$

are imaginary.

On comparison with the expressions obtained from Cardan's formulas we discover the rather curious fact that

$$\sqrt[3]{26 + 15\sqrt{3}} + \sqrt[3]{26 - 15\sqrt{3}} = 4.$$

although both cube roots are irrational numbers. The explanation of this follows from a comparison of the imaginary roots. This comparison gives for the difference of the same cube roots

$$\sqrt[3]{26 + 15\sqrt{3}} - \sqrt[3]{26 - 15\sqrt{3}} = 2\sqrt{3},$$

whence

$$\sqrt[3]{26 + 15\sqrt{3}} = 2 + \sqrt{3}, \qquad \sqrt[3]{26 - 15\sqrt{3}} = 2 - \sqrt{3}.$$

It follows that $26 + 15\sqrt{3}$ and $26 - 15\sqrt{3}$ are cubes of numbers $2 + \sqrt{3}$ and $2 - \sqrt{3}$. Such simplification of cube roots occurs always when the cubic equation has a rational root but not otherwise.

Example 2. To solve the equation

$$x^3 + 9x - 2 = 0.$$

Here the preliminary transformation is not necessary and Cardan's formulas can be applied directly. We have

$$p = 9, \qquad q = -2, \qquad \Delta = 3024, \qquad \frac{\Delta}{108} = 28,$$

$$A = 1 + \sqrt{28}, \qquad B = 1 - \sqrt{28}.$$

Consequently, the real root is

$$\sqrt[3]{\sqrt{28} + 1} - \sqrt[3]{\sqrt{28} - 1}$$

while the imaginary roots are

$$-\frac{1}{2}\left(\sqrt[3]{\sqrt{28} + 1} - \sqrt[3]{\sqrt{28} - 1}\right) \pm \frac{i\sqrt{3}}{2}\left(\sqrt[3]{\sqrt{28} + 1} + \sqrt[3]{\sqrt{28} - 1}\right).$$

To calculate these roots approximately, use can be made of any of the well-known handbooks that contain tables of squares, cubes, square roots, and cube roots. From such tables it is found that

$$\sqrt{28} + 1 = 6.2915026, \qquad \sqrt{28} - 1 = 4.2915026,$$

and further

$$\sqrt[3]{\sqrt{28} + 1} = 1.8460840, \qquad \sqrt[3]{\sqrt{28} - 1} = 1.6250615.$$

Hence, the real root of the proposed equation is approximately

$$1.8460840 - 1.6250615 = 0.2210225$$

with the degree of approximation allowed by seven-place tables.

In case $\Delta = 0$

$$A = B = -\frac{q}{2},$$

and the roots of the equation

$$y^3 + py + q = 0$$

are

$$y = 2\sqrt[3]{-\frac{q}{2}}, \qquad y_2 = \sqrt[3]{\frac{q}{2}}, \qquad y_3 = \sqrt[3]{\frac{q}{2}}.$$

Thus, $y_2 = y_3$ is a double root unless $q = 0$, which implies $p = 0$, when all three roots are equal to 0 and the equation

$$y^3 = 0$$

is trivial.

Problems

Solve the cubic equations:

1. $x^3 - 6x - 6 = 0$.
2. $x^3 - 12x - 34 = 0$.
3. $x^3 + 9x - 6 = 0$.
4. $x^3 + 18x - 6 = 0$.
5. $2x^3 + 6x + 3 = 0$.
6. $2x^3 - 3x + 5 = 0$.
7. $3x^3 - 6x^2 - 2 = 0$.
8. $x^3 + 6x^2 - 36 = 0$.
9. $x^3 + 3x^2 + 9x + 14 = 0$.
10. $x^3 + 6x^2 + 6x + 5 = 0$.
11. $x^3 - 15x^2 + 105x - 245 = 0$.
12. $8x^3 + 12x^2 + 102x - 47 = 0$.
13. $x^3 - 2x + 2 = 0$.
14. $x^3 + 3x - 2 = 0$.
15. $x^3 + 6x^2 + 9x + 8 = 0$.
16. $8x^3 + 12x^2 + 30x - 3 = 0$.

Show that

17. $\sqrt[3]{\sqrt{5} + 2} - \sqrt[3]{\sqrt{5} - 2} = 1$.
18. $\sqrt[3]{7 + \sqrt{50}} + \sqrt[3]{7 - \sqrt{50}} = 2$.
19. $\sqrt[3]{\sqrt{108} + 10} - \sqrt[3]{\sqrt{108} - 10} = 2$.
20. $\sqrt[3]{\sqrt{243} + \sqrt{242}} - \sqrt[3]{\sqrt{243} - \sqrt{242}} = 2\sqrt{2}$.

21. What is the outer radius of a spherical shell of thickness 1 in. if the volume of the shell is equal to the volume of the hollow space inside?

22. Solve Prob. 21 if the volume of the hollow space is twice the volume of the shell.

23. An open box has the shape of a cube with each edge 10 in. long. If the capacity of the box is 500 cu. in., what is the thickness of the walls? The walls are supposed to be uniformly thick.

4. Irreducible Case. We return now to the discussion of the general solution and consider what happens when $\Delta < 0$. A curious phenomenon occurs here, for in this case

$$\sqrt{\frac{q^2}{4} + \frac{p^3}{27}} = i \sqrt{\frac{-\Delta}{108}}$$

is purely imaginary and both numbers

$$A = -\frac{q}{2} + i \sqrt{\frac{-\Delta}{108}}, \qquad B = -\frac{q}{2} - i \sqrt{\frac{-\Delta}{108}}$$

are imaginary, so that the roots of equation (1) in Sec. 2 are expressed through the cube roots of imaginary numbers, and yet all three of them are real. To see this let

$$\sqrt[3]{A} = a + bi$$

be one of the cube roots of A. Since B is conjugate to A, the number $a - bi$ will be one of the cube roots of B, and it must be taken equal to $\sqrt[3]{B}$ in order to satisfy the condition

$$\sqrt[3]{A} \cdot \sqrt[3]{B} = -\frac{p}{3}.$$

Thus,

$$\sqrt[3]{A} = a + bi, \qquad \sqrt[3]{B} = a - bi,$$

and from Cardan's formulas it follows that the roots

$$\begin{aligned} y_1 &= 2a, \\ y_2 &= (a + bi)\omega + (a - bi)\omega^2 = -a - b\sqrt{3}, \\ y_3 &= (a + bi)\omega^2 + (a - bi)\omega = -a + b\sqrt{3} \end{aligned}$$

are real and, moreover, unequal. It is clear that $y_2 \neq y_3$. If $y_1 = y_2$, we should have

$$b = -a\sqrt{3},$$

so that

$$\sqrt[3]{A} = a(1 - i\sqrt{3}).$$

But then

$$A = a^3(1 - i\sqrt{3})^3 = -8a^3$$

would be real, which is not true. Similarly, it is shown that $y_1 \neq y_3$.

Example. To solve the equation

$$y^3 - 3y + 1 = 0.$$

In this case

$$p = -3, \qquad q = 1, \qquad \Delta = -81, \qquad \sqrt{\frac{-\Delta}{108}} = \frac{\sqrt{3}}{2},$$

$$A = -\frac{1}{2} + i\frac{\sqrt{3}}{2} = \omega, \qquad B = -\frac{1}{2} - i\frac{\sqrt{3}}{2} = \omega^2,$$

and the real roots are presented in the form

$$y_1 = \sqrt[3]{\omega} + \sqrt[3]{\omega^2},$$
$$y_2 = \omega\sqrt[3]{\omega} + \omega^2\sqrt[3]{\omega^2},$$
$$y_3 = \omega^2\sqrt[3]{\omega} + \omega\sqrt[3]{\omega^2}.$$

These expressions are not suitable for direct calculation because of the cube roots of imaginary numbers. If we try to find

$$\sqrt[3]{\omega} = a + bi$$

algebraically, we are led to the solution of the two simultaneous equations

$$a^3 - 3ab^2 = -\frac{1}{2}, \qquad 3a^2b - b^3 = \frac{\sqrt{3}}{2}.$$

Solving for b^2 in the first and substituting

$$b^2 = \frac{2a^3 + 1}{6a}$$

into the second, we find

$$b\left(3a^2 - \frac{2a^3 + 1}{6a}\right) = \frac{\sqrt{3}}{2},$$

whence

$$b = \frac{3\sqrt{3}a}{16a^3 - 1}, \qquad b^2 = \frac{27a^2}{(16a^3 - 1)^2}.$$

Equating the two expressions for b^2, we have the equation

$$\frac{2a^3 + 1}{6a} = \frac{27a^2}{(16a^3 - 1)^2},$$

which after due simplifications becomes

$$(2a)^9 + 3(2a)^6 - 24(2a)^3 + 1 = 0.$$

Setting $x = 8a^3$, we have for x a cubic equation

$$x^3 + 3x^2 - 24x + 1 = 0.$$

which by the substitution $x = y - 1$ is transformed into

$$y^3 - 27y - 27 = 0;$$

or, setting $y = -3z$, into

$$z^3 - 3z + 1 = 0.$$

But this is the same equation that we wanted to solve. Consequently, we did not advance a step in trying to find a and b by an algebraic process. The fact that the real roots of a cubic equation

$$y^3 + py + q = 0$$

in case

$$4p^3 + 27q^2 < 0$$

are presented in a form involving the cube roots of imaginary numbers puzzled the old algebraists for a long time, and this case was called by them *casus irreducibilis*, irreducible case. We know now that, for instance, when p and q are rational numbers, but among the three real roots of an equation

$$y^3 + py + q = 0$$

none is rational, it is absolutely impossible to express any of these roots in a form involving only real radicals of any kind.

5. Trigonometric Solution. In spite of the algebraic difficulties inherent in the irreducible case, it is possible to present the roots in a form suitable for numerical computation by extracting the cube root of

$$A = -\frac{q}{2} + i\sqrt{-\frac{q^2}{4} - \frac{p^3}{27}}$$

trigonometrically. The square of the modulus of A is

$$\rho^2 = \left(-\frac{q}{2}\right)^2 - \left(\frac{q}{2}\right)^2 - \frac{p^3}{27} = \left(\frac{-p}{3}\right)^3$$

whence

$$\rho = \left(\frac{-p}{3}\right)^{3/2} = \frac{-p\sqrt{-p}}{\sqrt{27}}.$$

The argument of A can be determined either by its cosine

$$\cos \phi = \frac{\sqrt{27}q}{2p\sqrt{-p}}$$

or by its tangent

$$\tan \phi = \frac{-\sqrt{-\Delta}}{q\sqrt{27}}$$

on the condition that ϕ is taken in the first or second quadrant according as q is negative or positive. Having found ρ and ϕ, we can take

$$\sqrt[3]{A} = \sqrt[3]{\rho}\left(\cos\frac{\phi}{3} + i\sin\frac{\phi}{3}\right) = \sqrt{\frac{-p}{3}}\left(\cos\frac{\phi}{3} + i\sin\frac{\phi}{3}\right),$$

and

$$\sqrt[3]{B} = \sqrt{\frac{-p}{3}}\left(\cos\frac{\phi}{3} - i\sin\frac{\phi}{3}\right).$$

Then, since

$$\omega = \cos 120° + i\sin 120°,$$

the roots y_1, y_2, y_3 will be given by

$$y_1 = 2\sqrt{\frac{-p}{3}}\,\cos\frac{\phi}{3},$$

$$y_2 = 2\sqrt{\frac{-p}{3}}\,\cos\left(\frac{\phi}{3} + 120°\right),$$

$$y_3 = 2\sqrt{\frac{-p}{3}}\,\cos\left(\frac{\phi}{3} + 240°\right).$$

In practice it is more convenient to present y_2 and y_3 thus:

$$y_2 = -2\sqrt{\frac{-p}{3}}\cos\left(60° - \frac{\phi}{3}\right),$$

$$y_3 = -2\sqrt{\frac{-p}{3}}\cos\left(60° + \frac{\phi}{3}\right).$$

Example 1. For the equation

$$y^3 - 3y + 1 = 0$$

we have

$$\sqrt{\frac{-p}{3}} = 1, \qquad \cos\phi = -\frac{1}{2};$$

hence, $\phi = 120°$ and

$$y_1 = 2\cos 40°, \qquad y_2 = -2\cos 20°, \qquad y_3 = 2\cos 80°.$$

Approximate values of the roots can be found directly from the trigonometric tables and are

$$y_1 = 1.5320888862,$$
$$y_2 = -1.8793852416,$$
$$y_3 = 0.3472963554.$$

Example 2. To solve the equation

$$y^3 - 7y - 7 = 0.$$

For this equation

$$p = -7, \qquad q = -7, \qquad -\Delta = 49$$

and

$$\tan\phi = \frac{1}{\sqrt{27}}.$$

Further computation was made with six-place tables of logarithms. The calculations and results can be presented as follows:

$\log 27 = 1.431364 \qquad\qquad \log 7 = 0.845098 \quad \log\cos\dfrac{\phi}{3} = 9.999128 - 10$

$\log\sqrt{27} = 0.715682 \qquad\qquad \log 3 = 0.477121 \qquad\qquad 0.485018$

$\log\tan\phi = 9.284318 \qquad \overline{\log \tfrac{7}{3} = 0.367977} \quad \log y_1 = \overline{0.484146}$

$\phi = 10°53'36''.2 \quad \log\sqrt{\tfrac{7}{3}} = 0.183988 \qquad\quad y_1 = 3.04892$

$\tfrac{1}{3}\phi = 3°37'52''.0 \qquad\quad \log 2 = 0.301030 \quad \log\cos\left(60° + \dfrac{\phi}{3}\right) = 9.647528 - 10$

$\qquad\qquad\qquad \log 2\sqrt{\tfrac{7}{3}} = 0.485018 \qquad\qquad 0.485018$

$\qquad\qquad\qquad\qquad\qquad\qquad\qquad\qquad \log(-y_2) = \overline{0.132546}$

$\qquad\qquad\qquad\qquad\qquad\qquad\qquad\qquad\quad -y_2 = 1.35689$

$y_1 = \quad 3.04892 \qquad\qquad\qquad \log\cos\left(60° - \dfrac{\phi}{3}\right) = 9.743387 - 10$

$y_2 = -1.35689$

$\underline{y_3 = -1.69202} \qquad\qquad\qquad\qquad\qquad 0.485018$

$\qquad 0.00001 \qquad\qquad\qquad\qquad\quad \log(-y_3) = \overline{0.228405}$

$\qquad\qquad\qquad\qquad\qquad\qquad\qquad\qquad\quad -y_3 = 1.69202$

The roots have been calculated independently, and the sum of their approximate values turns out to be 0.00001 instead of 0, which serves as a control to show that the values found are correct within the limits of approximation attainable with six-place tables.

Problems

Solve trigonometrically:

1. $x^3 - 3x^2 + 1 = 0$.   **2.** $x^3 + 3x^2 - 3 = 0$.

3. $x^3 + x^2 - 2x - 1 = 0$.   **4.** $x^3 - 6x + 2 = 0$.

5. $x^3 + 6x^2 + 10x + 3 = 0$.   **6.** $x^3 + x^2 - 4x + 1 = 0$.

7. $x^3 + 3x^2 - 2x - 3 = 0$.   **8.** $x^3 + 6x^2 + 8x - 1 = 0$.

9. Locate a plane parallel to the base of a hemispherical solid that divides it into two parts of equal volume.

10. If it is required to divide the same solid into three parts of equal volume by two planes parallel to the base, show how to choose these planes.

11. Assuming the specific gravity of cork to be 0.25, to what depth will a sphere of radius 10 in. made out of cork sink in the water? By Archimedes' principle the cork will displace a volume of water equal in weight to the weight of the cork.

12. Solve the equation

$$(x^2 - x + 1)^3 = 9x^2(x - 1)^2.$$

Set $y = x - \frac{1}{2}$.

13. Solve the equation

$$(x^2 - x + 1)^3 = 8x(x - 1).$$

6. Solution of Biquadratic Equations.

The algebraic solution of biquadratic equations was discovered by Ferrari, a pupil of Cardan. Write the equation

$$x^4 + ax^3 + bx^2 + cx + d = 0$$

in the form

$$x^4 + ax^3 = -bx^2 - cx - d$$

and add $\dfrac{a^2}{4} x^2$ to both sides; then,

$$\left(x^2 + \frac{a}{2}x\right)^2 = \left(\frac{a^2}{4} - b\right)x^2 - cx - d \qquad (1)$$

is an equation equivalent to the original equation. If the right-hand member of (1) were a perfect square, the solution of this equation would be immediate. But in general it is not so. The basic idea in Ferrari's method consists in adding to both sides of (1)

$$y\left(x^2 + \frac{a}{2}x\right) + \frac{y^2}{4}$$

so as to have a perfect square in the left-hand side for an indeterminate y. Equation (1) is then transformed into

$$\left(x^2 + \frac{a}{2}x + \frac{y}{2}\right)^2 = \left(\frac{a^2}{4} - b + y\right)x^2 + \left(-c + \frac{1}{2}ay\right)x + \left(-d + \frac{1}{4}y^2\right). \quad (2)$$

Now we can seek to determine y so that

$$\left(\frac{a^2}{4} - b + y\right)x^2 + \left(-c + \frac{1}{2}ay\right)x + \left(-d + \frac{1}{4}y^2\right) \tag{3}$$

becomes the square of a linear expression $ex + f$.

In general, if

$$Ax^2 + Bx + C = (ex + f)^2, \tag{4}$$

then

$$B^2 - 4AC = 0, \tag{5}$$

and conversely. In fact, equation (4) is equivalent to the three relations

$$A = e^2, \qquad B = 2ef, \qquad C = f^2 \tag{6}$$

so that (5) is satisfied. Conversely, suppose that (5) holds. Then, if both $A = 0$ and $C = 0$, we have also $B = 0$, and the relations (6) will hold for $e = f = 0$. If both A and C are not zero let, for example, $A \neq 0$. Then, we take

$$e = \sqrt{A}, \qquad f = \frac{B}{2e},$$

and by virtue of (5) we shall have

$$C = f^2.$$

Thus, the right-hand side of (2) will be the square of a linear expression $ex + f$ if y satisfies the equation

$$\left(\frac{1}{2}ay - c\right)^2 = 4\left(y + \frac{a^2}{4} - b\right)\left(\frac{1}{4}y^2 - d\right),$$

or, in expanded form,

$$y^3 - by^2 + (ac - 4d)y + 4bd - a^2d - c^2 = 0. \tag{7}$$

It suffices to take for y any root of this cubic equation, called the *resolvent* of the biquadratic equation, in order to have

$$\left(\frac{a^2}{4} - b + y\right)x^2 + \left(\frac{1}{2}ay - c\right)x + \frac{1}{4}y^2 - d = (ex + f)^2$$

with properly chosen e and f. The biquadratic equation appears then in the form

$$\left(x^2 + \frac{a}{2}x + \frac{1}{2}y\right)^2 = (ex + f)^2$$

and splits into two quadratic equations

$$x^2 + \frac{a}{2}x + \frac{1}{2}y = ex + f, \qquad x^2 + \frac{a}{2}x + \frac{1}{2}y = -ex - f,$$

which, being solved in turn, supply the four requested roots. The solution is simplified in case the resolvent (7) has a root rationally expressible through a, b, c, d. In such a case this root may be chosen for y, and the roots of the biquadratic equation are expressible through quadratic radicals. But in general the expression for the roots will involve quadratic and cubic radicals.

Example 1. Let us apply this method to the equation

$$x^4 + 4x - 1 = 0.$$

In this case

$$a = 0, \qquad b = 0, \qquad c = 4, \qquad d = -1,$$

and the corresponding cubic resolvent

$$y^3 + 4y - 16 = 0$$

has a rational root 2. Taking $y = 2$, expression (3) becomes

$$2x^2 - 4x + 2 = (\sqrt{2}x - \sqrt{2})^2,$$

and the solution will be achieved by solving two quadratic equations

$$x^2 + 1 = \sqrt{2}x - \sqrt{2}, \qquad x^2 + 1 = -\sqrt{2}x + \sqrt{2}.$$

The four roots of the proposed equation are then

$$\frac{1 \pm i\sqrt{\sqrt{8}+1}}{\sqrt{2}}, \qquad -\frac{1 \pm \sqrt{\sqrt{8}-1}}{\sqrt{2}}$$

Example 2. For the second example we take the following geometric problem: Through a point P taken on the bisector of the angle between two perpendicular lines OX and OY, to draw a line so that its intercept QR between OX and OY shall have a given length. Taking OX and OY for coordinate axes, let the coordinates of P be (a, a). Let $OQ = x$ and $OR = y$. The equation of the line making intercepts x and y on OX and OY is

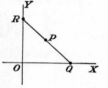

$$\frac{X}{x} + \frac{Y}{y} = 1.$$

The condition that this line goes through the point $P(a, a)$ gives one relation between x and y:

$$\frac{a}{x} + \frac{a}{y} = 1.$$

On the other hand, if l is the given length of QR, the second relation

$$x^2 + y^2 = l^2$$

is supplied by the Pythagorean theorem. Substituting into it the expression

$$y = \frac{ax}{x-a}$$

resulting from the first relation, we have to determine x from the equation

$$x^2 + \frac{a^2 x^2}{(x-a)^2} = l^2$$

or

$$x^4 - 2ax^3 + (2a^2 - l^2)x^2 + 2al^2 x - a^2 l^2 = 0.$$

The problem is thus reduced to the solution of this biquadratic equation. Its cubic resolvent

$$y^3 + (l^2 - 2a^2)y^2 - 4a^4 l^2 = (y - 2a^2)(y^2 + ly + 2a^2 l^2) = 0$$

has a root $y = 2a^2$, and with this y the expression (3) becomes

$$(a^2 + l^2)x^2 - 2a(a^2 + l^2)x + a^2 + l^2 = [\sqrt{l^2 + a^2}\,(x - a)]^2.$$

Our quartic equation now splits into two quadratic equations

$$x^2 - ax + a^2 = \sqrt{l^2 + a^2}\,(x - a),$$
$$x^2 - ax + a^2 = -\sqrt{l^2 + a^2}\,(x - a),$$

or

$$x^2 - (\sqrt{l^2 + a^2} + a)x + a(a + \sqrt{l^2 + a^2}) = 0,$$
$$x^2 + (\sqrt{l^2 + a^2} - a)x + a(a - \sqrt{l^2 + a^2}) = 0.$$

The same equations determine abscissas of the points of intersection of two circles

$$\left(x - \frac{\sqrt{l^2 + a^2} + a}{2}\right)^2 + (y - a)^2 = \left(\frac{\sqrt{l^2 + a^2} - a}{2}\right)^2$$

$$\left(x + \frac{\sqrt{l^2 + a^2} - a}{2}\right)^2 + (y - a)^2 = \left(\frac{\sqrt{l^2 + a^2} + a}{2}\right)^2$$

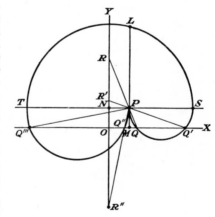

with the line $y = 0$. Hence, the following construction is easily derived: Through P draw two lines PM and PN respectively parallel to OY and OX so that $ON = PN = a$. On the extension of MP lay off the segment $PL = l$, and, taking N as a center, describe a circle with the radius NL, which intersects PN at S and T. Then, on PS and PT as diameters describe circles that intersect OX, respectively, at Q, Q' and Q'', Q'''. The requested lines are obtained by joining P to Q, Q', Q'', Q''', so that the problem may have four solutions and always has two. Points Q'' and Q''' exist always, but Q and Q' may not exist. The latter case happens if the circle with diameter PS does not intersect OX, that is, if

$$\frac{\sqrt{l^2 + a^2} - a}{2} < a,$$

or

$$l^2 < 8a^2.$$

Problems

Solve the following biquadratic equations:

1. $x^4 - 8x^2 - 4x + 3 = 0.$ **2.** $x^4 - 4x^2 + x + 2 = 0.$

3. $x^4 + x^3 - 5x^2 + 2 = 0.$ **4.** $2x^4 + 5x^3 - 8x^2 - 17x - 6 = 0.$

5. $x^4 - 3x^2 + 6x - 2 = 0.$ **6.** $x^4 - x^2 - 2x - 1 = 0.$

7. $x^4 + 3x^3 - 2x^2 - 10x - 12 = 0.$ **8.** $x^4 + 5x^3 + x^2 - 13x + 6 = 0.$

9. $x^4 + 2x^2 + x + 2 = 0.$ **10.** $x^4 + 5x^2 + 2x + 8 = 0.$

11. $2x^4 + x^3 + 2x^2 - 3x - 2 = 0.$ **12.** $x^4 + x^3 + 5x^2 + 5x + 12 = 0.$

13. Show that a reciprocal equation

$$x^4 + px^3 + qx^2 + px + 1 = 0$$

is solvable by quadratic radicals.

14. Show that $x = 2 \cos (2\pi/15)$ satisfies the equation

$$x^4 - x^3 - 4x^2 + 4x + 1 = 0.$$

What are the other roots of this equation?

15. Solve the equation

$$[(x + 2)^2 + x^2]^3 = 8x^4(x + 2)^2.$$

HINT: Set $x + 1 = y.$

CHAPTER VI

SEPARATION OF ROOTS

1. Object of This Chapter. In this and the following two chapters we shall deal with equations whose coefficients are given numbers. Such equations are called *numerical* equations to distinguish them from *literal* equations whose coefficients are letters capable of representing any numbers. The direct methods for solving literal cubic and biquadratic equations were dealt with in the preceding chapter. For equations of higher degree such direct methods are not available, but there are indirect procedures for the computation of roots of numerical equations equally applicable to equations of any degree. Often these indirect methods are far more advantageous even in case of cubic and biquadratic equations, and their exposition constitutes an important chapter of algebra. In what follows we shall consider only equations with real coefficients and concentrate our attention on real roots. In regard to real roots, the first problem is to *separate* or *isolate* them. A real root is isolated if an interval is found containing this root and no others. Real roots are separated if each of them is included in such an interval. Thus, for instance, the roots of the equation

$$x^3 + x^2 - 2x - 1 = 0$$

are all real and located in the following intervals, each containing one root: $(-2, -1)$, $(-1, 0)$, $(1, 2)$. The roots are therefore separated. Clearly the separation of real roots requires, in the first place, the solution of the following fundamental question: How many real roots has a proposed equation? In turn, this question will be answered if we find the solution of the more general problem: How many real roots of a given equation are contained between two given numbers? Mainly with these problems we shall deal in this and the next chapters.

2. The Sign of a Polynomial for Small and Large Values of the Variable. Consider a polynomial

$$f(x) = c_1 x + c_2 x^2 + \cdots + c_n x^n$$

with real coefficients, and suppose that x assumes real values. For simplicity we shall write $|x| = r$. Then,

$$|f(x)| \leqq |c_1|r + |c_2|r^2 + \cdots + |c_n|r^n,$$

and a fortiori

$$|f(x)| \leqq c(r + r^2 + \cdots + r^n),$$

c being the greatest of the numbers

$$|c_1|, |c_2|, \ldots, |c_n|.$$

Supposing $r < 1$, we have

$$r + r^2 + \cdots + r^n = \frac{r - r^{n+1}}{1 - r} < \frac{r}{1 - r},$$

and so

$$|f(x)| < \frac{cr}{1 - r}$$

provided $r < 1$. On the other hand, ϵ being a given positive number, the inequality

$$\frac{cr}{1 - r} < \epsilon$$

is satisfied if

$$r < \frac{\epsilon}{c + \epsilon} < 1.$$

Hence,

$$|f(x)| < \epsilon$$

provided

$$|x| < \frac{\epsilon}{c + \epsilon}.$$

In other words, we can state the following result: The numerical value of the polynomial

$$c_1x + c_2x^2 + \cdots + c_nx^n$$

will be less than any given positive number ϵ for sufficiently small x. It suffices to take

$$|x| < \frac{\epsilon}{c + \epsilon}$$

where c is the greatest of the numbers

$$|c_1|, |c_2|, \ldots, |c_n|.$$

This proposition leads to the following conclusions:

1. For x sufficiently small in absolute value, the sign of the polynomial

$$\phi(x) = k_0 + k_1x + \cdots + k_nx^n$$

is the same as that of k_0 provided $k_0 \neq 0$. We can write

$$\phi(x) = k_0(1 + c_1x + \cdots + c_nx^n),$$

where

$$c_i = \frac{k_i}{k_0}; \qquad i = 1, 2, \ldots, n.$$

Now for sufficiently small x

$$c_1 x + c_2 x^2 + \cdots + c_n x^n$$

will be numerically less than a given number, say $\frac{1}{2}$. Then, for such small x,

$$1 + c_1 x + \cdots + c_n x^n$$

will be certainly greater than $\frac{1}{2}$ and therefore positive, and hence $\phi(x)$ will have the same sign as k_0.

2. If the coefficients a, b, c, $\ldots$ are not all equal to 0, the sign of the polynomial

$$a x^m + b x^n + c x^p + \cdots$$

arranged according to ascending powers of x will be, for sufficiently small x, the same as the sign of its lowest term $a x^m$. In fact,

$$a x^m + b x^n + c x^p + \cdots = a x^m \left(1 + \frac{b}{a} x^{n-m} + \frac{c}{a} x^{p-m} + \cdots \right)$$

and by conclusion 1

$$1 + \frac{b}{a} x^{n-m} + \frac{c}{a} x^{p-m} + \cdots$$

will be positive for sufficiently small x. Thus, for instance,

$$- 2x^3 + 3x^5 - 100x^6$$

for small x will be negative or positive according as x is positive or negative.

3. The sign of the polynomial

$$a_0 x^n + a_1 x^{n-1} + \cdots + a_n,$$

arranged in descending powers of x, is the same as that of its leading term $a_0 x^n$ for all sufficiently large values of x. In fact,

$$a_0 x^n + a_1 x^{n-1} + \cdots + a_n = a_0 x^n \left(1 + \frac{a_1}{a_0} x^{-1} + \cdots + \frac{a_n}{a_0} x^{-n} \right),$$

and for sufficiently large x (or sufficiently small x^{-1})

$$1 + \frac{a_1}{a_0} x^{-1} + \cdots + \frac{a_n}{a_0} x^{-n}$$

is positive. Thus, for large x

$$- 3x^4 + 100x^3 + 1000x - 100000$$

will be negative, while

$$x^5 - 1000x^4 + 20000x^3 - 1000000$$

will be positive or negative according as x is positive or negative.

3. Theorem.   *If a real polynomial $f(x)$ for $x = a$ and $x = b$ takes values $f(a)$ and $f(b)$ of opposite signs, so that, for instance, $f(a) < 0$, $f(b) > 0$, then there is at least one root of the equation $f(x) = 0$ in the interval (a,b).*

PROOF.   From a certain point of view the theorem is intuitively evident.   For two points of the curve $y = f(x)$ corresponding to $x = a$ and $x = b$ are on the opposite sides of the axis OX; hence, the curve, being continuous, must cross OX at some point between $x = a$ and $x = b$.

★The more rigorous reasoning is the following: In the first place, without restriction of generality, it can be assumed that a and b are integers. For otherwise it suffices to make a linear transformation of the variable

$$x = a + (b - a)t.$$

Then, $f(x)$ will be transformed into another real polynomial $\phi(t)$, and $\phi(0) = f(a)$, $\phi(1) = f(b)$ are numbers of opposite signs.   If, therefore, it is true that $\phi(t) = 0$ has a root between 0 and 1, it will follow that the original equation $f(x) = 0$ has a root between a and b.   Thus, it is no restriction of generality to assume that a and b are integers.   Then, among the values

$$f(a), \qquad f(a + 1), \qquad \ldots, \qquad f(b)$$

of which the first is negative and the last positive, there will be the last that is not positive, say $f(c_0)$, so that

$$f(c_0) \leqq 0, \qquad f(c_0 + 1) > 0.$$

In case $f(c_0) = 0$ the equation $f(x) = 0$ has an integral root c_0 between a and b, and nothing remains to be proved.   In the contrary case divide the interval from c_0 to $c_0 + 1$ in 10 equal parts and consider the numbers

$$f(c_0), \qquad f(c_0 + \tfrac{1}{10}), \qquad f(c + \tfrac{2}{10}), \qquad \ldots, \qquad f(c_0 + 1)$$

of which the first is negative and the last positive.   Among them there will be the last that is not positive, say

$$f\!\left(c_0 + \frac{c_1}{10}\right); \qquad 0 \leqq c_1 \leqq 9,$$

so that

$$f\!\left(c_0 + \frac{c_1}{10}\right) \leqq 0, \qquad f\!\left(c_0 + \frac{c_1 + 1}{10}\right) > 0.$$

If

$$f\!\left(c_0 + \frac{c_1}{10}\right) = 0,$$

the equation $f(x) = 0$ has a rational root $c_0 + (c_1/10)$ between a and b, and nothing remains to prove. In the contrary case divide the interval between

$$c_0 + \frac{c_1}{10} \quad \text{and} \quad c_0 + \frac{c_1 + 1}{10}$$

in 10 equal parts and consider the numbers

$$f\left(c_0 + \frac{c_1}{10}\right), \quad f\left(c_0 + \frac{c_1}{10} + \frac{1}{10^2}\right), \quad \cdots, \quad f\left(c_0 + \frac{c_1 + 1}{10}\right);$$

among them will be found the last, say

$$f\left(c_0 + \frac{c_1}{10} + \frac{c_2}{10^3}\right); \quad 0 \leqq c_2 \leqq 9$$

which is not positive. If

$$f\left(c_0 + \frac{c_1}{10} + \frac{c_2}{10^2}\right) = 0,$$

a rational root

$$c_0 + \frac{c_1}{10} + \frac{c_2}{10^2}$$

between a and b is found. In the contrary case, divide the interval between

$$c_0 + \frac{c_1}{10} + \frac{c_2}{10^2} \quad \text{and} \quad c_0 + \frac{c_1}{10} + \frac{c_2 + 1}{10^2}$$

again in 10 parts and proceed as before. The process will end if the equation $f(x) = 0$ has a rational root of the form

$$c_0 + \frac{c_1}{10} + \frac{c_2}{10^2} + \cdots + \frac{c_m}{10^m}$$

represented by a finite decimal fraction, between a and b, and in the contrary case can be continued indefinitely, producing an infinite decimal

$$\xi = c_0 + \frac{c_1}{10} + \frac{c_2}{10^2} + \cdots .$$

From the way this decimal is obtained it is clear that, setting

$$\xi_m = c_0 + \frac{c_1}{10} + \cdots + \frac{c_m}{10^m},$$

$$\eta_m = c_0 + \frac{c_1}{10} + \cdots + \frac{c_m + 1}{10^m},$$

we shall have

$$f(\xi_m) < 0, \quad f(\eta_m) > 0$$

for $m = 1, 2, 3, \ldots$ On the other hand, by Taylor's formula

$$f(\xi_m) = f(\xi) + (\xi_m - \xi)f'(\xi) + (\xi_m - \xi)^2\frac{f''(\xi)}{1 \cdot 2} + \cdots < 0,$$

$$f(\eta_m) = f(\xi) + (\eta_m - \xi)f'(\xi) + (\eta_m - \xi)^2\frac{f''(\xi)}{1 \cdot 2} + \cdots > 0,$$

and so

$$f(\xi) < (\xi - \xi_m)f'(\xi) - (\xi - \xi_m)^2\frac{f''(\xi)}{1 \cdot 2} + \cdots, \tag{1}$$

$$f(\xi) > (\xi - \eta_m)f'(\xi) - (\xi - \eta_m)^2\frac{f''(\xi)}{1 \cdot 2} + \cdots. \tag{2}$$

But according to the result established in Sec. 2 the polynomial in h

$$hf'(\xi) - h^2\frac{f''(\xi)}{1 \cdot 2} + \cdots$$

will be numerically less than any given positive number ϵ provided the absolute value of h is less than some number δ, which can be assigned as soon as ϵ is given. Considering that the differences $\xi - \xi_m$ and $\xi - \eta_m$ are numerically less than $1/10^m$ and taking m so large as to have $1/10^m < \delta$, the right-hand members of the inequalities (1) and (2) will be numerically less than ϵ, that is to say, greater than $-\epsilon$ and less than ϵ. Consequently,

$$f(\xi) < \epsilon \qquad \text{and} \qquad f(\xi) > -\epsilon$$

no matter how small ϵ is taken, but this implies $f(\xi) = 0$. Thus, it is proved that the number

$$\xi = c_0 + \frac{c_1}{10} + \frac{c_2}{10^2} + \cdots,$$

which certainly belongs to the interval (a,b), is a root of the equation $f(\xi) = 0$.

Notice that this proof also gives the procedure to evaluate approximately, and to any desired degree of approximation, that root ξ whose existence is established.★

The theorem just proved is only a particular case of a more general theorem concerning continuous functions: *If a function continuous in a closed interval $a \leq x \leq b$ at the end points of this interval takes values of opposite signs, it vanishes at some point interior to it.* We shall refer occasionally to this general property of continuous functions, the proof of which is very similar to the one developed for polynomials.

4. Corollaries. Among the immediate consequences of the theorem of Sec. 3 are the following:

1. A real equation

$$f(x) = a_0 x^{2n+1} + a_1 x^{2n} + \cdots + a_{2n+1} = 0$$

of an odd degree has at least one real root. This is evident in case $a_{2n+1} = 0$, for then $x = 0$ is a root. In case a_{2n+1} is not 0, it will be either positive or negative. Now without restricting the generality the leading coefficient a_0 may be assumed to be positive. Then, for sufficiently large c (Sec. 2, conclusion 3) $f(c)$ will be positive and $f(-c)$ negative. Supposing $a_{2n+1} < 0$, we have

$$f(0) < 0, \qquad f(c) > 0,$$

and so there is a positive root of the equation $f(x) = 0$. If $a_{2n+1} > 0$, then

$$f(-c) < 0, \qquad f(0) > 0$$

and in this case the equation has a negative root.

2. An equation

$$f(x) = a_0 x^{2n} + a_1 x^{2n-1} + \cdots + a_{2n} = 0$$

of an even degree, in which the leading coefficient a_0 and the constant term a_{2n} have opposite signs, has at least two real roots, one positive and one negative. Supposing $a_0 > 0$ and c sufficiently large, both $f(c)$ and $f(-c)$ will be positive; then, since

$$f(-c) > 0, \qquad f(0) < 0, \qquad f(c) > 0,$$

each of the intervals $(-c, 0)$ and $(0, c)$ contains at least one root of the equation $f(x) = 0$.

3. A real polynomial without real roots in an interval $a \leq x \leq b$ keeps a constant sign in this interval; that is, $f(x)$ is either positive or negative no matter what value x is taken in the interval (a, b). Let x' and x'' be any two values taken in (a, b). Then, neither $f(x')$ nor $f(x'')$ is zero, since the interval does not contain roots of $f(x)$, and so $f(x')$ and $f(x'')$ cannot have opposite signs, for otherwise $f(x)$ would have a root between x' and x'' and this root would belong to (a, b) contrary to hypothesis.

4. The number of roots of $f(x) = 0$ between a and b counted according to their multiplicity is odd or even according as $f(a)$ and $f(b)$ have opposite or like signs. Let $c, d, \ldots, l$ be distinct roots of $f(x)$ between a and b and $\gamma, \delta, \ldots, \lambda$ their multiplicities. Then,

$$f(x) = (x - c)^\gamma (x - d)^\delta \cdots (x - l)^\lambda \phi(x),$$

and $\phi(x)$ has no root between a and b. On substituting $x = b$ and $x = a$ and taking the quotient, we have

$$\frac{f(b)}{f(a)} = \left(\frac{b - c}{a - c}\right)^\gamma \left(\frac{b - d}{a - d}\right)^\delta \cdots \left(\frac{b - l}{a - l}\right)^\lambda \frac{\phi(b)}{\phi(a)}.$$

Now by corollary 3 the quotient

$$\frac{\phi(b)}{\phi(a)}$$

is positive while

$$\frac{b-c}{a-c}, \qquad \frac{b-d}{a-d}, \qquad \cdots, \qquad \frac{b-l}{a-l}$$

are negative numbers. The sign of $f(b)/f(a)$ is the same as the sign of

$$(-1)^{\gamma+\delta+\cdots+\lambda};$$

hence, $\gamma + \delta + \cdots + \lambda$—the number of roots of $f(x)$ in (a,b) counted according to their multiplicity—is odd if $f(b)$ and $f(a)$ have opposite signs, and even if $f(b)$ and $f(a)$ have like signs.

5. Examples. Before considering the examples intended to illustrate the use of the results established thus far, it is convenient to introduce certain symbols and to explain their meaning. When we write $f(+\infty) = +\infty$ or $f(+\infty) = -\infty$, we mean that for all sufficiently large positive values of x the polynomial $f(x)$ keeps the sign $+$ or $-$ and becomes numerically larger than any given positive number. Similarly, symbols $f(-\infty) = +\infty$ or $f(-\infty) = -\infty$ mean that for any negative x, sufficiently large numerically, $f(x)$ keeps the sign $+$ or $-$, surpassing numerically any given positive number.

Example 1. Consider the equation

$$f(x) = (x-1)(x-3)(x-5)(x-7) + \lambda(x-2)(x-4)(x-6) = 0,$$

λ being an arbitrary real number. On substituting into $f(x)$, respectively, $-\infty$, 2, 4, 6, $-\infty$, we have

Values	$f(-\infty)$	$f(2)$	$f(4)$	$f(6)$	$f(+\infty)$
Signs	$+$	$-$	$+$	$-$	$+$

From this we conclude that the proposed equation has roots in each of the following intervals:

$$(-\infty, 2); \quad (2, 4); \quad (4, 6); \quad (6, +\infty).$$

The degree of this equation being 4, all its roots are real and simple. Denoting them by c, d, e, f, we see that

$$-\infty < c < 2 < d < 4 < e < 6 < f < \infty.$$

In other words, the roots of the equations

$$f(x) = 0 \qquad \text{and} \qquad (x-2)(x-4)(x-6) = 0$$

follow in alternate order or separate each other. In general, if two equations have all their roots real and simple and are so located that between any two consecutive roots of one there is just one root of another, we say that the roots of one separate the roots of the other. Clearly in such a case the degrees of the equations are either equal or differ just by one unit.

Example 2. Consider an equation of the form

$$F(x) = \frac{A}{x - a} + \frac{B}{x - b} + \frac{C}{x - c} - P = 0$$

in which A, B, C are positive numbers, $a < b < c$, and P is any number different from 0. Writing

$$g(x) = (x - a)(x - b)(x - c)$$

and

$$F(x) = \frac{f(x)}{g(x)},$$

it is clear that the roots of the equation in the original form and those of $f(x) = 0$ are the same. Now in $F(x)$ make the substitutions $x = a - \epsilon$, $x = a + \epsilon$; the results are

$$F(a - \epsilon) = -\frac{A}{\epsilon} + \frac{B}{a - b - \epsilon} + \frac{C}{a - c - \epsilon} - P,$$

$$F(a + \epsilon) = \frac{A}{\epsilon} + \frac{B}{a - b + \epsilon} + \frac{C}{a - c + \epsilon} - P.$$

If ϵ is positive and very small, the terms

$$-\frac{A}{\epsilon} \quad \text{and} \quad \frac{A}{\epsilon}$$

outweigh the other terms in the preceding expressions and therefore $F(a - \epsilon) < 0$, $F(a + \epsilon) > 0$ for small positive ϵ. In the same way we find

$$F(b - \epsilon) < 0, \qquad F(c - \epsilon) < 0,$$
$$F(b + \epsilon) > 0, \qquad F(c + \epsilon) > 0,$$

for small positive ϵ or, in other words,

$$\frac{f(a - \epsilon)}{g(a - \epsilon)} < 0, \qquad \frac{f(b - \epsilon)}{g(b - \epsilon)} < 0, \qquad \frac{f(c - \epsilon)}{g(c - \epsilon)} < 0,$$

$$\frac{f(a + \epsilon)}{g(a + \epsilon)} > 0, \qquad \frac{f(b + \epsilon)}{g(b + \epsilon)} > 0, \qquad \frac{f(c + \epsilon)}{g(c + \epsilon)} > 0,$$

for small positive ϵ. The signs of $g(a - \epsilon)$, $g(a + \epsilon)$, etc., are the following:

$$g(a - \epsilon), \quad g(a + \epsilon), \quad g(b - \epsilon), \quad g(b + \epsilon), \quad g(c - \epsilon), \quad g(c + \epsilon)$$
$$- \qquad + \qquad + \qquad - \qquad - \qquad + \quad ,$$

and consequently those of $f(a - \epsilon)$, $f(a + \epsilon)$, etc., are

$$f(a - \epsilon), \quad f(a + \epsilon), \quad f(b - \epsilon), \quad f(b + \epsilon), \quad f(c - \epsilon), \quad f(c + \epsilon)$$
$$+ \qquad + \qquad - \qquad - \qquad + \qquad + \quad .$$

Since, moreover,

$$F(-\infty) = F(+\infty) = -P,$$

the sign of $f(+\infty)$ is $-$ in the case of positive P, and the sign of $f(-\infty)$ is $-$ in the case of negative P. Thus, for $P > 0$

signs of	$f(a + \epsilon)$,	$f(b - \epsilon)$,	$f(b + \epsilon)$,	$f(c - \epsilon)$,	$f(c + \epsilon)$,	$f(+\infty)$
are	$+$	$-$	$-$	$+$	$+$	$-$

while for $P < 0$

signs of	$f(-\infty)$,	$f(a - \epsilon)$,	$f(a + \epsilon)$,	$f(b - \epsilon)$,	$f(b + \epsilon)$,	$f(c - \epsilon)$
are	$-$	$+$	$+$	$-$	$-$	$+$

In the first case there is a root in each of the intervals

$$(a,b); \quad (b,c); \quad (c,+\infty)$$

and in the second case in each of the intervals

$$(-\infty,a); \quad (a,b); \quad (b,c).$$

The roots of the proposed equation are therefore real, simple, and separate *a, b, c*.

Problems

Verify that the following equations have roots in the intervals indicated:

1. $x^3 - 7x + 7 = 0$. Roots in $(-4, -3)$, $(1, \frac{3}{2})$, $(\frac{3}{2}, 2)$.

2. $x^3 - 3x^2 - 4x + 13 = 0$. Roots in $(1, \frac{8}{3})$, $(\frac{8}{3}, 3)$, $(-3, -2)$.

3. $x^4 - 6x^3 + 5x^2 + 14x - 4 = 0$. Roots in $(-2, -1)$, $(0, 1)$, $(3, \frac{7}{2})$, $(\frac{7}{2}, 4)$.

4. $x^4 + 4x^3 - 3x^2 - 6x + 2 = 0$. Roots in $(-5, -4)$, $(-2, 0)$, $(0, 1)$, $(1, 2)$.

5. $5x^4 + 16x^3 - 9x^2 - 12x + 2 = 0$. Three roots in $(-1, 0)$, $(0, \frac{1}{2})$, $(\frac{1}{2}, 1)$. Isolate the fourth root.

6. Show that for all real values of λ the equation

$$(x - 2)(x - 5)(x - 7)(x - 9) + \lambda(x - 3)(x - 6)(x - 8)(x - 10) = 0$$

has all roots real and simple and separate them.

7. Solve Prob. 6 for

$$x(x^2 - 1)(x - 2) + \lambda(2x + 1)(3x - 2)(2x - 3) = 0.$$

*8. If $a_1 < b_1 < a_2 < b_2 < \cdots < a_{n-1} < b_{n-1} < a_n$ and λ is a real number, show that the roots of

$$(x - a_1)(x - a_2) \cdots (x - a_n) + \lambda(x - b_1)(x - b_2) \cdots (x - b_{n-1}) = 0$$

are real and simple. Separate them.

*9. If $a_1 < b_1 < a_2 < b_2 < \cdots < a_n < b_n$ and λ is real, what is the nature of the roots of the equation

$$(x - a_1)(x - a_2) \cdots (x - a_n) + \lambda(x - b_1)(x - b_2) \cdots (x - b_n) = 0?$$

Indicate intervals each containing just one root.

10. Show that the roots of the equation

$$\frac{1}{x + 1} + \frac{2}{x} + \frac{1}{x + 2} + \frac{3}{x + 3} - 10 = 0$$

are real and simple and isolate them.

*11. Prove in general that the roots of the equation

$$\frac{A_1}{x - a_1} + \frac{A_2}{x - a_2} + \cdots + \frac{A_n}{x - a_n} - P = 0$$

are real and simple if $A_1 > 0$, $A_2 > 0$, $\cdots$, $A_n > 0$. Indicate intervals each containing just one root.

6 An Important Identity and Lemma. Let $x_1, x_2, \ldots, x_n$ be the roots of a polynomial

$$f(x) = a_0 x^n + a_1 x^{n-1} + \cdots + a_n$$

and let

$$f(x) = a_0(x - x_1)(x - x_2) \cdots (x - x_n)$$

be its factorization. Replacing here x by $x + h$, we can present $f(x + h)$ thus:

$$f(x + h) = a_0(x + h - x_1)(x + h - x_2) \cdots (x + h - x_n)$$

$$= a_0(x - x_1) \cdots (x - x_n)\left(1 + \frac{h}{x - x_1}\right)\left(1 + \frac{h}{x - x_2}\right) \cdots \left(1 + \frac{h}{x - x_n}\right),$$

or

$$f(x + h) = f(x)\left(1 + \frac{h}{x - x_1}\right) \cdots \left(1 + \frac{h}{x - x_n}\right).$$

The product

$$\left(1 + \frac{h}{x - x_1}\right) \cdots \left(1 + \frac{h}{x - x_n}\right)$$

can be expanded in ascending powers of h, and the first two terms of this expansion are

$$1 + h\left(\frac{1}{x - x_1} + \frac{1}{x - x_2} + \cdots + \frac{1}{x - x_n}\right).$$

Hence,

$$f(x + h) = f(x) + h\left[\frac{f(x)}{x - x_1} + \frac{f(x)}{x - x_2} + \cdots + \frac{f(x)}{x - x_n}\right] + \cdots.$$

On the other hand, by Taylor's formula

$$f(x + h) = f(x) + hf'(x) + \cdots$$

so that, comparing coefficients of h in both expressions, we obtain an important identity

$$f'(x) = \frac{f(x)}{x - x_1} + \frac{f(x)}{x - x_2} + \cdots + \frac{f(x)}{x - x_n},$$

which also can be written thus:

$$\frac{f'(x)}{f(x)} = \frac{1}{x - x_1} + \frac{1}{x - x_2} + \cdots + \frac{1}{x - x_n}.$$

The roots $x_1, x_2, \ldots, x_n$ need not be all different; let the distinct ones among them be $a, b, \ldots, l$ and let $\alpha, \beta, \ldots, \lambda$ be their multiplicities. Then, the identity just derived can be presented in the form

$$\frac{f'(x)}{f(x)} = \frac{\alpha}{x - a} + \frac{\beta}{x - b} + \cdots + \frac{\lambda}{x - l}.$$

Writing

$$f(x) = (x - a)^\alpha g(x),$$

the polynomial $g(x)$ has roots $b, \ldots, l$ of multiplicities $\beta, \ldots, \lambda$. Hence, applying the same identity to $g(x)$, we have

$$\frac{g'(x)}{g(x)} = \frac{\beta}{x-b} + \cdots + \frac{\lambda}{x-l}$$

and so

$$\frac{f'(x)}{f(x)} = \frac{\alpha}{x-a} + \frac{g'(x)}{g(x)}. \tag{1}$$

From now on, we shall suppose that $f(x)$ has real coefficients; we suppose also that a is a real root. Then, $g(x)$ will be a real polynomial and $g(a) \neq 0$. Substituting in (1) $x = a - \epsilon$ and $x = a + \epsilon$, we have

$$\frac{f'(a-\epsilon)}{f(a-\epsilon)} = -\frac{\alpha}{\epsilon} + \frac{g'(\alpha-\epsilon)}{g(\alpha-\epsilon)}, \qquad \frac{f'(a+\epsilon)}{f(a+\epsilon)} = \frac{\alpha}{\epsilon} + \frac{g'(a+\epsilon)}{g(a+\epsilon)}$$

Now, if ϵ is positive and small, the terms

$$-\frac{\alpha}{\epsilon} \qquad \text{and} \qquad \frac{\alpha}{\epsilon}$$

will be large negative and positive numbers and will outweigh

$$\frac{g'(a-\epsilon)}{g(a-\epsilon)} \qquad \text{and} \qquad \frac{g'(a+\epsilon)}{g(a+\epsilon)},$$

which for small ϵ are nearly equal to the finite quantity

$$\frac{g'(a)}{g(a)}.$$

Therefore, for small positive ϵ the quotient

$$\frac{f'(a-\epsilon)}{f(a-\epsilon)}$$

will be negative, and

$$\frac{f'(a+\epsilon)}{f(a+\epsilon)}$$

will be positive, and both expressions will be large in absolute value. It is this important property that we state as a *lemma*. *When x increases and passes through a real root of $f(x)$, the quotient*

$$\frac{f'(x)}{f(x)}$$

on becoming infinite changes sign from $-$ *to* $+$ *or passes from* $-\infty$ *to* $+\infty$.

7. Rolle's Theorem. It will be easy now to prove an important theorem known as the *theorem of Rolle*. *Between two consecutive (real)* roots a and b of a polynomial $f(x)$ there is at least one and at any rate an odd number of roots of its derivative $f'(x)$.

PROOF. Since a and b are two consecutive roots of $f(x)$, in the interval $a + \epsilon \leqq x \leqq b - \epsilon$ the sign of $f(x)$ cannot change, and so

$$f(a + \epsilon) \qquad \text{and} \qquad f(b - \epsilon)$$

will be numbers of the same sign. But if ϵ is positive and small,

$$\frac{f'(a + \epsilon)}{f(a + \epsilon)} > 0, \qquad \frac{f'(b - \epsilon)}{f(b - \epsilon)} < 0$$

by the lemma of Sec. 6; consequently,

$$f'(a + \epsilon) \qquad \text{and} \qquad f'(b - \epsilon)$$

are of opposite signs, and between a and b there will be at least one root of the derivative $f'(x)$ (Sec. 4, corollary 4). At any rate, the number of such roots, each counted according to its multiplicity, will be odd.

COROLLARY. *Between two consecutive roots c and d of the derivative $f'(x)$ lies at most one root of $f(x)$.* In the first place, any such root must be simple. Further, suppose that there are two roots α and β of the equation $f(x) = 0$ between c and d, so that $c < \alpha < \beta < d$. Then, by Rolle's theorem the derivative $f'(x)$ will have at least one root between α and β, and c and d would not be two consecutive roots of $f'(x)$. Similarly, if f is the smallest root of $f'(x)$ and g the greatest, each of the intervals $(-\infty, f)$ and $(g, +\infty)$ may contain only one root of $f(x)$ in its interior. It is clear that between c and d there will be no root of $f(x)$ if

$$f(c)f(d) > 0,$$

and just one if

$$f(c)f(d) < 0.$$

Similarly, the intervals $(-\infty, f)$ and $(g, +\infty)$ contain none or just one root of $f(x)$ if at their end points the signs of $f(x)$ are like or unlike.

On these remarks can be based a method for separating the roots of a polynomial in case they are simple, provided the roots of the derivative are known. Let all the distinct roots of $f'(x)$ be

$$c_1 < c_2 < \cdots < c_r.$$

Write one after another the signs of

$$f(-\infty), \qquad f(c_1), \qquad f(c_2), \qquad \ldots, \qquad f(c_r), \qquad f(+\infty) \qquad (1)$$

In a sequence of signs $+$ or $-$ there is a *variation* if two consecutive signs are unlike and a *permanence* if they are like. For instance, the sequence

$$+ \quad + \quad - \quad - \quad + \quad - \quad + \quad -$$

presents five variations and two permanences. This terminology being adopted, the number of real roots of the equation $f(x) = 0$ is equal to the number of variations in the sequence (1). Moreover, the roots will be separated and assigned to intervals each containing just one root.

Example 1. Separate the roots of the equation
$$f(x) = 2x^5 - 5x^4 + 10x^2 - 10x + 1.$$
Since
$$f'(x) = 10(x^4 - 2x^3 + 2x - 1) = 10(x - 1)^3(x + 1)$$
has two distinct roots -1 and 1, consider
$$f(-\infty), \qquad f(-1), \qquad f(1), \qquad f(+\infty)$$
and write the corresponding sequence of signs
$$- \quad + \quad - \quad +.$$
It presents three variations that indicate three real and simple roots, one in each of the intervals:
$$(-\infty, -1); \ (-1, 1); \ (1, +\infty).$$

Example 2. How many real roots has the equation
$$f(x) = x^4 - 4ax + b = 0?$$
The derivative
$$f'(x) = 4(x^3 - a)$$
has only one real root
$$\sqrt[3]{a}$$
and
$$f(\sqrt[3]{a}) = -3a\sqrt[3]{a} + b.$$
If
$$b = 3a\sqrt[3]{a} \qquad \text{or} \qquad b^3 = 27a^4,$$
the only real root of the proposed equation will be the multiple root $\sqrt[3]{a}$. If $b^3 > 27a^4$, the signs of
$$f(-\infty), \qquad (f\sqrt[3]{a}), \qquad f(+\infty)$$
are
$$+ \quad + \quad +$$
so that in this case all roots are imaginary. Finally, in case $b^3 < 27a^4$ the sequence of signs
$$+ \quad - \quad +$$
indicates two real and simple roots, the other being imaginary. The use of this method for separating roots based on Rolle's theorem is rather limited since only rarely are the roots of the derivative known. The importance of this theorem lies in other applications, some of which will be considered presently.

Problems

Separate the roots of the following equations:

1. $3x^4 - 4x^3 - 6x^2 + 12x - 1 = 0.$ **2.** $3x^4 - x^3 - 6x^2 + 3x + 1 = 0.$
3. $3x^4 - 2x^3 - 6x^2 + 6x - 2 = 0.$ **4.** $x^4 - 4x^3 + 4x^2 - 24x - 1 = 0.$

5. $2x^5 + 5x^4 - 10x^3 - 20x^2 + 40x + 5 = 0$.

6. $3x^5 - 25x^3 + 60x - 10 = 0$.   **7.** $x^{10} - 10x^3 + 5 = 0$.

8. $x^5 - 80x + 35 = 0$.   **9.** $x^6 - 6x^5 + 4 = 0$.

10. $5x^6 + 24x^5 + 30x^4 - 20x^3 - 75x^2 - 60x + 3 = 0$.

11. For what values of A has the equation

$$(x + 3)^3 - A(x - 1) = 0$$

three real roots?

12. For what values of A has the equation

$$(x + 3)^3 - A(x - 1)^2 = 0$$

three real roots?

13. Show that the necessary and sufficient condition for an equation

$$x^3 + px + q = 0$$

to have three real and distinct roots is $4p^3 + 27q^2 < 0$.

*14. In a trinomial equation

$$x^n + px^m + q = 0$$

of an odd degree, the exponent m can be taken odd. Show that the number of real roots is one or three, and that there are three real and distinct roots only if

$$\left(\frac{mp}{n}\right)^n + \left(\frac{mq}{n - m}\right)^{n-m} < 0.$$

*15. In a trinomial equation

$$x^n + px^m + q = 0$$

of an even degree, the exponent m may be odd or even and p may be taken positive if m is odd. Then, there will be two distinct real roots only if

$$\left(\frac{mq}{n - m}\right)^{n-m} < \left(\frac{mp}{n}\right)^n,$$

and none if

$$\left(\frac{mq}{n - m}\right)^{n-m} > \left(\frac{mp}{n}\right)^n.$$

If m is even, the number of real roots according to these two cases is four or zero.

*16. If the roots of $f(x) = 0$ are real and simple, prove that the roots of

$$f'(x)^2 - f(x)f''(x) = 0$$

are all imaginary.

*17. The roots of equations $f(x) = 0$ and $g(x) = 0$ are real and simple and separate each other; that is, between any two consecutive roots of one there is just one root of the other. Denoting the roots of $g(x) = 0$ by $x_1 < x_2 < \cdots < x_m$, show that the quotients

$$\frac{f(x_1)}{g'(x_1)}, \qquad \frac{f(x_2)}{g'(x_2)}, \qquad \cdots, \qquad \frac{f(x_m)}{g'(x_m)}$$

are of the same sign.

*18. The conditions being the same as in the preceding problem, show that all the roots of the equation

$$f(x)g'(x) - f'(x)g(x) = 0$$

are imaginary.

★8. Other Applications of Rolle's Theorem. Let a polynomial $f(x)$ have the distinct real roots

$$b_1 < b_2 < \cdots < b_s$$

with respective multiplicities β_1, β_2, . . . , β_s so that altogether we count

$$r = \beta_1 + \beta_2 + \cdots + \beta_s$$

real roots. By Rolle's theorem, each of the intervals

$$(b_1, b_2); \qquad (b_2, b_3); \qquad \ldots ; \qquad (b_{s-1}, b_s)$$

contains at least one root of the derivative $f'(x)$. Since the number of intervals is $s - 1$, we have not less than $s - 1$ distinct roots of $f'(x)$. Moreover, b_i will be a root of multiplicity $\beta_i - 1$ of $f'(x)$ (in case $\beta_i = 1$ no root at all). Summing, we therefore have

$$\beta_1 - 1 + \beta_2 - 1 + \cdots + \beta_s - 1 = r - s$$

roots different from those enumerated earlier. Altogether, the equation $f'(x) = 0$ will have at least

$$r - s + s - 1 = r - 1$$

real roots. Hence, we may state the conclusion:

If an equation $f(x) = 0$ has r real roots, the number of real roots of $f'(x) = 0$ is at least $r - 1$; or, what is the same, the number of imaginary roots of the derivative is not greater than the number of imaginary roots of $f(x)$. In fact, let $2k$ be the number of imaginary roots of $f(x)$ and n the degree of this polynomial; then,

$$n = r + 2k.$$

Similarly, if r' is the number of real roots of $f'(x)$ and $2k'$ that of imaginary roots,

$$n - 1 = r' + 2k'.$$

But $r' \geq r - 1$ and so

$$n = r + 2k \geq r + 2k'$$

whence $2k' \leq 2k$. In particular, if all the roots of the equation $f(x) = 0$ are real, the derived equation $f'(x) = 0$ cannot have imaginary roots; that is, all its roots are also real. In this particular case something more can be said about the roots of the derivative. Since $k' = 0$, we must have $r' = r - 1$. This means that each of the intervals

$$(b_1, b_2); \qquad (b_2, b_3); \qquad \ldots ; \qquad (b_{s-1}, b_s)$$

will contain just one root of $f'(x)$, necessarily simple, and the other roots of $f'(x)$ will be multiple roots of $f(x)$. Hence, the multiple roots of $f'(x)$

are necessarily multiple roots of $f(x)$, provided this polynomial has only real roots, and $f'(x)$ will have simple roots if $f(x)$ has at least two distinct roots. Moreover, the roots of $f'(x)$ are contained between the smallest and the largest root of $f(x)$. Applying the same reasoning to $f'(x)$, then to $f''(x)$, etc., we can finally state the following proposition: If all the roots of an equation $f(x) = 0$ are real, the same will be true of the equations

$$f'(x) = 0, \qquad f''(x) = 0, \qquad f'''(x) = 0, \ldots$$

Multiple roots of each of these will be multiple roots of $f(x)$, and each of them will have simple roots if not all the roots of $f(x) = 0$ are equal. No root will be outside of the interval between the smallest and the largest roots of $f(x) = 0$.

The following two examples will show what applications can be made of this theorem:

Example 1. Let $a < b$ be two real numbers and let

$$f(x) = (x - a)^n (x - b)^n.$$

The equation $f(x) = 0$ has only the real roots a and b, both of multiplicity n. Hence, the equation

$$\frac{d^n}{dx^n} [(x - a)^n (x - b)^n] = 0$$

of degree n will have only real roots. These roots will be simple, since neither a nor b occurs among them, and will be contained between a and b.

Example 2. Let

$$f(x) = (x + a_1)(x + a_2) \cdots (x + a_n).$$

On expanding $f(x)$ in descending powers of x, the coefficient of x^{n-i} will be the sum

$$s_i = \Sigma a_1 a_2 \cdots a_i$$

of all products of i quantities taken among $a_1, a_2, \ldots, a_n$. The number of the terms in this sum is

$$\frac{n(n - 1) \cdots (n - i + 1)}{1 \cdot 2 \cdots i} = \binom{n}{i}.$$

Denoting therefore by p_i the arithmetic mean of all products of i factors taken among $a_1, a_2, \ldots, a_n$, we have

$$s_i = \binom{n}{i} p_i,$$

and we can write

$$f(x) = x^n + \binom{n}{1} p_1 x^{n-1} + \binom{n}{2} p_2 x^{n-2} + \cdots + p_n,$$

whence it follows that

$$f'(x) = n \left[x^{n-1} + \binom{n-1}{1} p_1 x^{n-2} + \binom{n-1}{2} p_2 x^{n-3} + \cdots + p_{n-1} \right].$$

By repeated application of this result we can conclude that the derivative of order s differs only by a constant factor from

$$x^{n-s} + \binom{n-s}{1}p_1 x^{n-s-1} + \cdots + p_{n-s}.$$

Suppose now that $a_1, a_2, \ldots, a_n$ are real numbers and not all equal. Then, the equation

$$x^{n-s} + \binom{n-s}{1}p_1 x^{n-s-1} + \cdots + p_{n-s} = 0$$

has only real roots, and they are not all equal. Replace s by $n - k - 1$ and substitute $x = y^{-1}$; the transformed equation

$$p_{k+1}y^{k+1} + \binom{k+1}{1}p_k y^k + \binom{k+1}{2}p_{k-1}y^{k-1} + \cdots + 1 = 0$$

likewise has only real roots and not all equal. Taking the $(k-1)$st derivative and removing the constant factor, we come to the equation

$$p_{k+1}y^2 + 2p_k y + p_{k-1} = 0$$

with real and unequal roots, which implies the inequality

$$p_k^2 > p_{k-1}p_{k+1}$$

holding for $k = 1, 2, \ldots, n-1$. In particular, if $a_1, a_2, \ldots, a_n$ are positive numbers, $p_1, p_2, \ldots, p_n$ will be positive. Taking $k = 1$, we have

$$p_1^2 > p_2$$

or

$$p_2^{1/2} < p_1.$$

For $k = 2$ we have

$$p_2^2 > p_1 p_3 > p_2^{1/2}p_3$$

whence

$$p_3^{1/3} < p_2^{1/2}.$$

Taking again $k = 3$, we have

$$p_3^2 > p_2 p_4 > p_3^{2/3}p_4$$

whence

$$p_4^{1/4} < p_3^{1/3};$$

and, continuing in the same way, it is easy to verify that in general

$$p_{\nu+1}^{\frac{1}{\nu+1}} < p_\nu^{\frac{1}{\nu}}.$$

Thus, we establish the series of remarkable inequalities

$$p_1 > p_2^{1/2} > p_3^{1/3} > \cdots > p_n^{1/n},$$

holding for any positive quantities $a_1, a_2, \ldots, a_n$ provided they are not all equal. In particular,

$$p_1 = \frac{a_1 + a_2 + \cdots + a_n}{n}$$

and

$$p_n = a_1 a_2 \cdots a_n,$$

so that

$$\frac{a_1 + a_2 + \cdots + a_n}{n} > \sqrt[n]{a_1 a_2 \cdots a_n},$$

and this is the classical Cauchy inequality expressing the fact that the arithmetic mean of positive numbers is greater than their geometric mean provided not all the numbers are equal.★

Problems

*1. Show that the roots of the equation

$$1 + \left(\frac{n}{1}\right)^2 x + \left(\frac{n(n-1)}{1 \cdot 2}\right)^2 x^2 + \left(\frac{n(n-1)(n-2)}{1 \cdot 2 \cdot 3}\right)^2 x^3 + \cdots + x^n = 0$$

are real, simple, and contained between 0 and 1.

*2. If the roots of the equation

$$a_0 x^n + a_1 x^{n-1} + \cdots + a_n = 0$$

are real, then

$$a_i^2 \geqq a_{i-1} a_{i+1}$$

for $i = 1, 2, \ldots, n-1$

*3. If the roots of the equation $f(x) = 0$ are real and simple, show that the roots of the equation

$$(n-1)f'^2 - nff'' = 0$$

are imaginary. Note that for an arbitrary real x the roots of the equation in z

$$f(x)z^n + f'(x)z^{n-1} + \frac{f''(x)}{1 \cdot 2} z^{n-2} + \cdots = 0$$

are real and distinct.

*4. The roots of $f(x) = 0$ being real, show that the same is true of the equations

$$xf'(x) + f(x) = 0, \qquad xf'(x) + 2f(x) = 0, \qquad xf'(x) + 3f(x) = 0, \qquad \cdots .$$

*5. The roots of an equation

$$a_0 x^n + a_1 x^{n-1} + \cdots + a_n = 0$$

being real, show that the same is true of the equation

$$(n+1)^k a_0 x^n + n^k a_1 x^{n-1} + (n-1)^k a_2 x^{n-2} + \cdots + a_n = 0$$

where k is a positive integer.

*6. Show that the roots of the equation

$$(n+1)^{n-1} x^n + n^n x^{n-1} + \frac{n(n-1)^n}{1 \cdot 2} x^{n-2} + \cdots + 1 = 0$$

are real.

*7. The roots of the equations $f(x) = 0$ and $g(x) = 0$ are real and simple and separare each other. Show that the same property holds for $f'(x) = 0$ and $g'(x) = 0$. See Prob. 18, Sec. 7.

*8. Show that the roots of the equation

$$1 + x + \frac{x^2}{2} + \cdots + \frac{x^n}{n} = 0$$

are imaginary if n is even, and all but one of the roots are imaginary if n is odd.

*9. Prove the same for the equation

$$1 + \frac{x}{1} + \frac{x^2}{1 \cdot 2} + \cdots + \frac{x^n}{1 \cdot 2 \cdots n} = 0.$$

9. A Theorem of de Gua. We shall consider now an equation of the form

$$F(x) = xf'(x) + \alpha f(x)$$

in which α is an arbitrary constant positive or negative. Let

$$b_1 < b_2 < \cdots < b_s$$

be distinct positive roots of

$$f(x) = 0$$

and let $\beta_1, \beta_2, \ldots, \beta_s$ be their multiplicities so that altogether we have

$$r = \beta_1 + \beta_2 + \cdots + \beta_s$$

positive roots. The root b_i in case $\beta_i > 1$ will also be a root of the equation $F(x) = 0$, but of multiplicity $\beta_i - 1$. For

$$f(x) = (x - b_i)^{\beta_i} f_1(x),$$
$$f'(x) = (x - b_i)^{\beta_i - 1} f_2(x),$$

whence

$$F(x) = (x - b_i)^{\beta_i - 1}[xf_2(x) + \alpha(x - b_i)f_1(x)],$$

but

$$f_3(x) = xf_2(x) + \alpha(x - b_i)f_1(x),$$

for $x = b_i$, reduces to

$$b_i f_2(b_i)$$

—a number that is different from 0—and this proves that b_i is a root of $F(x)$ of multiplicity $\beta_i - 1$. Thus, the equation $F(x) = 0$ has for certain

$$\beta_1 - 1 + \beta_2 - 1 + \cdots + \beta_s - 1 = r - s$$

roots, and, besides, it has at least one root in each of the intervals

$$(b_1, b_2); \qquad (b_2, b_3); \qquad \ldots; \qquad (b_{s-1}, b_s).$$

To prove this let ϵ be a small positive number. Then,

$$x \frac{f'(x)}{f(x)}$$

for $x = b_i + \epsilon$ will be a very large positive number, while for $x = b_{i+1} - \epsilon$ it will be negative and large numerically (Sec. 6). It follows that

$$\frac{F(b_i + \epsilon)}{f(b_i + \epsilon)} > 0, \qquad \frac{F(b_{i+1} - \epsilon)}{f(b_{i+1} - \epsilon)} < 0,$$

for sufficiently small ϵ, while

$$f(b_i + \epsilon) \qquad \text{and} \qquad f(b_{i+1} - \epsilon)$$

have the same sign. Therefore,

$$F(b_i + \epsilon) \qquad \text{and} \qquad F(b_{i+1} - \epsilon)$$

have opposite signs, and the equation

$$F(x) = 0$$

has at least one root in the interval (b_i, b_{i+1}). To the previously counted $r - s$ roots at least $s - 1$ roots different from them may be added so that altogether the equation

$$F(x) = xf'(x) + \alpha f(x) = 0$$

has at least $r - 1$ positive roots if the polynomial $f(x)$ has r positive roots, an important conclusion that will be used in the next section.

★It can be proved in a similar manner that the number of negative roots of $F(x) = 0$ is not less than $r' - 1$ if r' denotes the number of negative roots of $f(x) = 0$. The total number of real roots of the last equation is

$$m = r + r'$$

if $f(0) \neq 0$, and

$$m = r + r' + \nu$$

if $f(x)$ has 0 as a root of multiplicity ν. In this case it is easy to verify that $F(x)$ is divisible by x^ν so that 0 is a root of multiplicity at least ν for the equation $F(x) = 0$. Since the number of positive and negative roots of this equation is at least $r + r' - 2$, the number of all real roots of it is never less than $m - 2$. In particular, if all the roots of $f(x)$ are real, the equation

$$xf'(x) + \alpha f(x) = 0$$

cannot have more than two imaginary roots since its degree is not higher than that of $f(x)$. That it may have imaginary roots even when all the roots of $f(x)$ are real may be seen from an example. Take $f(x) = x^3 - x$; then,

$$F(x) = (\alpha + 3)x^3 - (\alpha + 1)x,$$

and for $\alpha = -2$

$$F(x) = x^3 + x = x(x^2 + 1).$$

This polynomial has two imaginary roots. However, all the roots of

$$xf'(x) + \alpha f(x) = 0$$

will be real if α is positive. To prove this, consider the quotient

$$\frac{F(\epsilon)}{f(\epsilon)} = \epsilon \frac{f'(\epsilon)}{f(\epsilon)} + \alpha.$$

This quotient is evidently positive if $f(0) \neq 0$ and ϵ is sufficiently small. On the other hand, in case $f(0) = 0$

$$\frac{f'(\epsilon)}{f(\epsilon)}$$

is positive for small positive ϵ. Hence, for small positive ϵ the quotient

$$\frac{F(\epsilon)}{f(\epsilon)}$$

is positive while the quotient

$$\frac{F(b_1 - \epsilon)}{f(b_1 - \epsilon)}$$

is negative. Since $f(\epsilon)$ and $f(b_1 - \epsilon)$ have the same sign, $F(\epsilon)$ and $F(b_1 - \epsilon)$ have opposite signs and there is at least one root of $F(x)$ between 0 and b_1. Thus, if n is the degree of $f(x)$, the equation $F(x) = 0$ has at least $n - 1$ real roots, so that all its roots are real.★

Problems

*1. The roots of $f(x) = 0$ being real, show that the same is true of the equation

$$xf'(x) + \alpha f(x) = 0$$

not only for positive α but also for $\alpha < -n$, where n is the degree of $f(x)$.

*2. The roots of $f(x) = 0$ being real, show that the same is true of the equation

$$f(x) + cxf'(x) + x^2f''(x) = 0$$

if $c \geqq 3$.

*3. Let $g(x) = (x + \alpha_1)(x + \alpha_2) \cdots (x + \alpha_k)$ be a polynomial with real and negative roots. The roots of

$$f(x) = a_0x^n + a_1x^{n-1} + \cdots + a_n = 0$$

being real, show that the same holds for

$$a_0g(n)x^n + a_1g(n - 1)x^{n-1} + \cdots + g(0)a_n = 0.$$

*4. For an arbitrary real λ the equation

$$\lambda f(x) + f'(x) = 0$$

has not more imaginary roots than $f(x) = 0$.

*5. Show that the same is true of the equation

$$af(x) + bf'(x) + cf''(x) = 0,$$

if the roots of

$$ax^2 + bx + c = 0$$

are real.

*6. Hermite's polynomial $H_n(x)$ is defined by

$$\frac{d^ne^{-x^2}}{dx^n} = H_n(x)e^{-x^2}.$$

Show that

$$H_{n+1}(x) = H_n'(x) - 2xH_n(x)$$

and hence deduce that the roots of $H_n(x)$ are real and simple. Prove this by induction.

10. Descartes' Rule of Signs. In a sequence of numbers

$$a_0, \quad a_1, \quad a_2, \quad \ldots, \quad a_n$$

none of which is zero, two consecutive terms

$$a_{i-1} \quad \text{and} \quad a_i$$

may have the same sign or opposite signs. In the first case we say that the terms a_{i-1}, a_i present a *permanence* of signs and in the second that they present a *variation* of signs. For example, in the sequence

$$-2, \quad -3, \quad 4, \quad 4, \quad -1, \quad 7, \quad 7, \quad 7, \quad -5, \quad -4, \quad 1$$

there are five variations and five permanences. If some terms in a sequence are 0, they are simply disregarded in counting the number of variations and permanences. Thus, in the sequence

$$1, \quad 0, \quad 0, -1, \quad -1, \quad 0, \quad 0, \quad 0, \quad 2, \quad 3, -1, \quad 0, \quad 0$$

there are three variations and two permanences. With this terminology adopted we can state the following classical theorem known as

Descartes' Rule of Signs. *The number of positive real roots of an equation with real coefficients,*

$$f(x) = a_0 x^n + a_1 x^{n-1} + \cdots + a_n = 0,$$

is never greater than the number of variations in the sequence of its coefficients

$$a_0, a_1, \ldots, a_n$$

and, if less, then always by an even number.

PROOF. Let V denote the number of variations and r the number of positive roots, each root counted according to its multiplicity. We want to prove that

$$V = r + 2h$$

where h is a nonnegative integer. Now the theorem is evident in case $V = 0$. For, if all coefficients that are different from 0 are of the same sign, then the equation has no positive roots so that $r = 0$. Assuming that the theorem is true for $V - 1$ variations, we shall prove that it is true in case of V variations, and this suffices to conclude the generality of the theorem by induction.

Let a_α and a_β, $\beta > \alpha$, be two coefficients of opposite signs, the intermediate coefficients (if there are any) being 0. The number of variations V is composed of three parts: the number of variations v_1 in the section

$$a_0, a_1, \ldots, a_\alpha,$$

one variation in the section

$$a_\alpha, \ldots, a_\beta,$$

and the number of variations v_2 in the section

$$a_\beta, \ldots, a_n,$$

so that

$$V = v_1 + v_2 + 1.$$

Now consider a new equation

$$F(x) = xf'(x) - \lambda f(x) = 0$$

whose coefficients are

$$(n - \lambda)a_0, \quad (n - 1 - \lambda)a_1, \quad \ldots, \quad (n - \alpha - \lambda)a_\alpha, \quad \ldots,$$
$$(n - \beta - \lambda)a_\beta, \quad \ldots, \quad - \lambda a_n,$$

and choose λ so that

$$n - \alpha - \lambda > 0, \quad n - \beta - \lambda < 0$$

or

$$n - \beta < \lambda < n - \alpha,$$

which is possible since $\beta > \alpha$. Noticing that the factors

$$n - \lambda, \quad n - 1 - \lambda, \quad \ldots, \quad n - \alpha - \lambda$$

are positive while the factors

$$n - \beta - \lambda, \quad n - \beta - 1 - \lambda, \quad \ldots, \quad n - n - \lambda = -\lambda$$

are negative, in the sections

$$(n - \lambda)a_0, \quad \ldots, \quad (n - \alpha - \lambda)a_\alpha$$

and

$$(n - \beta - \lambda)a_\beta, \quad \ldots, \quad - \lambda a_n$$

we count, respectively, v_1 and v_2 variations, but in the section

$$(n - \alpha - \lambda)a_\alpha, \quad \ldots, \quad (n - \beta - \lambda)a_\beta$$

there is no variation since the extreme terms are of the same sign and the intermediate ones are 0. Thus, for the equation

$$F(x) = xf'(x) - \lambda f(x) = 0$$

the number of variations is $v_1 + v_2 = V - 1$. By de Gua's theorem (Sec. 9) the number of positive roots of this equation is not less than $r - 1$. Assuming that the theorem is true in the case of $V - 1$ variations, we have

$$r - 1 \leqq V - 1$$

whence

$$r \leqq V.$$

It remains to show that the difference $V - r$ is an even number. In the sequence

$$a_0, a_1, \ldots, a_n$$

let a_ν be the last term that is different from 0. Then, if V is even, a_ν and a_0 have like signs, and opposite signs if V is odd. The polynomial

$$f(x) = a_0 x^n + \cdots + a_\nu x^{n-\nu}$$

for small positive x has the sign of a_ν, and for large positive x the sign of a_0. Hence, if a_0 and a_ν have the same sign, the number of positive roots is even, the same as V, and if a_0 and a_ν have opposite signs, the number of such roots is odd, the same as V. Thus, r and V are together even or odd, and their difference $V - r$ is an even number and that is what remained to be proved.

Changing x into $- x$ in the equation $f(x) = 0$, we get another equation $f(- x) = 0$, which obviously has as many positive roots as the proposed equation has negative roots. Hence, if r' is the number of negative roots of the proposed equation, and V' the number of variations corresponding to $f(- x)$, then

$$V' = r' + 2h'$$

where h' is a nonnegative integer.

The rule of signs indicates the exact number of positive real roots in two cases: $V = 0$ and $V = 1$. In the first case obviously $r = 0$, and in the second the relation

$$r + 2h = 1$$

with h an integer and nonnegative requires $h = 0$ and $r = 1$. This particular result can be proved independently as follows: If there is only one variation in the sequence

$$a_0, a_1, \ldots, a_n,$$

it can be divided into two parts: the first

$$a_0, a_1, \ldots, a_{\mu-1}$$

consisting of, say, positive or zero terms and the second

$$a_\mu = - b_0, \qquad a_{\mu+1} = - b_1, \qquad \ldots, \qquad a_n = - b_{n-\mu}$$

starting with a negative term a_μ and consisting of negative and zero terms. The polynomial $f(x)$ can be presented thus:

$$f(x) = x^{n-\mu+1} \left[a_0 x^{\mu-1} + \cdots + a_{\mu-1} - \left(\frac{b_0}{x} + \cdots + \frac{b_{n-\mu}}{x^{n-\mu+1}} \right) \right].$$

The expression in the brackets, being the difference between an increasing and a decreasing function, is itself an increasing function, which, from very large negative value (for x small and positive), passes to a very large positive value (for x large and positive) and therefore goes through 0 only once. Hence, there is only one positive root of the equation $f(x) = 0$.

In the case $V > 1$, Descartes' rule of signs indicates only an upper limit to the number of positive and negative roots and sometimes reveals unmistakably the presence of imaginary roots, as we shall see by examples.

Example 1. Let the equation be

$$f(x) = x^4 + x^2 - x - 3 = 0.$$

Since both $f(x)$ and $f(-x)$ present one variation, there is one positive and one negative root. The remaining two roots are imaginary.

Example 2. Consider the equation

$$f(x) = x^6 - x^3 + 2x^2 - 3x - 1 = 0.$$

For this equation $V = 3$, so that there may be one or three positive roots. Changing x into $-x$, for the transformed equation

$$f(-x) = x^6 + x^3 + 2x^2 + 3x - 1 = 0,$$

$V' = 1$, and so there is just one negative root of the proposed equation. The total number of real roots is not greater than four, and so two roots at least are imaginary. The exact number of imaginary roots in this example can be found as follows: Multiplying $f(x)$ by $(x + 1)^2$, we do not change the number of positive roots, but

$$(x + 1)^2 f(x) = x^8 + 2x^7 + x^6 - x^5 - 5x^2 - 5x - 1$$

presents only one variation. Hence, there is only one positive root, and four roots are imaginary.

Problems

How many real roots do the following equations have?

1. $x^6 + x^4 - x^3 - 2x - 1 = 0.$
2. $x^4 - x^2 + x - 2 = 0.$ Multiply by $x + 2$.
3. $x^5 + x^3 - 2x^2 + x - 2 = 0.$ Multiply by $x + 1$.
4. $x^5 + 2x^3 - x^2 + x - 1 = 0.$ Multiply by $x + 1$.
5. $1 - 2x + 3x^2 - 4x^3 + 5x^4 = 0.$ Multiply by $(1 + x)^2$.
6. $1 - 2x + 3x^2 - 4x^3 + 5x^4 - 6x^5 = 0.$ Multiply by $(1 + x)^2$.
7. $1 - 2x + 3x^2 - \cdots + (2n + 1)x^{2n} = 0.$
8. $1 - 2x + 3x^2 - \cdots - 2nx^{2n-1} = 0.$

★11. Equations with Real Roots. The rule of signs indicates the number of positive and negative roots exactly in case all the roots of an equation are real. As before, let V and V' represent the number of variations in the sequence of coefficients

$$a_0, \ a_1, \ a_2, \ \ldots, \ a_n \tag{1a}$$

of $f(x)$, and in the sequence

$$a_0, \ -a_1, \ a_2, \ \ldots, \ (-1)^n a_n \tag{2a}$$

corresponding to $(-1)^n f(-x)$. Then, if $f(x)$ is a complete polynomial, so that all members of the sequence $(1a)$ are different from zero, we have

$$V + V' = n.$$

In fact, to each permanence in $(1a)$ corresponds a variation in $(2a)$, and vice versa. Hence, $V + V'$ is the number of variations and permanences in the sequence $(1a)$, which is n.

If $f(x)$ is not a complete polynomial, we shall prove that always

$$V + V' \leqq n.$$

Let

$$a_0, a_\alpha, a_\beta, \ldots, a_\mu, a_\nu \tag{1b}$$

be those terms of the sequence $(1a)$ that are different from 0. Then, terms different from 0 in the sequence $(2a)$ are

$$a_0, (-1)^\alpha a_\alpha, (-1)^\beta a_\beta, \ldots, (-1)^\mu a_\mu, (-1)^\nu a_\nu \tag{2b}$$

Clearly, the sequences $(1b)$ and $(2b)$ present V and V' variations, respectively. Now we replace sequence $(1b)$ by

$$a_0, a_0, \ldots, a_\alpha, a_\alpha, \ldots, a_\beta, \ldots, a_\mu, \ldots, a_\mu, a_\nu, a_\nu, \ldots, a_\nu \tag{1c}$$

so as to have a complete sequence of $n + 1$ terms, all different from 0. It is evident that the number of variations in $(1c)$ is again V. From $(1c)$ we derive another sequence of $n + 1$ terms by changing the signs of alternate terms of $(1c)$:

$$a_0, -a_0, \ldots, (-1)^\alpha a_\alpha, (-1)^{\alpha+1} a_\alpha, \ldots, (-1)^\beta a_\beta, \ldots,$$
$$(-1)^\nu a_\nu, \ldots, (-1)^n a_\nu \tag{2c}$$

Let V'' be the number of variations in this sequence $(2c)$. Then, on the one hand,

$$V + V'' = n,$$

since $(1c)$ is a complete sequence of $n + 1$ terms; and, on the other hand, $V' \leqq V''$. This we can see by splitting the sequences $(2c)$ and $(2b)$ into corresponding sections as follows:

$$a_0, -a_0, \ldots, (-1)^\alpha a_\alpha \text{ corresponds to } a_0, (-1)^\alpha a_\alpha,$$
$$(-1)^\alpha a_\alpha, \ldots, (-1)^\beta a_\beta \text{ corresponds to } (-1)^\alpha a_\alpha, (-1)^\beta a_\beta,$$
$$\cdots \cdots \cdots \cdots \cdots \cdots \cdots \cdots \cdots \cdots \cdots \cdots \cdots \cdots$$
$$(-1)^\mu a_\mu, \ldots, (-1)^\nu a_\nu \text{ corresponds to } (-1)^\mu a_\mu, (-1)^\nu a_\nu,$$

and the section

$$(-1)^\nu a_\nu, \ldots, (-1)^n a_\nu$$

has no correspondence in $(2b)$. Clearly each section of $(2c)$ has no less variations than the corresponding section of $(2b)$, and hence it follows that $V' \leqq V''$, and further

$$V + V'' \leqq n,$$

as we wanted to prove. Now, if r and r' are the numbers of the positive and negative roots of the equation $f(x) = 0$ of degree n, whose roots are real and different from 0, we have

$$r + r' = n.$$

On the other hand,

$$V = r + 2h, \qquad V' = r' + 2h'$$

so that

$$V + V' = r + r' + 2h + 2h' = n + 2h + 2h' \leqq n$$

or

$$h + h' \leqq 0$$

which is possible only if $h = h' = 0$, and then

$$r = V, \qquad r' = V'.$$

It was supposed that 0 is not a root of the equation, but it is almost evident that the conclusion holds even if there are roots equal to 0.

Example. Given that all roots of the equation

$$f(x) = x^6 - 15x^4 + 40x^3 - 45x^2 + 24x - 5 = 0$$

are real, to find how many roots it has between 0 and 2 and between 2 and 3. In general, to find the number of roots in the interval $a < x \leqq b$, it suffices to find the number of roots $> a$ and subtract it from the number of roots $> b$. The number of roots $> a$ is the same as the number of positive roots of the transformed equation obtained from the original equation by the substitution $x = a + y$. In our example we make two transformations: $x = 2 + y$ and $x = 3 + y$ by Horner's process.

2) 1	0	−15	40	−45	24	− 5		3) 1	0	−15	40	−45	24	−5
	2	4	−22	36	−18	12			3	9	−18	66	63	261
	2	−11	18	−9	6	7			3	−6	22	21	87	256
	2	8	−6	24	30				3	18	36	174	585	
	4	−3	12	15	36				6	12	58	195	672	
	2	12	18	60					3	27	117	525		
	6	9	30	75					9	39	175	720		
	2	16	50						3	36	225			
	8	25	80						12	75	400			
	2	20							3	45				
	10	45							15	120				
	2								3					
	12								18					
1								1						

The underlined numbers read from the bottom up and represent the coefficients of the transformed equations. Since they are all positive, there are no roots greater than 2; hence, the number of roots between 2 and 3 is 0. On the other hand, the original equation has five variations, which indicates the presence of five positive roots. Consequently, between 0 and 2 there are just five roots, and, the degree of the equation being 6, the sixth root is negative. In fact,

$$f(x) = (x - 1)^5(x + 5).\star$$

Problems

The roots of the following equations are real. Find the number of roots in the specified intervals.

    1. $2x^3 - 9x^2 + 6 = 0$; $(0,1)$ and $(4,5)$.

    2. $5x^3 - 9x + 2 = 0$; $(0,1)$ and $(1,2)$.

    3. $x^3 - 27x + 5 = 0$; $(4,5)$.

    4. $24x^4 - 96x^3 + 72x^2 - 16x + 1 = 0$; $(1,2)$ and $(2,3)$.

    5. $x^4 - 5x^2 - 2x + 1 = 0$; $(-3, -1)$ and $(0,1)$.

    6. $x^5 - 10x^3 + 6x + 1 = 0$; $(0,1)$ and $(-1,0)$.

    7. Show that an equation has imaginary roots if there are terms missing between two terms of the same sign, or more than one term missing between two terms of opposite signs.

    *8. An equation has imaginary roots if three consecutive coefficients are in geometric progression.

    *9. An equation has imaginary roots if the coefficients of four consecutive terms are in arithmetic progression. Multiply by $(x - 1)^2$.

    *10. Show that the equation

$$x^n + (a + b)x^{n-1} + (a^2 + ab + b^2)x^{n-2} + \cdots + (a^n + a^{n-1}b + \cdots + b^n) = 0$$

cannot have more than one real root provided a and b are real. Multiply by $(x - a)(x - b)$.

12. A Complete Method of Separating Roots. The theorem of Rolle and the rule of signs, though they may often help in separating the roots of an equation, do not provide by themselves a complete and exhaustive solution of this important problem. The application of Rolle's theorem for this purpose requires a knowledge of the real roots of the derivative, which is, for the most part, lacking. The rule of signs is a rather weak proposition and, applied to an equation the nature of whose roots is not known, does not give the exact number of positive (or negative) roots except when the number of variations is zero or one. But exactly these two particular cases, when combined with a remarkable theorem published by Vincent in 1836 and hinted at earlier by Fourier, supply the most efficient method, not only to determine the exact number of positive and negative roots but also to effect their separation, in case the proposed equation has no multiple real roots. We have seen (Chap. III, Sec. 6) that the solution of equations with multiple roots can be reduced to the solution of a number of equations with simple roots. Therefore, there is no essential limitation to assume that the proposed equation has no multiple roots.

The theorem of Vincent can be stated as follows: *Let a, b, c, . . . be an arbitrary sequence of positive integers. Transforming an equation without multiple roots by a series of successive substitutions*

$$x = a + \frac{1}{y}, \qquad y = b + \frac{1}{z}, \qquad z = c + \frac{1}{t}, \text{ etc.},$$

after a number of such substitutions independent of the choice of integers a, b, c, . . . , we come to a transformed equation with not more than one variation. The proof of this theorem will be found in Appendix II. The idea of the method, which presently will be illustrated by several examples, is very simple. To find the exact number of positive roots (and to this case we can confine ourselves) we notice that the positive roots may be > 1 or < 1 excluding the case when 1 is a root. The positive roots > 1 may be written in the form $x = 1 + y$ with $y > 0$, while those < 1 may be presented in the form $x = 1/(1 + y)$ where again $y > 0$. Accordingly, the proposed equation is transformed by the substitutions $x = 1 + y$ and $x = 1/(1 + y)$, and these transformations can be very conveniently made using only additions as will be seen in the examples. If the transformed equations have no variations or just one, the question is settled. For if, for instance, the equation obtained by the transformation $x = 1 + y$ has no variations, it means that the original equation has no roots > 1; and the presence of just one variation in the transformed equation indicates just one root > 1 of the proposed equation. Similar conclusions hold for the equation resulting from the transformation $x = 1/(1 + y)$.

If one or both of the transformed equations have more than one variation, we transform them again by the substitutions $y = 1 + z$, $y = 1/(1 + z)$, and if necessary continue the transformations by substitutions of the same type until the transformed equations obtained by this process have no more than one variation. This necessarily must happen after a finite number of steps. For transformations of the form $x = 1 + y$, $y = 1 + z$, . . . followed by a transformation of the form $v = 1/(1 + w)$ are equivalent to two transformations: one of the type

$$x = a + \frac{1}{y},$$

where a is a positive integer, followed by another of the type $y = 1 + z$.

It follows from this remark that any transformed equation results from the proposed equation either by a series of transformations

$$x = a + \frac{1}{y}, \qquad y = b + \frac{1}{z}, \qquad \dots, \qquad u = l + \frac{1}{v}, \qquad v = \frac{1}{w}$$

or by the transformations

$$x = \frac{1}{y}, \qquad y = a + \frac{1}{z}, \qquad z = b + \frac{1}{t}, \qquad \ldots, \qquad u = l + \frac{1}{v}, \qquad v = \frac{1}{w},$$

$a, b, \ldots, l$ being positive integers. The second case does not differ from the first because the number of variations is not changed by the substitution $x = 1/y$. By Vincent's theorem the transformations

$$x = a + \frac{1}{y}, \qquad y = b + \frac{1}{z}, \qquad \ldots, \qquad u = l + \frac{1}{v}$$

in sufficient number lead to an equation with not more than one variation, and an additional transformation of the type $v = 1/w$ does not change the number of variations. Thus, it is certain that the above described process will lead to equations with not more than one variation. These general considerations will be better understood by examples to which we now turn.

Example 1. To separate the roots of the equation

$$x^3 - 7x + 7 = 0.$$

Let us examine first the positive roots. Now, if 1 is not a root, the positive roots are either greater than 1 or less than 1. The positive roots > 1 are of the form $x = 1 + y$, and the positive roots < 1 are of the form $x = 1/(1 + y)$ with positive y. Hence, to find the number of positive roots > 1 we transform the equation by the substitution $x = 1 + y$ and seek the number of positive roots of the transformed equation. Only additions are required to effect this transformation. In our example the necessary operations are as follows:

$$
\begin{array}{rrrr}
1 & 0 & -7 & 7 \\
\hline
1 & 1 & -6 & 1 \\
1 & 2 & -4 & \\
1 & 3 & & \\
1 & & &
\end{array}
$$

so that the transformed equation is

$$y^3 + 3y^2 - 4y + 1 = 0$$

and the number of positive roots of it may be zero or two. To perform the transformation

$$x = \frac{1}{1 + y}$$

we make two steps. First, x is replaced by $1/x$, which leads to

$$7x^3 - 7x^2 + 1 = 0.$$

The effect of this preliminary transformation is the reversal of the order of the coefficients. Next, we set $x = 1 + y$ in the new equation and perform the operations as indicated:

$$
\begin{array}{rrrr}
7 & -7 & 0 & 1 \\
\hline
7 & 0 & 0 & \underline{1} \\
7 & 7 & \underline{7} & \\
7 & \underline{14} & & \\
\underline{7} & & &
\end{array}
$$

The final transformed equation is

$$7y^3 + 14y^2 + 7y + 1 = 0.$$

Instead of reversing the order of coefficients first and then making the substitution $x = 1 + y$, we can proceed directly, starting additions from the right and going upward as shown:

$$
\begin{array}{rrrr}
 & & & \dfrac{7}{} \\
 & & \dfrac{14}{} & 7 \\
 & \dfrac{7}{} & 7 & 7 \\
\underline{1} & 0 & 0 & 7 \\
\hline
1 & 0 & -7 & 7
\end{array}
$$

The underlined numbers, when read down, supply the coefficients 7, 14, 7, 1 of the transformed equation by means of the substitution

$$x = \frac{1}{1+y}.$$

Both transformations $x = 1 + y$ and $x = 1/(1 + y)$ can be performed in the same scheme, shown below with the parallelogrammatic arrangement of numbers:

$$
\begin{array}{lrrr}
 & & & \dfrac{7}{} \\
[\text{Read down}] & & \dfrac{14}{} & 7 \\
 & \dfrac{7}{} & 7 & 7 \\
\underline{1} & 0 & 0 & 7 \\
\hline
1 & 0 & -7 & 7 \\
1 & 1 & -6 & \underline{1} \\
1 & 2 & -\underline{4} & \\
1 & \underline{3} & & [\text{Read up}] \\
\underline{1} & & &
\end{array}
$$

Since the equation

$$7y^3 + 14y^2 + 7y + 1 = 0$$

has no variations, it has no positive roots, and hence the proposed equation has no root of the form

$$x = \frac{1}{1+y}$$

with $y > 0$, that is, no root between 0 and 1. But the equation

$$y^3 + 3y^2 - 4y + 1 = 0$$

resulting from the substitution $x = 1 + y$ has two variations, and we have to treat it further by making the two substitutions

$$y = 1 + z \qquad \text{and} \qquad y = \frac{1}{1+z}.$$

The necessary calculations are shown in the parallelogrammatic scheme:

$$
\begin{array}{rrrc}
 & & & \tfrac{1}{1} \\
\text{[Read down]} & -\,1 & 1 \\
 & -\,2 & -\,2 & 1 \\
1 & 0 & -\,3 & 1 \\
\hline
1 & 3 & -\,4 & 1 \\
1 & 4 & 0 & 1 \\
1 & 5 & 5 \\
1 & 6 & \text{[Read up]} \\
1
\end{array}
$$

The equation resulting from the substitution $y = 1 + z$

$$z^3 + 6z^2 + 5z + 1 = 0$$

has no variation and no positive roots, but the equation resulting from the substitution $y = 1/(1 + z)$, namely,

$$z^3 - z^2 - 2z + 1 = 0$$

still has two variations and must be subjected again to the transformations

$$z = 1 + t \quad\text{and}\quad z = \frac{1}{1+t}.$$

The necessary calculations are

$$
\begin{array}{rrrc}
 & & & \tfrac{1}{1} \\
\text{[Read down]} & 1 & 1 \\
 & -\,2 & 0 & 1 \\
-\,1 & -\,2 & -\,1 & 1 \\
\hline
1 & -\,1 & -\,2 & 1 \\
1 & 0 & -\,2 & -\,1 \\
1 & 1 & -\,1 \\
1 & 2 & \text{[Read up]} \\
1
\end{array}
$$

so that the transformed equations are

$$t^3 + 2t^2 - t - 1 = 0 \quad\text{and}\quad t^3 + t^2 - 2t - 1 = 0$$

and have each only one variation and hence only one positive root. Now the first equation results from the original one by the substitutions

$$x = 1 + y, \qquad y = \frac{1}{1+z}, \qquad z = 1 + t,$$

which can be combined into one:

$$x = 1 + \frac{1}{2+t}.$$

The substitutions leading to the second equation:

$$x = 1 + y, \qquad y = \frac{1}{1+z}, \qquad z = \frac{1}{1+t}$$

are combined into one:

$$x = 1 + \cfrac{1}{1 + \cfrac{1}{1 + t}} \cdot$$

Each of the transformed equations having just one positive root, there are two positive roots to the proposed equation, and intervals within which they lie are obtained by taking the extreme values $t = 0$ and $t = \infty$ in the formulas expressing x. Thus, we find two intervals

$$(1, \tfrac{3}{2}) \quad \text{and} \quad (\tfrac{3}{2}, 2),$$

each containing one root of the equation

$$x^3 - 7x + 7 = 0.$$

The successive substitutions that serve to pass from x to t are immediately seen if the results of the transformations applied are arranged in a scheme resembling a genealogical tree:

$$\begin{array}{cccc} 1 & 0 & -7 & 7 \end{array}$$

$$x = (1 + y)^{-1} \mid x = 1 + y$$

$$\begin{array}{cccccccc} 7 & 14 & & 7 & 1 & & 1 & 3 & -4 & 1 \\ \text{(no var.)} & & & & & & \text{(2 var.)} \end{array}$$

$$y = (1 + z)^{-1} \mid y = 1 + z$$

$$\begin{array}{cccccccc} 1 & -1 & -2 & 1 & & 1 & 6 & 5 & 1 \\ & \text{(2 var.)} & & & & & \text{(no var.)} \end{array}$$

$$z = (1 + t)^{-1} \mid z = 1 + t$$

$$\begin{array}{cccccccc} 1 & 1 & -2 & -1 & & 1 & 2 & -1 & -1 \\ & \text{(1 var.)} & & & & & \text{(1 var.)} \end{array}$$

Such schemes will henceforth be called "genealogical trees" for abbreviation. To find the number of negative roots it suffices to substitute $- x$ for x; there will be as many negative roots as there are positive roots of the transformed equation. In our example this transformed equation

$$x^3 - 7x - 7 = 0$$

has one variation and consequently one positive root which, by inserting $x = 1, 2, 3, \ldots$ and observing the signs of the results, is found to be contained between 3 and 4. Hence, the proposed equation has one negative root in the interval $(-4, -3)$.

Example 2. To separate the roots of the equation

$$x^4 - 4x^3 + 12x^2 - 24x + 24 = 0.$$

The equation obtained in replacing x by $- x$ has no variations; hence, the proposed equation has no negative roots. Since it has four variations, it is necessary to start its genealogical tree by making the two transformations

$$x = 1 + y \quad \text{and} \quad x = \frac{1}{1 + y} \cdot$$

The necessary calculations are as follows:

$$
\begin{array}{rrrrr}
 & & \cdot & & \\
 & & \cdot & & \\
 & & \cdot & & \\
9 & 8 & 12 & 0 & 24 \\
\hline
1 & -4 & 12 & -24 & 24 \\
\hline
1 & -3 & 9 & -15 & 9 \\
1 & -2 & 7 & -8 & \\
1 & -1 & 6 & & \\
1 & 0 & & & \text{[Read up]} \\
1 & & & & \\
\hline
\end{array}
$$

It was not necessary to continue the scheme upward since the first line has no variations and the coefficients of the transformed equation evidently will be of the same sign. The sequence

$$1, \qquad 0, \qquad 6, \qquad -8, \qquad 9$$

having two variations, the same process is repeated:

$$
\begin{array}{rrrrr}
 & & \cdot & & \\
 & & \cdot & & \\
 & & \cdot & & \\
8 & 7 & 7 & 1 & 9 \\
\hline
1 & 0 & 6 & -8 & 9 \\
\hline
1 & 1 & 7 & -1 & 8 \\
1 & 2 & 9 & 8 & \\
 & & \cdot & & \\
 & & \cdot & & \\
\end{array}
$$

Now it is useless to continue either upward or downward since in both cases clearly there will be no variations. The genealogical tree is therefore

$$
\begin{array}{ccccc}
1 & -4 & 12 & -24 & 24
\end{array}
$$

$$x = (1+y)^{-1} \mid x = 1 + y$$

(no var.) $\qquad \begin{array}{ccccc} 1 & 0 & 6 & -8 & 9 \end{array}$

$$y = (1+z)^{-1} \mid y = 1 + z$$

(no var.) $\qquad\qquad$ (no var.)

This means that, transforming the given equation by any of the substitutions

$$x = \frac{1}{1+y}, \qquad x = 2 + z, \qquad x = 1 + \frac{1}{1+z},$$

we come to equations without variations and therefore without positive roots. Hence, the proposed equation has no real roots.

Example 3. To separate the roots of the equation

$$x^6 + x^5 - x^4 - x^3 + x^2 - x + 1 = 0.$$

We examine first the positive roots. Since the equation has four variations, we start building up the genealogical tree as we explained in the two preceding examples:

	1	1	2	2	1	1
<u>1</u>	0	-1	0	1	0	1
1	1	-1	-1	1	-1	1
1	2	1	0	1	0	<u>1</u>

Without continuing further we see that the genealogical tree is

$$1 \quad 1 \quad -1 \quad -1 \quad \quad 1 \quad -1 \quad 1$$

$$x = (1 + y)^{-1} \mid x = 1 + y$$

(no var.)   (no var.)

Hence, there are no positive roots. To examine the negative roots, substitute $-x$ for x. The resulting equation

$$x^6 - x^5 - x^4 + x^3 + x^2 + x + 1 = 0$$

has two variations, and so we continue as before:

<u>3</u>	2	3	4	3	2	1
1	-1	-1	1	1	1	1
1	0	-1	0	1	2	<u>3</u>
1	1	0	0	1	<u>3</u>	

and draw the conclusion that neither of the transformed equations has positive roots, whence it follows that the proposed equation has no negative roots. Thus, all six roots are imaginary.

Example 4. Separate the roots of the equation

$$6x^7 - 5x^6 + 4x^5 - 3x^4 - 2x^2 + 1 = 0.$$

1. Investigation of positive roots. Having four variations, we begin with the usual transformations:

[Read down]

						7	1
							1
					19	6	1
				25	13	5	1
			12	12	8	4	1
		− 4	0	4	4	3	1
	− 9	− 4	− 4	0	1	2	1
1	− 5	0	− 4	− 1	− 1	1	1
6	− 5	4	− 3	0	− 2	0	1
6	1	5	2	2	0	0	1

Next, the same transformations are applied to the equation whose coefficients are 1, 7, 19, 25, 12, − 4, − 9, 1:

[Read down]

						− 2	1
							1
					− 37	− 3	1
				− 108	− 34	− 4	1
			− 112	− 74	− 30	− 5	1
		12	− 38	− 44	− 25	− 6	1
	101	50	6	− 19	− 19	− 7	1
52	51	44	25	0	− 12	− 8	1
1	7	10	25	12	− 4	− 9	1
1	8	27	52	64	60	51	52

The same process is repeated again:

− 93	− 94	− 92	− 55	53	165	153	52
1	− 2	− 37	− 108	− 112	12	101	52
1	− 1	− 38	− 146	− 258	− 246	− 145	− 93

Since in each of the horizontal lines we have one variation, it is useless to continue. for it can readily be seen that the transformed equations will have one variation The genealogical tree is the following:

$$6 \quad -5 \quad 4 \quad -3 \quad 0 \quad -2 \quad 0 \quad 1$$

$$x = (1+y)^{-1} \mid x = 1 + y$$

$$1 \qquad 7 \qquad 19 \qquad 25 \qquad 12 \qquad -4 \qquad -9 \qquad 1$$

(2 var.)   (no var.)

$$y = (1+z)^{-1} \mid y = 1 + z$$

$$1 \quad -2 \quad -37 \quad -108 \quad -112 \quad 12 \quad 101 \quad 52$$

(2 var.)   (no var.)

$$z = (1+t)^{-1} \mid z = 1 + t$$

(one var.)   (one var.)

Thus, by the transformations

$$x = \cfrac{1}{1 + \cfrac{1}{2+t}} \qquad \text{and} \qquad x = \cfrac{1}{1 + \cfrac{1}{1 + \cfrac{1}{1+t}}}$$

the proposed equation is transformed into equations with one variation and one positive root. The intervals for the corresponding roots x are obtained by taking $t = 0$ and $t = \infty$ and are

$$(\tfrac{2}{3}, 1) \quad \text{and} \quad (\tfrac{1}{2}, \tfrac{2}{3}).$$

2. Investigation of negative roots. When x is replaced by $-x$, the transformed equation

$$6x^7 + 5x^6 + 4x^5 + 3x^4 + 2x^2 - 1 = 0$$

has one variation and one positive root, which indicates one negative root of the proposed equation. It has three real roots located in the intervals

$$(-\infty, 0); \quad (\tfrac{1}{2}, \tfrac{2}{3}); \quad (\tfrac{2}{3}, 1)$$

and the remaining four roots are imaginary.

The number of operations required to separate the roots by this method depends on how close the roots are and naturally will be large if there are roots with small difference. Large roots, say > 10, should be considered as "close" to ∞ and the separation of them may be shortened by using substitutions of the type

$$x = 10 \, (1 + y), \qquad x = \frac{10}{1 + y}$$

or

$$x = 100 \, (1 + y), \qquad x = \frac{100}{1 + y}, \qquad \text{etc.}$$

Problems

Separate the roots of the following equations:

1. $2x^3 - 3x^2 + 4x - 1 = 0.$ **2.** $6x^3 - 10x^2 + 5x + 3 = 0.$

3. $x^3 - 9x^2 + 20x + 1 = 0.$ **4.** $x^3 + 11x^2 - 102x + 181 = 0.$

5. $2x^4 - 6x^3 + x^2 - 10x + 2 = 0.$ **6.** $x^4 - 2x^3 + 3x^2 - 7x + 1 = 0.$

7. $x^4 - x^3 + 3x^2 - 2x + 1 = 0.$ **8.** $x^4 + 6x^3 - 7x^2 - 4x + 8 = 0.$

9. $x^4 + 6x^3 + 5x^2 - 4x - 2 = 0.$ **10.** $x^4 + 10x^3 + 23x^2 + 6x - 2 = 0.$

11. $x^5 - 3x^4 + 6x^3 - 2x + 2 = 0.$ **12.** $2x^5 - 3x^4 + x^3 - 2x^2 + x - 1 = 0.$

13. $x^5 + x^4 - 2x^3 - 2x^2 + 4 = 0.$ **14.** $3x^5 - 7x^4 + x^3 + x^2 - 7x + 5 = 0.$

15. $x^6 - 4x^4 + 9x^3 - 7x^2 + 3x - 3 = 0.$

16. $2x^6 + 3x^4 - 7x^3 - 5x^2 + 8x - 7 = 0.$

17. $x^6 - 6x^5 + 30x^4 - 120x^3 + 360x^2 - 720x + 720 = 0.$

18. $x^6 - x^5 + 2x^4 - x^3 + x^2 - x + 3 = 0.$

19. $x^7 - x^5 - 8x^2 + 3 = 0.$

20. $x^7 - 3x^6 + 5x^5 + 6x^3 - 8x + 1 = 0.$

21. $2x^8 - 3x^7 + 6x^4 - 2x^2 + x - 1 = 0.$

22. $x^8 - 3x^7 + x^6 - x^5 + 3x^3 - 2x^2 + 3 = 0.$

CHAPTER VII

THE THEOREM OF STURM

1. Sturm's Functions. Another method to effect the separation of real roots is based on a remarkable theorem discovered by C. Sturm (1803–1855) and published by him in 1829. It allows one to find the exact number of real roots contained between two given numbers for equations without multiple roots. Let $V = 0$ be an equation without multiple roots so that the polynomial V has no repeated factors. Starting with V, it is possible, and in many ways, to form a sequence of polynomials

$$V, V_1, V_2, \ldots, V_s \tag{1}$$

that in a given interval (a,b), where a is less than b, possesses the following four properties:

1. When x increases from a to b and passes through a root of the equation $V = 0$, the quotient V/V_1 changes sign from $-$ to $+$.

2. No two consecutive terms of the sequence (1) vanish for the same value x in the interval (a,b).

3. If for some x in this interval a term $V_i(i = 1, 2, \ldots, s - 1)$ vanishes, two adjoining terms V_{i-1} and V_{i+1} for the same x have opposite signs.

4. The last term V_s does not vanish in the interval (a,b) and hence keeps constant sign when x increases from a to b.

Any sequence of polynomials satisfying these four conditions is called a Sturm's series relative to the interval (a,b).

2. A Method for Constructing Sturm's Series. One way of constructing a Sturm's series relative to the interval $(-\infty, \infty)$, and consequently also relative to any other interval, is the following: For the second term V_1 of Sturm's series we can always take the derivative V' of V or this derivative multiplied by any positive constant. Then, condition 1 will be satisfied. The other terms $V_2, V_3, \ldots$ are determined by a uniform process essentially the same that serves to find the highest common divisor of V and V_1. The first step in this process is to divide V by V_1 until a remainder of degree less than that of V_1 is obtained. This remainder with signs of all its coefficients changed is taken for V_2, so that

$$V = V_1 Q_1 - V_2,$$

138

Q_1 being the quotient. If V_2 actually involves x, the second step consists in dividing V_1 by V_2 until a remainder of degree less than that of V_2 is obtained; this remainder with changed sign is taken for V_3 so that

$$V_1 = V_2 Q_2 - V_3,$$

and unless V_3 is a constant, the same process is continued. In the sequence of polynomials thus obtained,

$$V_2, V_3, \ldots,$$

we find of necessity a polynomial V_s that is a constant different from 0, since the polynomial V has no repeated factors and hence V and V' can have no common divisors save constants. It is easy to see now that the sequence

$$V, V_1, V_2, \ldots, V_s$$

is a Sturm's series. The property 1 is assured by the choice $V_1 = V'$. To see that the property 2 holds we notice that in general

$$V_{i-1} = V_i Q_i - V_{i+1}.$$

Assume now that for some real $x = \xi$

$$V_i(\xi) = 0, \qquad V_{i+1}(\xi) = 0.$$

Then, also $V_{i-1}(\xi) = 0$, similarly $V_{i-2}(\xi) = 0$ and thus retroceding finally $V_1(\xi) = V'(\xi) = 0$ and $V(\xi) = 0$. But this is impossible since it would imply that ξ is a multiple root of V. Now if for $x = \xi$

$$V_i(\xi) = 0,$$

then

$$V_{i-1}(\xi) = -V_{i+1}(\xi),$$

which shows that the numbers

$$V_{i-1}(\xi) \qquad \text{and} \qquad V_{i+1}(\xi),$$

which are different from 0, have opposite signs, so that property 3 holds. Finally, property 4 holds because V_s is a constant different from 0.

Theoretically therefore it is always possible to find at least one Sturm's series relative to any interval, although in a given case, and especially when the degree of V is somewhat high, the calculations may involve large numbers. To avoid the appearance of fractions it is well to bear in mind that, multiplying members of one Sturm's series by arbitrary positive factors, we obtain another Sturm's series. The introduction of proper positive factors to avoid fractions can be done in the process of division by multiplying by appropriate positive factors the coefficients of each partial remainder. The effect of this on the final remainder is simply the multiplication of it by a positive factor. If in forming a Sturm's series we come to a term, say V_m, that either has no real roots or has real roots but outside of the interval (a, b) with which we are concerned, we can terminate the series with V_m since members of the truncated series

$$V, V_1, V_2, \ldots, V_m$$

relative to the interval (a, b) possess all the properties 1, 2, 3, and 4. Such simplification occurs, for instance, when V_m is of the second degree and has imaginary roots, a circumstance that may be verified almost at a glance.

Example 1. Let

$$V = x^3 - 7x + 7.$$

Then,

$$V_1 = V' = 3x^2 - 7,$$

and to proceed further we have to divide V by V_1, but to avoid fractions it is well to multiply V by 3 and start the division thus:

$$
\begin{array}{rrr|rrr}
3 & 0 \ -21 & 21 & 3 & 0 & -7 \\
3 & 0 \ -7 & & 1 & & \\
\hline
& -14 & 21 & & &
\end{array}
$$

Hence, the remainder is $-14x + 21$. Changing sign and suppressing the positive factor 7, we can take

$$V_2 = 2x - 3.$$

Next, we divide $3x^2 - 7$ by $2x - 3$ in the manner indicated:

$$
\begin{array}{lrrr|rr}
\text{Multiply by 2:} & 3 & 0 & -7 & 2 & -3 \\
 & 6 & 0 & -14 & 3 & 9 \\
 & 6 & -9 & & & \\
\cline{1-4}
\text{Multiply by 2:} & & 9 & -14 & & \\
 & & 18 & -28 & & \\
 & & 18 & -27 & & \\
\cline{3-4}
 & & & -1 & &
\end{array}
$$

Changing the sign of the remainder, we take

$$V_3 = +1$$

and thus have Sturm's series for the equation $V = 0$:

$$
\begin{aligned}
V &= x^3 - 7x + 7, \\
V_1 &= 3x^2 - 7, \\
V_2 &= 2x - 3, \\
V_3 &= 1.
\end{aligned}
$$

Notice that what is written in the place of coefficients of the quotient are numbers that differ from them. But this has no importance since we are interested only in remainders or these remainders multiplied by positive factors.

Example 2. Let

$$V = x^4 - 6x^3 + x^2 - 1.$$

We can take
$$V_1 = \tfrac{1}{2}V' = 2x^3 - 9x^2 + x$$

and proceed, finding V_2 as shown:

2	-12	2	0	-2	2	-9	1	0
2	-9	1	0		1	-3

Multiply by 2:

	-3	1	0	-2
	-6	2	0	-4
	-6	27	-3	0
	-25	3	-4	

Hence,
$$V_2 = 25x^2 - 3x + 4.$$

Since roots of this polynomial are imaginary, it is not necessary to continue, so that in this example Sturm's series consists of three terms,

$$V = x^4 - 6x^3 + x^2 - 1,$$
$$V_1 = 2x^3 - 9x^2 + x,$$
$$V_2 = 25x^2 - 3x + 4.$$

Example 3. Let
$$V = x^5 - 10x^4 + 15x^3 - x^2 + 3x - 7.$$

We take
$$V_1 = V' = 5x^4 - 40x^3 + 45x^2 - 2x + 3$$

and seek V_2 as shown:

5	-50	75	-5	15	-35	5	-40	45	-2	3
5	-40	45	-2	3		1	-2			
	-10	30	-3	12	-35					
	-10	80	-90	4	-6					
	-50	87	8	-29						

whence
$$V_2 = 50x^3 - 87x^2 - 8x + 29.$$

Before proceeding with the next division, all coefficients of V_1 are multiplied by 10:

50	-400	450	-20	30	50	-87	-8	29
50	-87	-8	29		1	-313		

Multiply by 50:

	-313	458	-49	30
	-15650	22900	-2450	1500
	-15650	27231	2504	-9077
	-4331	-4954	10577	

whence
$$V_3 = 4331x^2 + 4954x - 10577.$$

Roots of this polynomial are real, and so it is necessary to continue; but as the numbers grow large, it is better, having multiplied V_2 by 200, to find the coefficients of the remainder only approximately, retaining four decimals in the coefficients of the quotient. Here is the entire calculation made with the help of a calculating machine:

10000	-17400	-1600	5800	4331	4954	-10577
10000	11438	-24421		2.3089	-6.6585	
	-28838	22821	5800			
	-28838	-32986	70427			
	55807	-64627				

whence, approximately,

$$V_4 = -5580.7x + 6462.7.$$

Finally, V_3 is divided by V_4 in the same approximate manner:

$$
\begin{array}{ll}
\begin{array}{r}
4331 \quad 4954 \ -10577 \\
4331 \ -5016 \\
\hline
9970 \ -10577 \\
9970 \ -11546 \\
\hline
+\,969
\end{array}
&
\left|
\begin{array}{rr}
-5580.7 & 6462.7 \\
\hline
-0.7761 & -1.7865
\end{array}
\right.
\end{array}
$$

whence it follows that V_5 is a negative constant so that we can take $V_5 = -1$. Thus, Sturm's series in this example is

$$
\begin{aligned}
V &= x^5 - 10x^4 + 15x^3 - x^2 + 3x - 7, \\
V_1 &= 5x^4 - 40x^3 + 45x^2 - 2x + 3, \\
V_2 &= 50x^3 - 87x^2 - 8x + 29, \\
V_3 &= 4331x^2 + 4954x - 10577, \\
V_4 &= -5580.7x + 6462.7, \\
V_5 &= -1.
\end{aligned}
$$

Appearance of very large numbers in the process of the formation of a Sturm's series is a practical disadvantage of Sturm's method for the separation of roots, but this inconvenience can be avoided by resorting to approximations, as in the last example, since we are only interested in the signs of Sturm's functions for particular values of x and even roughly approximated coefficients suffice for this purpose.

3. The Theorem of Sturm. Suppose that

$$V, V_1, V_2, \ldots, V_s$$

is some sequence of Sturm's functions relative to a given interval (a,b). Let ξ be any number belonging to (a,b), and suppose that $x = \xi$ is substituted in all Sturm's functions. This gives the numbers

$$V(\xi), \quad V_1(\xi), \quad V_2(\xi), \quad \ldots, \quad V_s(\xi)$$

the last of which is different from 0. Suppose also that the first $V(\xi)$ is not 0 and disregard the intermediate numbers that might be equal to 0. Replacing the remaining ones by the signs $+$ or $-$ according as they are positive or negative, we get a sequence of signs $+$ or $-$ in which we count the number of variations. This number of variations $v(\xi)$ is called the number of variations in the Sturm's series for $x = \xi$ and may be considered as a function of ξ defined for all values of ξ in the interval (a,b), except the roots of the equation $V = 0$ that may happen to be in this interval. After these preliminary explanations the famous theorem with which we deal can be stated as follows:

The Theorem of Sturm. Supposing that neither a nor b is a root of the equation $V(x) = 0$, the number of its roots contained between a and b is equal to the difference

$$v(a) - v(b)$$

or the number of variations lost in the Sturm's series when x passes from a to b.

PROOF. We shall suppose at first that none of the numbers

$$V(a), V_1(a), \ldots, V_{s-1}(a)$$
$$V(b), V_1(b), \ldots, V_{s-1}(b)$$

is zero. Let the real roots of all the equations

$$V(x) = 0, \qquad V_1(x) = 0, \qquad \ldots, \qquad V_{s-1}(x) = 0$$

falling in the interval (a, b) and arranged according to their magnitude be

$$c_1 < c_2 < c_3 < \cdots < c_m$$

so that $c_1 > a$ and $c_m < b$. Between c_1 and c_2; c_2 and c_3; $\ldots$; c_{m-1} and c_m let us choose arbitrarily the numbers $\xi_1, \xi_2, \ldots, \xi_{m-1}$ so that

$$c_1 < \xi_1 < c_2 < \xi_2 < c_3 < \cdots < c_{m-1} < \xi_{m-1} < c_m.$$

Then, we can write

$$v(a) - v(b) = [v(a) - v(\xi_1)] + [v(\xi_1) - v(\xi_2)] + \cdots + [v(\xi_{m-1}) - v(b)],$$ or

in a more condensed way

$$v(a) - v(b) = \sum_{i=1}^{m} [v(\xi_{i-1}) - v(\xi_i)],$$

taking for uniformity of notation $\xi_0 = a$, $\xi_m = b$. Note that between ξ_{i-1} and ξ_i there is just one of the numbers $c_1, c_2, c_3, \ldots$, namely, c_i, so that

$$\xi_{i-1} < c_i < \xi_i.$$

This number c_i is a root of some one or more of the functions $V(x)$, $V_1(x), \ldots, V_{s-1}(x)$. Suppose at first that it is not a root of the first of them, $V(x)$, but of

$$V_\alpha(x), V_\beta(x), \ldots, V_\kappa(x), V_\lambda(x),$$

where, by property 2 of the Sturm's series, $\beta > \alpha + 1, \gamma > \beta + 1, \ldots$, $\lambda > \kappa + 1$, and $\lambda < s$ by property 4. To compare $v(\xi_{i-1})$ and $v(\xi_i)$ the whole Sturm's series is divided in sections as follows:

$$
\begin{array}{llllll}
V, & V_1, & \ldots, & V_{\alpha-1}, & V_\alpha, & V_{\alpha+1} \\
V_{\alpha+1}, & V_{\alpha+2}, & \ldots, & V_{\beta-1}, & V_\beta, & V_{\beta+1} \\
\cdots & \cdots & \cdots & \cdots & \cdots & \\
V_{\kappa+1}, & V_{\kappa+2}, & \ldots, & V_{\lambda-1}, & V_\lambda, & V_{\lambda+1} \\
V_{\lambda+1}, & & \ldots, & V_s & &
\end{array}
$$

The terms $V, V_1, \ldots, V_{\alpha-1}$, not having roots in the interval (ξ_{i-1}, ξ_i), have the same sign for $x = \xi_{i-1}$ and $x = \xi_i$. For this reason the number of variations in both sequences

$$V(\xi_{i-1}), \quad V_1(\xi_{i-1}), \quad \ldots, \quad V_{\alpha-1}(\xi_{i-1})$$

and
$$V(\xi_i), \quad V_1(\xi_i), \quad \ldots, \quad V_{\alpha-1}(\xi_i)$$

is the same, say A. As to
$$V_{\alpha-1}(x), \quad V_\alpha(x), \quad V_{\alpha+1}(x),$$

they present one variation for $x = \xi_{i-1}$ and one for $x = \xi_i$. In fact, the signs of
$$V_{\alpha-1}(\xi_{i-1}), \quad V_{\alpha-1}(c_i), \quad V_{\alpha-1}(\xi_i)$$

are the same. Similarly,
$$V_{\alpha+1}(\xi_i), \quad V_{\alpha+1}(c_i), \quad V_{\alpha+1}(\xi_i)$$

have the same sign, but $V_{\alpha-1}(c_i)$ and $V_{\alpha+1}(c_i)$ have opposite signs by property 3 since $V_\alpha(c_i) = 0$. Let, for example, the sign of $V_{\alpha-1}(c_i)$ be $+$ and that of $V_{\alpha+1}(c_i)$ be $-$. Then, there are only the following possibilities as to the signs of

$$V_{\alpha-1}(\xi_{i-1}), \quad V_\alpha(\xi_{i-1}), \quad V_{\alpha+1}(\xi_{i-1})$$

$+$	$+$	$-$
$+$	$-$	$-$

which gives in both cases just one variation. Again the signs of
$$V_{\alpha-1}(\xi_i), \quad V_\alpha(\xi_i), \quad V_{\alpha+1}(\xi_i)$$

can be only the following:

$+$	$+$	$-$
$+$	$-$	$-$

which gives again one variation. Hence, the first section both for $x = \xi_{i-1}$ and $x = \xi_i$ presents the same number of variations, namely, $A + 1$. What is proved for the first section is equally applicable to other sections, whence it follows that
$$v(\xi_{i-1}) - v(\xi_i) = 0$$

if $V(c_i) \neq 0$.

Suppose now that $V(c_i) = 0$. As before, it can be proved that for $x = \xi_{i-1}$ and $x = \xi_i$ the sequence
$$V_1(x), \quad V_2(x), \quad \ldots, \quad V_s(x)$$

presents the same number of variations since $V_1(c_i) \neq 0$. As to the part
$$V(x), \quad V_1(x)$$

it has, by property 1, one variation for $x = \xi_{i-1}$ and no variation for $x = \xi_i$ and therefore
$$v(\xi_{i-1}) - v(\xi_i) = 1$$

if c_i is a root of the equation $V(x) = 0$. The sum

$$\sum_{i=1}^{m} [v(\xi_{i-1}) - v(\xi_i)]$$

consists of as many terms each equal to 1 as there are roots of the equation $V(x) = 0$ between a and b, the other terms being equal to 0. But this sum is

$$v(a) - v(b),$$

which proves the theorem in case no term of Sturm's series vanishes for $x = a$ or $x = b$. It remains to remove this restriction supposing, however, that $V(a) \neq 0$, $V(b) \neq 0$. For sufficiently small positive ϵ, the number of roots of $V(x) = 0$ between a and b is the same as the number of roots between $a + \epsilon$ and $b - \epsilon$, and since no term in the Sturm's series vanishes for $x = a + \epsilon$ or $x = b - \epsilon$, the requested number of roots is

$$v(a + \epsilon) - v(b - \epsilon).$$

Now, suppose that $V_\alpha(a) = 0$; then, counting the number of variations in the section

$$V_{\alpha-1}(a), \quad V_\alpha(a), \quad V_{\alpha+1}(a),$$

we disregard the middle term 0; the extreme terms being of opposite signs, just one variation will be counted. Similarly, no matter what is the sign of $V_\alpha(a + \epsilon)$ the section

$$V_{\alpha-1}(a + \epsilon), \quad V_\alpha(a + \epsilon), \quad V_{\alpha+1}(a + \epsilon)$$

presents just one variation, and from these observations it follows that $v(a) = v(a + \epsilon)$; similarly, it can be proved that $v(b) = v(b - \epsilon)$. Hence, the number of roots of V contained between a and b is always given by

$$v(a) - v(b)$$

provided a and b are not among the roots.

4. Examples. A few examples will suffice to show how Sturm's theorem can be used for the purpose of the separation of roots.

Example 1. To separate the roots of the equation

$$V = x^3 - 7x + 7 = 0.$$

Sturm's functions are in this case

$$\begin{aligned}
V &= x^3 - 7x + 7, \\
V_1 &= 3x^2 - 7, \\
V_2 &= 2x - 3, \\
V_3 &= 1.
\end{aligned}$$

Substituting for x the integers 0, 1, 2, . . . and $-1, -2, . . .$, we form the following table exhibiting the signs of Sturm's functions:

x	V	V_1	V_2	V_3	
2	$+$	$+$	$+$	$+$	$\Big\}$ 2 var. lost
1	$+$	$-$	$-$	$+$	
0	$+$	$-$	$-$	$+$	
-1	$+$	$-$	$-$	$+$	
-2	$+$	$+$	$-$	$+$	
-3	$+$	$+$	$-$	$+$	$\Big\}$ 1 var. lost
-4	$-$	$+$	$-$	$+$	

We see at a glance that there is one root between -4 and -3 and two roots between 1 and 2. To separate them the interval $(1, 2)$ must be subdivided by inserting some intermediate number, for instance, $\frac{3}{2}$. The signs corresponding to $x = 1, \frac{3}{2}, 2$ are

x	V	V_1	V_2	V_3	
2	$+$	$+$	$+$	$+$	0 var.
$\frac{3}{2}$	$-$	$-$	0	$+$	1 var.
1	$+$	$-$	$-$	$+$	2 var.

Hence, there is one root between 1 and $\frac{3}{2}$ and one root between $\frac{3}{2}$ and 2. The roots are now separated and assigned to the intervals

$$(-4, -3); \quad (1, \tfrac{3}{2}); \quad (\tfrac{3}{2}, 2).$$

Example 2. To separate the roots of the equation

$$V = x^4 - 6x^3 + x^2 - 1 = 0.$$

In this case Sturm's functions are

$$V = x^4 - 6x^3 + x^2 - 1,$$
$$V_1 = 2x^3 - 9x^2 + x,$$
$$V_2 = 25x^2 - 3x + 4,$$

and the following table gives their signs for various values of x:

x	V	V_1	V_2	
6	$+$	$+$	$+$	$\Big\}$ 1 var. lost
5	$-$	$+$	$+$	
0	$-$	0	$+$	$\Big\}$ 1 var. lost
-1	$+$	$-$	$+$	

Inspection of the signs shows that there is one root between 5 and 6 and one root between -1 and 0. The other two roots are imaginary.

Example 3. To show how a proper choice of Sturm's functions can facilitate the investigation of the nature of roots, let us consider the equation

$$f = 1 + \frac{a}{1}x + \frac{a(a + 1)}{1 \cdot 2}x^2 + \cdots + \frac{a(a + 1) \cdots (a + n - 1)}{1 \cdot 2 \cdots n}x^n = 0.$$

Taking the derivative and multiplying it by $1 - x$, we get

$$(1 - x)f' = af - \frac{a(a + 1) \cdots (a + n - 1)}{1 \cdot 2 \cdots n}(a + n)x^n.$$

The verification of this identity is straightforward and does not present any difficulty. If we take now

$$V = f, \qquad V_1 = f', \qquad V_2 = -\frac{a(a + 1) \cdots (a + n - 1)}{1 \cdot 2 \cdots n}(a + n)x^n$$

and suppose that a is not 0 or a negative integer $\geqq -n$, it can be verified readily that these three functions form a Sturm's series for each of the intervals $(\epsilon, +\infty)$ and $(-\infty, -\epsilon)$, ϵ being an arbitrarily small positive number, and so we shall be able to find exactly the number of positive and negative roots of the proposed equation by Sturm's theorem. Suppose at first that a is a positive number or a negative number $< -n$. On substituting $x = \epsilon$, $x = +\infty$ and $x = -\infty$, $x = -\epsilon$ and supposing n even, we have

$a > 0$

x	V	V_1	V_2
$+\infty$	$+$	$+$	$-$
ϵ	$+$	$+$	$-$
$-\epsilon$	$+$	$+$	$-$
$-\infty$	$+$	$-$	$-$

$a < -n$

x	V	V_1	V_2
$+\infty$	$+$	$-$	$+$
ϵ	$+$	$-$	$+$
$-\epsilon$	$+$	$-$	$+$
$-\infty$	$-$	$-$	$+$

and since no variations are lost in the intervals $(-\infty, -\epsilon)$ and $(\epsilon, +\infty)$, the proposed equation has only imaginary roots.

In case of an odd n the results are

$a > 0$

x	V	V_1	V_2
$+\infty$	$+$	$+$	$-$
ϵ	$+$	$+$	$-$
$-\epsilon$	$+$	$+$	$+$
$-\infty$	$-$	$+$	$+$

$a < -n$

x	V	V_1	V_2
$+\infty$	$-$	$-$	$-$
ϵ	$+$	$-$	$-$
$-\epsilon$	$+$	$-$	$+$
$-\infty$	$+$	$-$	$+$

Examining them, we see that in case $a > 0$ just one variation is lost in the interval $(-\infty, -\epsilon)$, which indicates one real and negative root, while in the case $a < -n$ one variation is lost in the interval $(\epsilon, +\infty)$, which shows the existence of one positive root. Thus, in both the cases $a > 0$ and $a < -n$ the equation has no real roots if n is even, and one real root if n is odd. If a is negative and is contained between two consecutive integers $-k$ and $-k + 1$ and $k \leqq n$, the examination of the nature of roots can be made in an analogous manner and leads to the following result: For even n there are two real roots; for an odd n the number of real roots is one or three according as k is odd or even.

Problems

Separate the roots by means of Sturm's theorem:

1. $x^3 - 3x + 1 = 0$.
2. $x^3 + 6x^2 + 10x - 1 = 0$.
3. $x^3 - 4x + 2 = 0$.
4. $x^3 - 6x^2 + 8x + 40 = 0$.

5. $x^3 + x^2 - 2x - 1 = 0$. **6.** $x^3 - 4x^2 - 4x + 20 = 0$.

7. $6x^4 - 24x^3 + 42x^2 - 32x + 11 = 0$.

8. $16x^4 - 32x^3 + 88x^2 - 8x + 17 = 0$.

9. $x^4 - 4x^3 + 10x^2 - 8x + 3 = 0$. **10.** $x^4 - 4x^3 + 12x^2 - 12x + 5 = 0$.

11. $x^4 - 4x^3 + x^2 + 6x + 2 = 0$. **12.** $x^4 - 4x^3 + x^2 - 1 = 0$.

13. $x^4 + x^3 + x - 1 = 0$. **14.** $x^4 + 2x^2 - 4x + 10 = 0$.

15. $x^5 - 5x^4 + 10x^3 - 5x^2 + 1 = 0$. **16.** $x^5 + 5x^4 - 20x^2 - 10x + 2 = 0$.

17. $x^5 - 2x^4 + x^3 - 8x + 6 = 0$.

18. $x^6 - 6x^5 + 15x^4 - 20x^3 + 30x^2 - 24x + 14 = 0$.

19. $x^6 - 6x^5 + 16x^4 - 24x^3 + 22x^2 - 12x + 4 = 0$.

20. $5x^6 - 30x^5 + 75x^4 - 90x^3 + 60x^2 - 18x - 2 = 0$.

*21. Sturm's series for each of the intervals $(\epsilon, + \infty)$ and $(- \infty, - \epsilon)$, where ϵ is an arbitrary positive number, can be obtained in the following manner provided $f(x)$ has no multiple factors: Arrange $f(x)$, and other polynomials that will be used, in ascending powers of x. Let x^{ρ_1} be the highest power of x that divides $f'(x)$ so that

$$f'(x) = x^{\rho_1}f_1(x) \qquad \rho_1 \geqq 0$$

and $f_1(0) \neq 0$. Divide $f(x)$ by $f_1(x)$ of degree $n_1 = n - 1 - \rho_1$, retaining in the quotient exactly $n - n_1 + 1$ terms. The remainder will certainly be divisible by

$$x^{n-n_1+1}$$

but may occasionally be divisible by a higher power of x, say x^{ρ_2}, so that $\rho_2 \geqq n - n_1 + 1$. Representing this remainder by

$$- x^{\rho_2}f_2(x), \qquad f_2(0) \neq 0,$$

we shall have identically

$$f(x) = f_1(x)q_1(x) - x^{\rho_2}f_2(x)$$

and the degree of $f_2(x)$ will be $n_2 \leqq n - \rho_2 \leqq n_1 - 1$, that is, less than that of $f_1(x)$. If $f_2(x)$ is not a constant, divide in the same way $f_1(x)$ by $f_2(x)$, retaining in the quotient exactly $n_1 - n_2 + 1$ terms. The remainder will be divisible by $x^{n_1-n_2+1}$; representing it by

$$- x^{\rho_3}f_3(x), \qquad \rho_3 \geqq n_1 - n_2 + 1, \qquad f_3(0) \neq 0$$

we shall have

$$f_1(x) = f_2(x)q(x) - x^{\rho_3}f_3(x).$$

The continuation of this process leads to a sequence of polynomials

$$f, f_1, f_2, \ldots, f_s$$

the last of which is a constant different from 0. Show that

$$V = f, \qquad V_1 = f_1, \qquad \cdots, \qquad V_s = f_s$$

is a Sturm's series for the interval $(\epsilon, + \infty)$. Choosing the sign $\pm$ according as ρ_i is even or odd and setting

$$V = f, \qquad V_i = \pm f_i \qquad (i = 1, 2, \ldots, s)$$

show that

$$V, V_1, \ldots, V_s$$

is a Sturm's series for the interval $(- \infty, - \epsilon)$. To avoid fractions it is permissible to multiply intermediate remainders by properly chosen positive factors.

Apply this process to separate roots of the equations:

*22. $x^3 - 3x^2 + 1 = 0$. ***23.** $x^4 - 4x^3 + x^2 - 1 = 0$.

*24. $x^4 + x^3 + x - 1 = 0.$ ***25.** $x^5 + x^4 + 2x^3 - 1 = 0.$

*26. Show that an equation $V = 0$ of degree n has all its roots real and distinct if and only if Sturm's series consists of n functions with leading coefficients of the same sign. Apply this criterion to the cubic equation $x^3 + px + q = 0.$

*27. What is the condition that all of the roots of the following equation are real:

$$x^5 - 5px^3 + 5p^2x + 2q = 0?$$

*28. Examine the nature of the roots of the equation

$$V = 1 + \frac{x}{1} + \frac{x^2}{1 \cdot 2} + \cdots + \frac{x^n}{1 \cdot 2 \cdots n} = 0.$$

Notice that

$$V - V' = \frac{x^n}{1 \cdot 2 \cdots n}$$

and show that

$$V, V', -x^n$$

form a Sturm's series in each of the intervals $(\epsilon, +\infty)$ and $(-\infty, -\epsilon)$, ϵ being an arbitrary positive number.

*29. If the functions

$$V, V_1, \ldots, V_s$$

satisfy conditions 2, 3, and 4 enumerated in Sec. 1, but not the condition 1, what does the difference $v(a) - v(b)$ represent?

*30. If conditions 2, 3, and 4 are satisfied and $v(-\infty) - v(+\infty) = n$, denoting by n the degree of the equation $V = 0$, show that all roots of the equation are real and distinct.

*31. Consider Hermite's polynomials $H_n = H_n(x)$ defined by the expression

$$\frac{d^n e^{-x^2}}{dx^n} = e^{-x^2} H_n.$$

Show that

$$H_{n+1} + 2xH_n + 2nH_{n-1} = 0$$

for $n = 1, 2, 3, \ldots$, assuming $H_0 = 1.$ Referring to Prob. 30, show that all roots of the equation $H_n = 0$ are real and distinct.

HINT: If $y = e^{-x^2}$, then

$$y' + 2xy = 0.$$

Take the derivative of order n.

*32. Let

$$\frac{d^{n+1} e^{\frac{1}{x}}}{dx^{n+1}} = \frac{(-1)^{n+1}}{x^{2n+2}} P_n(x) e^{\frac{1}{x}}$$

where $P_n = P_n(x)$ is a polynomial of degree $n.$ Show that

$$P_{n+1} - [(2n + 2)x + 1]P_n + n(n + 1)x^2 P_{n-1} = 0; \qquad n = 1, 2, 3, \ldots.$$

Hence, conclude by induction that the leading coefficient of P_n is $1 \cdot 2 \cdot 3 \cdots n$ and that $P_n(0) = 1.$ Referring to Prob. 30, deduce that all roots of the equation $P_n(x) = 0$ are real, negative, and distinct.

*33. Let the imaginary parts of the complex numbers $a_1 + b_1 i, a_2 + b_2 i, \ldots;$ $a_n + b_n i$ be of the same sign, for example, all positive. Let

$$(x - a_1 - b_1 i)(x - a_2 - b_2 i) \cdots (x - a_n - b_n i) = P_n(x) + iQ_n(x),$$

$P_n(x)$ and $Q_n(x)$ being real polynomials. Show that the roots of the equation

$$\alpha P_n + \beta Q_n = 0$$

are real for arbitrary real α and β. *Indication of Solution:* Let $V_k = \alpha P_k + \beta Q_k$ and $V_0 = \alpha$. Show that

$$V_k = \left[x - a_k + \frac{b_k}{b_{k-1}} (x - a_{k-1}) \right] V_{k-1} - \left[\frac{b_k}{b_{k-1}} (x - a_{k-1})^2 + b_k b_{k-1} \right] V_{k-2}$$

for $k = 2, 3, \ldots, n$. Then, refer to Prob. 30. This problem can also be solved by very simple geometric considerations.

CHAPTER VIII

APPROXIMATE EVALUATION OF ROOTS

1. Object of This Chapter. After isolation of a real root has been effected, a new question arises: how to evaluate this root to any desired degree of approximation? Approximate values of numbers are ordinarily represented by decimal fractions, and so the question of approximate evaluation of roots can be presented in this form: to calculate to any prescribed number of decimals the root that has been isolated. To attain this aim several methods can be employed. In this chapter we shall consider three principal methods: the method of Horner, the method of iteration, and the method of Newton. Horner's method applies only to algebraic equations, while the other two have the advantage of being applicable also to transcendental equations. On the other hand, in application to algebraic equations Horner's method has definite advantages over that of Newton and over the iteration method. First, in Horner's method the necessary calculations are arranged in a very convenient manner, and, second, the root can be computed to a greater number of decimals for a given expenditure of labor.

2. Basic Part of Horner's Method. The essential features of Horner's method will be best understood in application to particular examples.

Example 1. Let us take the equation

$$x^3 - 7x + 7 = 0$$

examined in Chap. VI, Sec. 12. It has three real roots: two positive roots in the intervals $(1, \frac{3}{2})$, $(\frac{3}{2}, 2)$ and one negative root between -4 and -3. Suppose we want to calculate approximately the root contained between $\frac{3}{2} = 1.5$ and 2. Since the integral part of it is 1, we can set

$$x = 1 + \frac{a}{10},$$

where a is a number contained between 5 and 10. The transformed equation in a can be obtained in two steps: First we set

$$x = 1 + x'$$

and then

$$x' = \frac{a}{10}.$$

The transformed equation in x' is found by Horner's process, invented by him precisely in connection with the approximate evaluation of roots. In our case the necessary calculations are arranged as follows:

$$
\begin{array}{r}
1) \quad 1 \quad\; 0 \quad -7 \quad\;\; 7 \\
1 \quad\;\; 1 \quad -6 \\
\hline
1 \quad -6 \quad\;\; 1 \\
1 \quad\;\; 2 \\
\hline
2 \quad -4 \\
1 \\
\hline
3 \\
\underline{\;} \\
1 \\
\underline{\;}
\end{array}
\qquad \text{(I)}
$$

and the underlined numbers are the coefficients of the equation in x':

$$x'^3 + 3x'^2 - 4x' + 1 = 0.$$

Here we must substitute

$$x' = \frac{a}{10}.$$

Cleared of fractions the resulting equation in a will be

$$a^3 + 30a^2 - 400a + 1000 = 0, \tag{1}$$

and its coefficients are obtained by multiplying the numbers $1, 3, -4, 1$ in Scheme (I) by $1, 10, 10^2, 10^3$. The root of equation (1) is known to be contained between 5 and 10. To find the integral part of a we must seek consecutive integers between which a is contained. To this end we substitute into the left-hand side of (1) successively $a = 5, 6, 7, 8, 9$ and look for two successive integers giving results of opposite sign; the smaller of them will be the desired integral part of a. We find that for

$$a = 5, \quad 6, \quad 7$$

the signs of the results are

$$- \quad - \quad + \, .$$

Hence,

$$6 < a < 7,$$

and so we write

$$a = 6 + \frac{b}{10},$$

where b is contained between 0 and 10. The transformed equation in b is found by another application of Horner's process as follows:

$$
\begin{array}{r}
6) \quad 1 \quad\;\; 30 \quad -400 \quad\;\; 1000 \\
6 \quad\;\; 216 \quad -1104 \\
\hline
36 \quad -184 \quad -104000 \\
6 \quad\;\; 252 \\
\hline
42 \quad\;\; 6800 \\
6 \\
\hline
480 \\
\underline{\;} \\
1
\end{array}
\qquad \text{(II)}
$$

The equation in b is

$$b^3 + 480b^2 + 6800b - 104000 = 0. \tag{2}$$

To find the integral part of b we substitute into the left-hand member of (2), $b = 0, 1, 2, \ldots, 9$, and look for two consecutive numbers among these that give results of opposite sign. The number of substitutions under unfavorable circumstances may amount to nine, but it can be reduced to a half by the following simple remark:

Substitute first $b = 5$; then, if the result is positive, it is useless to substitute $b = 6$, 7, 8, 9; and if the result is negative, it is useless to substitute $b = 1, 2, 3, 4$. Now the substitution $b = 5$ actually gives a negative number, so that it is necessary to continue substituting $b = 6, 7, 8, 9$. All these substitutions giving negative results, it is certain that the integral part of b is 9, and so we can set

$$b = 9 + \frac{c}{10}$$

where $0 < c < 10$. Another application of Horner's process

9)	1	480	6800	− 104000	
		9	4401	100809	
		489	11201	− 3191000	
		9	4482		
		498	1568300		(III)
		9			
		5070			
	1				

shows that the equation in c is

$$c^3 + 5070c^2 + 1568300c - 3191000 = 0, \tag{3}$$

and to find the integral part of c we could proceed by trials as before, but at this stage a circumstance occurs that facilitates the trials very much. On examining equation (3) it is obvious that the terms

$$c^3 + 5070c^2$$

for $0 < c < 10$ are considerably smaller than the terms

$$1568300c - 3191000.$$

Neglecting for this reason all terms except these two, and considering the equation of the first degree

$$1568300c - 3191000 = 0,$$

it is clear that the root of this equation will not differ much from the exact value of c and presumably will have the same integral part. Now the integral part of c, as determined from the abridged equation, is 2, and so we set

$$c = 2 + \frac{d}{10}; \qquad 0 < d < 10$$

and seek the transformed equation in d by Horner's process:

2)	1	5070	1568300	− 3191000	
		2	10144	3156888	
		5072	1578444	− 34112000	
		2	10148		
		5074	158859200		(IV)
		2			
		50760			
	1				

Since the underlined number in the last column is negative, 2 is certainly not greater than the integral part of c; neither is it less than it. For the sum of all underlined numbers

$$1 + 5076 + 1588592 - 34112$$

is the result of the substitution $c = 3$, and it is positive. As to the equation in d, it is

$$d^3 + 50760d^2 + 158859200d - 34112000 = 0,$$

and with still more reason we can assume that the integral part of d is equal to the integral part of the quotient obtained by dividing 34112000 by 158859200. This integral part being 0, we set

$$d = 0 + \frac{e}{10}; \qquad 0 < e < 10,$$

and the coefficients of the equation in e are obtained by appending to the numbers in columns 2, 3, and 4 of scheme (IV) one, two, and three zeros, respectively. Thus,

$$e^3 + 507600e^2 + 15885920000e - 34112000000 = 0.$$

The integral part of e, as found by the simplified method, is 2, and so we set further

$$e = 2 + \frac{f}{10}; \qquad 0 < f < 10.$$

The transformed equation in f is found as before:

2)	1	507600	15885920000	− 34112000000	
		2	1015204	31773870408	
		507602	15886935204	− 2338129592000	
		2	1015208		(V)
		507604	1588795041200		
		2			
		5076060			
	1				

The equation in f is

$$f^3 + 5076060f^2 + 1588795041200f - 2338129592000 = 0. \qquad (4)$$

At this stage we stop for a moment since the continuation of the process would lead to excessively large numbers. One of the most important features in Horner's method, and one that cannot be omitted, is the so-called "contraction" that will be explained presently. At this moment we must mention only that the separate calculations in (I), (II), (III), (IV), and (V) are combined into one compact scheme in a manner shown below and not requiring any detailed explanation.

1)	1	0	− 7	7	
		1	1	− 6	
		1	− 6	1000	
		1	2	− 1104	
		2	− 400	− 104000	
		1	216	100809	
		30	− 184	− 3191000	
6)	1	6	252	3156888	
		36	6800	− 34112000000	
		6	4401	31773870408	
		42	11201	− 2338129592	
		6	4482		
		480	1568300		
9)	1	9	10144		
		489	1578444		
		9	10148		
		498	15885920000		
		9	1015204		
		5070	15886935204		
2)	1	2	1015208		
		5072	15887950412		
		2			
		5074			
		2			
		507600			
2) 0)	1	2			
		507602			
		2			
		507604			
		2			
		507606			
	1				

As to the value of the root x, it results from the combination of the substitutions

$$x = 1 + \frac{a}{10}, \qquad a = 6 + \frac{b}{10}, \qquad b = 9 + \frac{c}{10}, \qquad c = 2 + \frac{d}{10},$$

$$d = 0 + \frac{e}{10}, \qquad e = 2 + \frac{f}{10}$$

and can be expressed thus

$$x = 1 + \frac{6}{10} + \frac{9}{10^2} + \frac{2}{10^3} + \frac{0}{10^4} + \frac{2}{10^5} + \frac{f}{10^6},$$

or

$$x = 1.69202 + \frac{f}{10^6},$$

so that the numbers 1, 6, 9, 2, 0, 2 appearing on the left of the above scheme are successive digits of x. Moreover, since $0 < f < 10$, it is clear that

$$1.69202$$

is an approximation to x in defect with five correct decimals.

Problems

Calculate by Horner's method:

1. $\sqrt[3]{7}$ to three decimals. **2.** $\sqrt[4]{18}$ to four decimals.
3. $\sqrt{3.895}$ to three decimals. **4.** $\sqrt[4]{20}$ to four decimals.
5. $\sqrt[5]{100}$ to three decimals.

Compute by Horner's method the roots of

6. $x^3 + x^2 + x - 100 = 0$ to three decimals.
7. $x^4 + 4x - 1 = 0$. Negative root to three decimals.
8. $x^3 - 3x + 1 = 0$. Positive roots to three decimals.
9. $x^4 - 4x + 1 = 0$. Positive roots to three decimals.

Compute to three decimals the coordinates of the points of intersection of the curves:

10. $x^2 + y = 1,\qquad x + y^2 = 2.$ **11.** $x^2 + y^2 = 5,\qquad y = x^2 + x.$

Compute to three significant figures the roots of

12. $2x^3 - 30x^2 + 7x + 10 = 0.$ **13.** $x^3 - 50x^2 + 120x + 30 = 0.$
14. $2x^3 - 250x^2 + 500x + 1000 = 0.$
15. $x^4 - 125x^3 + 200x^2 - 300x + 500 = 0.$

16. An iron spherical shell, empty inside, sinks in the water to the depth of its outer radius. If the thickness of the shell is 1 in. and the specific gravity of iron 7.5, what is the value of its outer radius?

17. Solve Prob. 16 if the shell completely submerged does not sink to the bottom.

18. A cubic open box has iron walls 1 in. thick and sinks in the water to half of its height. Compute to three significant figures the outer dimensions of the box in inches.

19. The figure represents the section through its axis of an iron conical funnel floating in the water, the portion XCY being submerged. If $AA' = BB' = 1$ in. and (a) $CY = \frac{1}{2}CB$, (b) $CY = CB$, what is the height CD in inches? Make the computation to three significant figures.

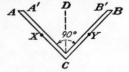

20. To what depth does a wooden ball sink in water if the specific gravity of wood is $\frac{3}{4}$?

3. Contraction. We return now to the method of "contraction" mentioned at the end of the last section. If the equation (4) for f is divided by 10^3, it takes the form

$$0.001f^3 + 5076.06f^2 + 1588795041.2f - 2338129592 = 0.$$

We may presume that the root of this equation will change but very little if fractional parts of the coefficients are dropped and the relation is replaced by

$$0f^3 + 5076f^2 + 1588795041f - 2338129592 = 0, \tag{1}$$

which is a quadratic equation. Its coefficients can be written immediately following the rule, which in essence is the contraction we wish to explain: We do not add zeros to numbers in columns 2, 3, 4 but cut off 1, 2, 3 digits on the right in the numbers found in columns 3, 2, 1, respectively. To calculate the root of equation (1), first we find its integral part, taking it equal to the integral part of the quotient obtained in dividing 2338129592 by 1588795041. This integral part being 1, we set

$$f = 1 + \frac{g}{10}; \qquad 0 < g < 10$$

and find the equation for g by Horner's process:

$$
\begin{array}{cccl}
1) & 5076 & 1588795041 & -\ 2338129592 \\
& & 5076 & 1588800117 \\
\cline{3-4}
& & 1588800117 & -\ 749329475 \qquad \text{(VI)} \\
& & 5076 & \\
\cline{3-3}
& & 1588805193 & \\
& 5076 & &
\end{array}
$$

This equation can be written thus:

$$50.76g^2 + 158880519.3g - 749329475 = 0.$$

When again fractional parts of the coefficients are neglected, it is replaced by another equation

$$50g^2 + 158880519g - 749329475 = 0,$$

whose root differs but little from the true value of g. The coefficients of the last equation are obtained by cutting off one and two digits on the right in the numbers of Scheme (VI) standing in columns 2 and 1, respectively. The integral part of g being 4, we set

$$g = 4 + \frac{h}{10}; \qquad 0 < h < 10$$

and calculate the coefficients of the equation in h in the usual way:

$$
\begin{array}{cccl}
4) & 50 & 158880519 & -\ 749329475 \\
& & 200 & 635522876 \\
\cline{3-4}
& & 158880719 & -\ 113806599 \qquad \text{(VII)} \\
& & 200 & \\
\cline{3-3}
& & 158880919 & \\
& 50 & &
\end{array}
$$

The equation in h can be written thus:

$$0.50h^2 + 15888091.9h - 113806599 = 0$$

and is replaced again, dropping fractional parts of the coefficients, by the equation of the first degree

$$15888091h - 113806599 = 0 \qquad (2)$$

whose coefficients are found from Scheme (VII) by cutting off one and two last digits in columns 2 and 1. Schemes (VI) and (VII) combined together

1)	5076	1588795041	− 2338129592
		5076	1588800117
		1588800117	− 749329475
		5076	635522876
		1588805193	− 113806599
4)	5076	200	
		158880719	
		200	
		158880919	
	50		

constitute the contracted part of Horner's process. Now the approximate value of h can be found from (2) by division. It is, however, necessary to know to how many digits the quotient should be computed. To this end the following general remark may serve as a guide to decide how many decimals in the root can be found by the contracted process. Examine how many digits there are in the number in the column preceding the last, obtained just before the contraction begins. Suppose there are N digits; in our example $N = 11$. If n is the degree of the equation, subtract $n + 1$ from N; the difference $N - (n + 1)$, which in our case is $11 - 4 = 7$, gives the number of digits that one might expect to find by the contracted process. According to this rule we can expect to find, in our example, seven digits of the root by contraction; hence, h must be determined to five digits by division. To this degree of approximation it is found

$$h = 7.1630,$$

and the corresponding value of the requested root is

$$x = 1.692021471630$$

to 12 decimals. In the regular part of Horner's process the numerical work increases with each new digit found, but in the contracted part, on the contrary, the work decreases with each step. The regular part is just like climbing a mountain: the exertion increases as we approach the summit; on the contrary, the contracted part is like a descent from the summit: the more we come down the less exertion we have to make.

We have treated our example with all the details necessary to explain the essential parts of Horner's method, but still it remains to examine two important points. In the first place, it remains to explain how to perform the last division with a large divisor in the most expedient manner and, in the second place, to examine more closely the question of the error committed by using the contracted process. In Sec. 4 rules of the remarkable method of division proposed by Fourier will be given, and in Sec. 5 we shall examine in detail the question of error.

Problems

Compute to the indicated number of decimals the roots of

1. $x^3 - 3 = 0$, six decimals.   **2.** $2x^3 - 5 = 0$, six decimals.
3. $x^3 - x - 1 = 0$, six decimals.
4. $3x^3 - 7x^2 + 2x + 5 = 0$, six decimals.
5. $6x^3 - 7x^2 + 10x + 5 = 0$, eight decimals.
6. $x^3 + 4x^2 - 7 = 0$. Positive root to six decimals.
7. $x^3 - 9x - 9 = 0$. Positive root to eight decimals.
8. $x^3 + 4x^2 - 10 = 0$. Positive root to eight decimals.
9. $x^3 - 7x + 7 = 0$. Negative root to eight decimals.
10. $x^4 - 4x + 1 = 0$. Smallest positive root to six decimals.
11. $x^4 - 6x + 2 = 0$. Largest positive root to six decimals.
12. $x^4 - 2x^3 + x^2 - 3 = 0$. Postive root to eight decimals.
13. $x^4 - 2x^3 + x^2 - x - 7 = 0$. Positive root to eight decimals.
14. $x^4 - 7x^3 + 6x^2 + 7x - 11 = 0$. Negative root to eight decimals.
15. $2x^4 - 3x^3 + 7x^2 - 12x + 3 = 0$. $\left.\right\}$ All real roots to six decimals.
16. $7x^4 - 16x^3 + x^2 - 9x + 7 = 0$.
17. $x^5 - x^3 + 2x^2 - 2x + 1 = 0$. $\left.\right\}$ All real roots to six decimals.
18. $x^5 + x^4 - 7x^3 - 22x^2 + x + 1 = 0$

★4. Fourier Division. When occasion arises to divide one number by another consisting of many digits, it is advantageous to employ the method of division proposed by Fourier. In Fourier's method of division a group or small number of first digits of the divisor consisting, say, of two or three digits is selected as an "abridged" divisor, and all the divisions are made with this abridged divisor, the remaining digits of the divisor being taken into consideration gradually to make the so-called "corrections." Suppose that B is the abridged divisor and the digits following it are $b_0, b_1, b_2, \ldots$ so that the complete divisor is

$$Bb_0b_1b_2 \ldots .$$

Let also

$$c_0c_1c_2 \ldots c_n$$

be the digits of the quotient found in the process of Fourier division
Then, the corresponding "correction" is computed by the formula

$$b_0c_n + b_1c_{n-1} + \cdots + b_nc_0$$

so that the corrections after finding one, two, three, etc., digits in the
quotient are

$$b_0c_0; \; b_0c_1 + b_1c_0; \; b_0c_2 + b_1c_1 + b_2c_0; \; \ldots .$$

These corrections at the beginning can be easily computed mentally;
but when several digits of the quotient have been found, the best way
to compute the corrections is the following: Write on a strip of paper
the digits of the divisor, following the abridged divisor in the reverse
order thus:

$$\overline{\ldots \quad b_3 \quad b_2 \quad b_1 \quad b_0}$$

Then, place the strip so that b_0 is immediately below the last digit
found in the quotient, and make the sum of products of numbers standing one above another. This can be done easily by adding first units
of these products, then tens if there are any, etc.

To find the first digit of the quotient divide the dividend, supplemented if necessary by as many zeros on the right as we wish, by the
abridged divisor in the ordinary way. Append to the remainder the next
digit of the dividend and from the number obtained subtract the
first correction, which gives the corrected partial dividend. To find the
second digit of the quotient divide this corrected partial dividend again
by the abridged divisor, append to the remainder the next digit of the
dividend, and subtract the second correction, which will give the second
corrected partial dividend. On dividing it by the abridged divisor the
third digit of the quotient is found. To the remainder obtained append
the corresponding digit of the dividend and subtract the third correction,
after which proceed as before. As long as after the corrections no negative numbers are encountered, this regular march of operation can be
continued. If, however, a negative number appears after the correction,
this is a sign that the last digit found in the quotient should be diminished by 1. This, when the last digit and some of the preceding are
0, causes a change in a number of the digits of the quotient. Suppose
that exactly i digits change. Then, to the negative number resulting
from the correction one must add

$$10B + b_0 + b_1 + \cdots + b_{i-1}$$

and then proceed as before. If it is desirable at a certain stage of the
Fourier division to make a change from the abridged divisor B to a

larger abridged divisor Bb_0, then a correction must be made as usual, but instead of proceeding with the division by B another digit from the dividend should be appended and another correction made but with a new abridged divisor. After this the regular procedure is followed with the new abridged divisor. The following examples will serve to illustrate Fourier's method of division.★

Example 1. It is required to find five digits in the quotient obtained in dividing 113806599 by 15888091. The operations are arranged as shown; explanatory remarks are given below.

$$
\begin{array}{l}
\quad\;\;\overset{\surd\surd\surd\surd\surd\surd}{113806599} \;\Big|\; \overline{\overline{15888091}} \\
\;\;\underline{105} \qquad\qquad\quad 1 \\
\;\;\;\;88 \qquad\qquad\; 7\overset{.}{2}630 \\
-\;56 \quad 7 \times 8 = 56 \text{ correction} \\
\;\;\underline{\;\;32} \\
\;\;\;\;30 \\
\;\;\underline{\;\;20} \\
-\;72 \quad 2 \times 8 + 7 \times 8 = 72 \text{ correction} \\
-\;52 \quad (\text{digit 2 too high}) \\
\;\;\underline{158} \\
\;\;106 \\
\;\;\underline{\;\;90} \\
\;\;166 \\
-\;112 \quad 6 \times 8 + 1 \times 8 + 7 \times 8 = 112 \text{ correction} \\
\;\;\underline{545} \\
-\;56 \quad 6 \times 8 + 1 \times 8 + 7 \times 0 = 56 \text{ correction} \\
\;\;\underline{489} \quad \text{Division continued with divisor 158} \\
\;\;474 \\
\;\;\underline{159} \\
-135 \quad 3 \times 8 + 6 \times 8 + 1 \times 0 + 7 \times 9 = 135 \text{ correction} \\
\;\;\underline{249} \\
-\;40 \quad 0 \times 8 + 3 \times 8 + 6 \times 0 + 1 \times 9 + 7 \times 1 = 40 \text{ correction} \\
\;\;209
\end{array}
$$

Explanations

The first abridged divisor is 15. Having found 2 after the second division and having made the correction a negative number, -52 is found. Therefore, 2 is changed into 1, and, since this is the only digit changed, we add

$$15 \times 10 + 8 = 158$$

to -52, which gives 106. Having found 6 as a third digit, we wish to take 158 for the new abridged divisor. To this end from 166 a correction amounting to 112 is subtracted, which leaves 54. Instead of dividing 54 by 15 the next digit 5 is appended, and from 545 a correction 56 corresponding to the abridged divisor 158 is subtracted. The resulting number now is divided by 158, and following corrections are made corresponding to the abridged divisor 158. The place of the decimal point *is* determined by inspection, and the quotient to five digits is found to be

$$7.1630+.$$

Example 2. To find seven digits in the quotient obtained dividing 26.73385 by 324.754813. Since the place of the decimal point is easily found by inspection, we disregard decimal points and arrange the operation as shown, taking at first 32 for the abridged divisor.

$$
\begin{array}{r|l}
\sqrt{}\sqrt{}\sqrt{}\sqrt{} & \\
2673385 & \overline{\overline{324754813}} \\
256 & 8232010 \\
\hline
113 & \\
\end{array}
$$

Correction	$-\ 32$
	$\overline{81}$
	64
	$\overline{173}$
Correction	$-\ 64$
	$\overline{109}$
	96
	$\overline{138}$
Correction	$-\ 66$
	$\overline{72}$
	64
	$\overline{85}$
Correction	$-\ 71$
	$\overline{140}$
Correction	$-\ 101$
	$\overline{390}$ Here we change to a new divisor
Correction	$-\ 46$
	$\overline{344}$
	324
	$\overline{200}$
Correction	$-\ 65$
	$\overline{135}$

Hence, the requested quotient is

$$0.08232010+.$$

The complete theory of Fourier division is somewhat complicated; moreover, it plays only an auxiliary role here and for this reason we shall not go into further details about it. However, because of its practical usefulness, it is impossible to omit one thing, and that is the application of Fourier's method of division to the extraction of square roots. To take an example suppose we want to extract the square root of the number 500. It is seen immediately that the integral part of the root is 22, and so it can be written as $22 + x$ with $0 < x < 1$. Then,

$$(22 + x)^2 = 484 + 44x + x^2 = 500$$

whence

$$44x + x^2 = 16$$

or

$$x = \frac{16}{44 + x}.$$

Dividing 16 by 44 + x by Fourier's method and taking 44 for the abridged divisor, we can determine the first digit of the quotient, which will then be written immediately after 44, and make the corresponding correction, after which the second digit of x will be determined and again written in the divisor, etc. In case the correction gives a negative result, the last digit both in the quotient and in the divisor should be diminished by 1. This may affect a certain number of digits, say i digits. Then to continue the operation, to the negative number we should add 10 times the abridged divisor plus *double* the sum of the i digits following it, or this number increased by 1, according as i is not greater than or is greater than one-half of the total number of digits written in the quotient. The other rules remain the same as in the ordinary division. The operations in our example are arranged as follows:

	$\overline{446}$ ₇₉ ₄₉

$$\frac{446}{160 \quad | \quad 143606807750}$$

	$\overline{446}$		
160	143606807750		7 9 4 9
132			7 9 4 9
$\overline{280}$	360680775Ø		

Correction	− 9		
	$\overline{271}$		
	264		Continuation
	70		3620
Correction	− 36	Correction	− 144
	$\overline{340}$		$\overline{3476}$
Correction	− 36		3122
	$\overline{304}$		3540
	264	Correction	− 168
	$\overline{400}$		$\overline{3372}$
Correction	− 36		3122
	$\overline{364}$		2500
	352	Correction	− 241
	$\overline{120}$		$\overline{2259}$
Correction	− 120		2230
	$\overline{00}$		290
	− 96	Correction	− 270
	$\overline{-96}$		200
$44 \times 10 + 2(3 + 6) =$	458	Correction	− 263
	$\overline{362}$		$\overline{-63}$

Here we change to
the larger divisor 446

Without continuing the division further we may conclude that
$$\sqrt{500} = 22.3606797749+.$$

Problems

By Fourier division find the quotient in the division of

1. 237.69 by 33.65489 to six decimals. Divisor 33 in three divisions and 336 after.

2. 63.9878 by 24.85397 to eight decimals. Divisor 24 in four divisions and 248 after.

3. 3.173563 by 334.856921 to 10 significant figures. Divisor 33 in five divisions and 334 after.

4. 180 by π to eight decimals; $\pi = 3.14159265358979 \cdots$.

Extract the square roots by Fourier's method:

5. $\sqrt{2.185}$ to eight decimals.

6. $\sqrt{3465.23}$ to eight decimals.

7. $\sqrt{0.031429}$ to eight decimals.

8. $\sqrt{\sqrt{2} - 1}$ to eight decimals.

★**5. Estimation of Error.** We must examine now the question of estimating the error that is caused by the process of contraction. The regular part of Horner's process led us to represent the root of the equation

$$x^3 - 7x + 7 = 0$$

with which we were concerned in the form

$$x = 1.69202 + \frac{f}{10^6}$$

where f is a root of the equation

$$0.001f^3 + 5076.06f^2 + 1588795041.2f - 2338129592 = 0$$

contained between 0 and 10. This equation was replaced, however, by

$$\phi(x) = 5076x^2 + 1588795041x - 2338129592 = 0$$

whose one root, which is close to f but not equal to it, we shall denote by f_0. It is necessary to estimate the difference $f - f_0$. To this end notice that the equation for f can be written thus:

$$\phi(f) = -0.001f^3 - 0.06f^2 - 0.2f.$$

On the other hand, $\phi(f_0) = 0$, and so

$$\phi(f) - \phi(f_0) = -0.001f^3 - 0.06f^2 - 0.2f.$$

But by the formula for the finite increment taught in courses of differential calculus

$$\phi(f) - \phi(f_0) = (f - f_0)\phi'(\xi),$$

where ξ is a certain number contained between f and f_0. Consequently,

$$f - f_0 = -\frac{0.001f^3 + 0.06f^2 + 0.2f}{\phi'(\xi)}$$

so that the difference $f - f_0$ is certainly negative because
$$\phi'(\xi) > 1.5 \times 10^9.$$
On the other hand, since $f < 10$,
$$0.001f^3 + 0.06f^2 + 0.2f < (0.001 + 0.006 + 0.002) \times 10^3$$
$$= 0.009 \times 10^3 < 0.01 \times 10^3,$$
and so
$$-\frac{0.01}{1.5} \times 10^{-6} < f - f_0 < 0. \tag{1}$$
At the same time
$$x = 1.69202 + \frac{f_0}{10^6} + \frac{f - f_0}{10^6}.$$
As to f_0, it was taken in the form
$$f_0 = 1 + \frac{g}{10}$$
where g is a root of the equation
$$50.76g^2 + 158880519.3g - 749329475 = 0$$
contained between 0 and 10. But this equation is replaced by
$$f_1(x) = 50x^2 + 158880519x - 749329475 = 0$$
whose root close to g may be denoted by g_0. The difference $g - g_0$ can be estimated in the same manner as the difference $f - f_0$; it turns out that
$$-\frac{0.79}{1.5} \times 10^{-6} < g - g_0 < 0. \tag{2}$$
At the same time we can write
$$f_0 = 1 + \frac{g_0}{10} + \frac{g - g_0}{10}.$$
Now
$$g_0 = 4 + \frac{h}{10}$$
where h, contained between 0 and 10, satisfies the equation
$$0.50h^2 + 15888091.9h - 113806599 = 0.$$
But for h we substitute the root h_0 of the equation
$$15888091h_0 - 113806599 = 0.$$
The difference $h - h_0$ can be estimated as before and it is found that
$$\frac{0.59}{1.5} \times 10^{-5} < h - h_0 < 0. \tag{3}$$

At the same time

$$g_0 = 4 + \frac{h_0}{10} + \frac{h - h_0}{10},$$

and

$$f_0 = 1 + \frac{4}{10} + \frac{h_0}{10^2} + \frac{h - h_0}{10^2} + \frac{g - g_0}{10},$$

$$x = 1.69202 + \frac{1}{10^6} + \frac{4}{10^7} + \frac{h_0}{10^8} + \frac{h - h_0}{10^8} + \frac{g - g_0}{10^7} + \frac{f - f_0}{10^6}.$$

By the inequalities (1), (2), and (3) the quantity

$$\Delta = \frac{f - f_0}{10^6} + \frac{g - g_0}{10^7} + \frac{h - h_0}{10^8},$$

which really represents the error caused by contraction, is negative and its absolute value is less than

$$\frac{0.01}{1.5} \times 10^{-12} + \frac{0.79}{1.5} \times 10^{-13} + \frac{0.59}{1.5} \times 10^{-13} < \frac{1.48}{1.5} \times 10^{-13} < 10^{-13}.$$

Since $h_0 \times 10^{-8}$ is contained between the limits

$$7.16300 \times 10^{-8} < h_0 \times 10^{-8} < 7.16301 \times 10^{-8},$$

it is clear that x will be contained between the limits

$$x < 1.6920214716301$$
$$x > 1.6920214716299$$

and by taking

$$x = 1.692021471630$$

we have an approximate value of the root differing from the true value by less than 10^{-13}. It goes without saying that the error caused by contraction can be estimated in much the same manner in every other case.★

6. Another Example. It is well to illustrate all parts of Horner's method by another example.

Example. Equation

$$x^4 - 4x + 1 = 0$$

has two real roots: one in the interval $(0, 1)$ and another in the interval $(1, 2)$. Let us compute approximately the smaller root. Writing

$$x = \frac{a}{10}, \qquad 0 < a < 10,$$

the equation to determine a is found immediately:

$$a^4 - 4000a + 10000 = 0,$$

and it is found by trials that the root, which is less than 10, is contained between 2 and 3 so that we write

$$a = 2 + \frac{b}{10}, \qquad 0 < b < 10.$$

The calculations necessary to find the transformed equation in b, as well as other equations occurring in Horner's method, are shown in the following scheme:

2)	1	0	0	− 4000	10000
		2	4	8	− 7984
		2	4	− 3992	20160000
		2	8	24	−19769375
		4	12	− 3968000	3906250000
		2	12	14125	
		6	2400	− 3953875	
		2	425	16375	
		8)	2825	− 3937500000	
5)	1	5	450		
		85	3275		
		5	475		
		90	375000		
		5			
		95			
		5			
		1000			
0)	1				

At this stage we begin the contraction, the various steps of which are presented in this scheme:

9)	1	3750	− 393750000	3906250000
		9	33831	− 3543445521
		3759	− 393716169	362804479
		9	33912	− 354311028
		3768	− 393682257	8493451
		9	333	
	9)	3777	− 30367892	
	1		333	
			-- 39367559	
		37		

Thus far the decimals found are

$$0.25099.$$

The remaining decimals will be found by Fourier division. Since the number obtained in column 4 before contraction had 10 digits, we can count to find $10 - 5 = 5$ digits by contraction, and so the division should be carried to three digits. The result is the following:

$$\begin{array}{c|c}
8493451 & \overline{3936755} \\
78 & 2157 \\ \hline
69 & \\
6 & \\ \hline
63 & \\
39 & \\ \hline
243 & \\
15 & \\ \hline
228 & \\
195 & \\ \hline
334 & \\
35 & \\ \hline
299 & \\
273 & \\ \hline
26 & \\
\end{array}$$

Hence, the requested root is

$$0.25099215,$$

and it remains only to examine more closely the degree of approximation. Setting

$$x = 0.250 + \frac{d}{10^4}$$

where $0 < d < 10$, it follows that d satisfies the equation

$$0.0001d^4 + d^3 + 3750d^2 - 393750000d + 3906250000 = 0.$$

Instead of d we actually substitute the root d_0 of the equation

$$d_0^3 + 3750d_0^2 - 393750000d_0 + 3906250000 = 0.$$

In the contracted part of the process it is found that

$$d_0 = 9 + \frac{e}{10}, \qquad 0 < e < 10$$

where e satisfies the equation

$$0.001e^3 + 37.77e^2 - 39368225.7e + 362804479 = 0,$$

but instead of e actually we take root e_0 of the equation

$$37e_0^2 - 39368225e_0 + 362804479 = 0.$$

Again,

$$e_0 = 9 + \frac{f}{10}, \qquad 0 < f < 10,$$

and f satisfies the equation

$$0.37f^2 - 3936755.7f + 8493451 = 0;$$

but instead of f the root f_0 of the equation

$$- 3936755f_0 + 8493451 = 0$$

is taken. Now we have

$$x = 0.25099 + \frac{f_0}{10^6} + \Delta,$$

where

$$\Delta = \frac{d - d_0}{10^4} + \frac{e - e_0}{10^5} + \frac{f - f_0}{10^6}.$$

The signs of the differences $d - d_0$, $e - e_0$, $f - f_0$ are now not known with certainty, and we may conclude only that

$$|d - d_0| < \frac{0.0001}{3.9} \times 10^{-4},$$

$$|e - e_0| < \frac{0.085}{3.9} \times 10^{-4},$$

$$|f - f_0| < \frac{0.46}{3.9} \times 10^{-4},$$

whence

$$|\Delta| < \frac{0.0132}{3.9} \times 10^{-8} < 3.4 \times 10^{-11};$$

so that the error caused by contraction can affect only the eleventh decimal place and the approximation turns out to be far better than suggested by the crude rule of Sec. 3. By Fourier division it is found that

$$0.00000215747 < \frac{f_0}{10^6} < 0.00000215748,$$

and so

$$0.250992157413 < x < 0.250992157582.$$

Taking

$$x = 0.2509921575,$$

we can be sure that the error does not exceed 10^{-10} in absolute value.

Problems

Having found four significant figures by the regular Horner's process, examine how many more can be found by contraction:

1. $x^3 - 3x + 1 = 0$. Root between 1 and 2.
2. $x^3 - 7x + 7 = 0$. Root between 1 and 1.5.
3. $6x^3 - 7x^2 + 10x + 5 = 0$.
4. $x^4 - 4x + 1 = 0$. Root between 1 and 2.
5. $x^4 - 2x^3 + x^2 - 3 = 0$. Positive root.
6. $x^5 - x^3 + 2x^2 - 2x + 1 = 0$.

7. Method of Iteration. It remains to consider two other methods that are often used in computing roots: the method of iteration (also called the method of successive approximations), and the method of Newton. The last method is a particular form of iteration, and so it is natural to examine the method of iteration first. The idea of this method is very general and may be applied to a given equation in a number of ways. The proposed equation can be written in the form

$$x = \theta(x)$$

in a variety of ways. We shall suppose as known that in some interval (a,b) the equation has a unique root ξ so that

$$\xi = \theta(\xi) \quad \text{and} \quad a < \xi < b.$$

As to the function $\theta(x)$ we shall assume at first that it is an increasing function of x in the interval (a,b) and that besides

$$a < \theta(a), \quad b > \theta(b)$$

or

$$a - \theta(a) < 0, \quad b - \theta(b) > 0.$$

The root ξ being the unique root in the interval (a,b), it follows that

$$x - \theta(x) < 0$$

if $a \leqq x < \xi$. and

$$x - \theta(x) > 0$$

if $\xi < x \leqq b$. Now let x_0 be an arbitrary number $\geqq a$ and $\leqq b$. Taking it for the first approximation to the root ξ and calculating by successive substitutions

$$x_1 = \theta(x_0), \quad x_2 = \theta(x_1), \quad x_3 = \theta(x_2), \quad \ldots$$

a sequence of numbers

$$x_0, x_1, x_2, \ldots$$

is generated. This sequence will be increasing with all its terms $< \xi$ if $x_0 < \xi$; and it will be decreasing with all its terms $> \xi$ if $x_0 > \xi$. To prove these assertions assume at first $x_0 < \xi$. Then,

$$x_1 = \theta(x_0) > x_0 \quad \text{or} \quad x_1 > x_0;$$

on the other hand,

$$\theta(\xi) > \theta(x_0)$$

because $\theta(x)$ is an increasing function. But

$$\theta(\xi) = \xi \quad \text{and} \quad \theta(x_0) = x_1$$

and so

$$x_1 < \xi.$$

By repetition of the same argument it is proved that

$$x_2 > x_1, \quad x_2 < \xi,$$

etc. Suppose next that $x_0 > \xi$. Then,

$$x_1 = \theta(x_0) < x_0 \quad \text{or} \quad x_1 < x_0;$$

and, on the other hand,

$$\xi = \theta(\xi) < \theta(x_0) = x_1$$

or
$$x_1 > \xi.$$

By repetition of the same reasoning it is shown that
$$\xi < x_2 < x_1,$$
etc. The sequence
$$x_0, x_1, x_2, \ldots$$
whether increasing or decreasing necessarily tends to a limit, and this limit, assuming continuity of $\theta(x)$, cannot be other than the root ξ to which it is thus possible to approach indefinitely and from either side.

Example 1. The equation
$$x^3 - 2x - 5 = 0$$
has one root between 2 and 3. It can be written in the form
$$x = \sqrt[3]{2x + 5} = \theta(x),$$
and all the conditions enumerated above are satisfied: $\theta(x)$ is an increasing function and
$$\theta(2) > 2, \qquad \theta(3) < 3.$$
Taking $x_0 = 2$ as a first approximation, and making use of tables, we calculate the sequence of approximations to the root ξ, each number listed below being taken less than its true value:

$x_1 = 2.08,$	$x_5 = 2.09452,$
$x_2 = 2.092,$	$x_6 = 2.094546,$
$x_3 = 2.0941,$	$x_7 = 2.0945506,$
$x_4 = 2.0944,$	$x_8 = 2.0945513.$

On the other hand, starting with $x_0 = 3$, we find the following approximations each a little greater than its true value:

$x_1 = 2.23,$	$x_5 = 2.09463,$
$x_2 = 2.115,$	$x_6 = 2.09457,$
$x_3 = 2.09777,$	$x_7 = 2.094555,$
$x_4 = 2.09503,$	$x_8 = 2.0945520,$

$$x_9 = 2.0945516.$$

Thus,
$$2.0945513 < \xi < 2.0945516$$
and so
$$\xi = 2.094551$$
with six true decimals.

Example 2. Let us consider the transcendental equation
$$2^x = 4x$$
which has one root between 0 and 1. When it is written in the form
$$x = 2^{x-2},$$

all the conditions for iteration are fulfilled. Starting with $x_0 = 0$ and using four-place logarithmic tables, we find the sequence of increasing approximations each being taken less than its true value:

$$x_1 = 0.25, \quad x_4 = 0.3088,$$
$$x_2 = 0.29, \quad x_5 = 0.3096,$$
$$x_3 = 0.305, \quad x_6 = 0.3098,$$
$$x_7 = 0.3099.$$

Similarly, we can find a sequence of decreasing approximations starting with $x_0 = 1$. On comparing results we find that the requested root is very nearly

$$\xi = 0.3099.$$

The last digit is naturally not quite certain since all calculations have been made with four-place tables.

An equation of the form

$$x = \theta(x),$$

having only one root between a and b, can always be solved by a convergent process of iteration if the derivative $\theta'(x)$ for $a \leqq x \leqq b$ remains in absolute value less than a fixed number $\rho < 1$, that is, if

$$|\theta'(x)| < \rho < 1$$

for $a \leqq x \leqq b$. In fact, the same equation can be written in an equivalent form

$$x = \omega(x)$$

where

$$\omega(x) = \frac{x + \theta(x)}{2}.$$

Now

$$\omega'(x) = \frac{1 + \theta'(x)}{2} > \frac{1 - \rho}{2} > 0$$

for $a \leqq x \leqq b$, and so $\omega(x)$ is an increasing function in the interval (a, b). On the other hand, the derivative of $f(x) = x - \omega(x)$ is

$$f'(x) = \frac{1 - \theta'(x)}{2} > \frac{1 - \rho}{2} > 0$$

and, being positive, $f(x)$ is an increasing function that takes the value 0 only once in the interval (a,b), namely, for $x = \xi$, ξ being the unique root of the equation

$$x = \theta(x)$$

between a and b. Hence, $f(a) < 0$, $f(b) > 0$ or

$$a < \omega(a), \quad b > \omega(b).$$

Thus, the iteration applied to the equation

$$x = \omega(x)$$

will converge whether we start with $x_0 = a$ or $x_0 = b$. Further remarks on the convergence of iteration will be found in the problems.

Problems

1. If $|\theta'(x)| < \rho < 1$ for $a \leqq x \leqq b$, the equation $x = \theta(x)$ cannot have more than one root between a and b. If it has one such root ξ and $x_1 = \theta(x_0)$, while $a \leqq x_0 \leqq b$, show that

$$|x_1 - \xi| < \rho|x_0 - \xi|.$$

HINT: $x_1 - \xi = (x_0 - \xi)\theta'(\eta)$, where η is contained between x_0 and ξ.

2. If $x = \theta(x)$ has a root ξ between a and b, $|\theta'(x)| < \rho < 1$ for $a \leqq x \leqq b$, and the first two approximations

$$x_0 \quad \text{and} \quad x_1 = \theta(x_0)$$

belong to (a,b), then all subsequent approximations $x_2, x_3, \ldots$ will belong to (a,b) and

$$|x_n - \xi| < \rho^n|x_0 - \xi|.$$

Moreover, the approximations will alternately be greater and smaller than ξ if $\theta'(x) < 0$ throughout the interval (a,b).

By iteration compute to four decimals the roots of the following:

3. $x^3 - 2x - 2 = 0$. Write $x = \sqrt[3]{2x + 2}$ or $x = \sqrt{2 + \dfrac{2}{x}}$. Which is more advantageous?

4. $x^3 - 10x - 5 = 0$. Write $x = \sqrt{10 + \dfrac{5}{x}}$.

5. $x^3 - 3x^2 - 1 = 0$. Write $x = 3 + \dfrac{1}{x^2}$.

6. $x^3 + x^2 - 100 = 0$. Write $x = \dfrac{10}{\sqrt{x + 1}}$ or $x = 4 + \dfrac{20}{x^2 + 5x + 20}$.

7. $x^3 + x^2 + x - 100 = 0$. Write $x = \sqrt[3]{100 - x^2 - x}$ or

$$x = 4 + \dfrac{16}{x^2 + 5x + 21}.$$

Compute to five decimals the smallest positive roots of the following:

8. $x^3 - 6x + 1 = 0$. Write $x = \dfrac{x^3 + 1}{6}$.

9. $x^4 - 3x + 1 = 0$. Write $x = \dfrac{x^4 + 1}{3}$.

10. $x^{10} + x^4 + x^3 - 10x + 2 = 0$. Write $x = \dfrac{2 + x^3 + x^4 + x^{10}}{10}$.

11. $x = 1 - \dfrac{x^3}{10}$. **12.** $x^3 - x^2 - 10x + 1 = 0$.

*When a root of an equation $F(x) = 0$ is enclosed in a sufficiently narrow interval (a, b) where the extreme values A and B of the derivative $F'(x)$ are not very different, the process of iteration can be considerably accelerated by writing the equation in the form

$$x = x - \dfrac{2}{A + B}F(x).$$

The factor $2(A + B)^{-1}$ can advantageously be replaced by a good approximation in small terms obtained by means of continued fractions. Apply this remark to compute to five decimals the roots of the following equations:

*13. $x^4 - 4x + 1 = 0$. Root between 1 and 2. Write

$$x = x - \tfrac{19}{4}(x - \sqrt[3]{4x - 1}).$$

*14. $x^3 - 4x^2 - x + 3 = 0$. Root between 4 and 5. Write

$$x = x - \tfrac{25}{8}(x - \sqrt[2]{4x^2 + x - 3}).$$

*15. $x^3 + x^2 - 100 = 0$. Write $x = x - \dfrac{5}{7}\Big(x - \dfrac{10}{\sqrt{x + 1}}\Big).$

*16. $x = \dfrac{1 - (1 + x)^{-15}}{10}.$ Write $x = x - \dfrac{8}{3}\Big(x - \dfrac{1 - (1 + x)^{-15}}{10}\Big).$

*17. $x = \dfrac{1 - (1 + x)^{-30}}{15}.$ Write $x = x - \dfrac{10}{7}\Big(x - \dfrac{1 - (1 + x)^{-30}}{15}\Big).$

18. $x = \tfrac{1}{2}e^{-x}$. Write $x = x - \tfrac{3}{4}(x - \tfrac{1}{2}e^{-x}).$

19. $x = e^{-x}$. Write $x = x - \tfrac{7}{11}(x - e^{-x}).$

20. $x + 1 = 10^x$. Write $x = x - \tfrac{19}{4}(x + 1 - 10^x).$

21. $x^x = 100$. Write $x = x + 2 - x \log x.$

22. $x = 1 - \tfrac{1}{10} \sin x.$

23. $x + \sin x = 1$. Write $x = \dfrac{x + 1 - \sin x}{2}.$

24. $x = \cos x$. Write $x = x - \tfrac{3}{5}(x - \cos x).$

25. $\sin x = \tfrac{1}{2}x$. Set $x = \dfrac{\pi}{2} + y$ and write

$$y = y - \frac{5}{8}\Big(y + \frac{\pi}{2} - 2 \cos y\Big).$$

26. $x = \tan x$. Smallest positive root. Write $x = \pi + \arctan x.$

27. $x \tan x = 1$. Smallest positive root. Write

$$x = \frac{x + \arctan \dfrac{1}{x}}{2}.$$

8. Newton's Method. Newton's method to approximate roots may be considered as a particular form of iteration. The proposed equation $f(x) = 0$ can be written in the form

$$x = x - \frac{f(x)}{f'(x)}$$

or

$$x = \theta(x)$$

with

$$\theta(x) = x - \frac{f(x)}{f'(x)}.$$

To examine under what conditions successive approximations starting with x_0:

$$x_1 = \theta(x_0), \qquad x_2 = \theta(x_1), \qquad x_3 = \theta(x_2), \qquad \dots$$

converge, we shall assume that:

1. In the interval (a,b) the equation $f(x) = 0$ has one root ξ.

2. Neither $f'(x)$ nor $f''(x)$ vanishes in that interval. Calculating the derivative $\theta'(x)$, we find

$$\theta'(x) = \frac{f(x)f''(x)}{f'(x)^2}.$$

Since $f'(x)$ does not change sign between a and b, the function $f(x)$ is either increasing or decreasing; and because it vanishes for $x = \xi$, the extreme values $f(a)$ and $f(b)$ must be of opposite signs. On the other hand, the sign of $f''(x)$ does not change so that $f''(a)$ and $f''(b)$ have the same sign; therefore, $f(a)f''(a)$ and $f(b)f''(b)$ have opposite signs: one of these numbers is positive and the other negative. Of the end points of the interval (a,b) we denote by α the one at which $f(x)$ and $f''(x)$ take the same sign, the other end point being denoted by β, so that $\alpha = a, \beta = b$ if

$$f(a)f''(a) > 0,$$

and $\alpha = b, \beta = a$ if

$$f(b)f''(b) > 0.$$

From the expression of the derivative $\theta'(x)$ it is evident that it changes sign only on passing through $x = \xi$. Consequently, between α and ξ the sign of $\theta'(x)$ is the same as that for $x = \alpha$, that is, positive by the choice of α. Now if $\alpha = a$, in the interval (a, ξ) the function $\theta(x)$ is increasing since its derivative is positive and at the same time

$$\theta(a) - a = -\frac{f(a)}{f'(a)} > 0.$$

Again, if $\alpha = b$, the function $\theta(x)$ is increasing in the interval (ξ, b) and

$$\theta(b) - b = -\frac{f(b)}{f'(b)} < 0.$$

Moreover, $\theta(x)$ is continuous in (a,b) so that the conditions of the convergence of an iteration process as enumerated in Sec. 7 are fulfilled if we start with the first approximation $x_0 = \alpha$. The successive approximations

$$\alpha_1 = \alpha - \frac{f(\alpha)}{f'(\alpha)}, \qquad \alpha_2 = \alpha_1 - \frac{f(\alpha_1)}{f'(\alpha_1)}, \qquad \dots$$

will be all on the same side of the root ξ, each nearer to it than the preceding one, and will converge to this root. These conclusions become intuitive by considering the curve $y = f(x)$.

Since the second derivative y'' does not vanish in the interval (a,b), this curve is concave upward or downward according as $y'' > 0$ or

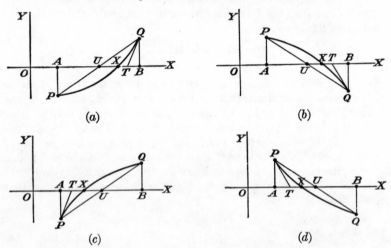

$$(a) \qquad (b)$$

$$(c) \qquad (d)$$

$y'' < 0$. In the figures (a), (b), (c), and (d) the points A and B on the x-axis have abscissas $x = a$ and $x = b$. For values of x contained between a and b the curve $y = f(x)$ is represented by the arc PXQ intersecting $y = 0$ at the point X whose abscissa is $OX = \xi$. In case of figures (a) and (b), y and y'' have the same sign at Q. The tangent at this point is represented by the equation

$$y - f(b) = f'(b)(x - b)$$

and intersects $y = 0$ at the point T with the abscissa

$$OT = b - \frac{f(b)}{f'(b)}.$$

Notice that T is on the same side of X as B but nearer to X than B. In case of figures (c) and (d), y and y'' have the same sign at P. The tangent at this point is represented by the equation

$$y - f(a) = f'(a)(x - a)$$

and intersects $y = 0$ at the point T with the abscissa

$$OT = a - \frac{f(a)}{f'(a)}.$$

Again, T is nearer to X than A and is located on the same side. These intuitive considerations confirm previously reached conclusions and also suggest the consideration of the point of intersection of the chord PQ with the x-axis. If this point is U, it is on the opposite side of T with respect to X and nearer to it than to A and B, respectively. Since the equation of PQ is

$$y = f(a) + \frac{f(b) - f(a)}{b - a}(x - a),$$

the abscissa of the point U will be

$$OU = a - \frac{f(a)}{f(b) - f(a)}(b - a) = b - \frac{f(b)}{f(a) - f(b)}(a - b)$$

and can also be represented by

$$\beta_1 = \beta - \frac{f(\beta)}{f(\beta) - f(\alpha)}(\beta - \alpha).$$

The double inequality

$$\alpha_1 < \xi < \beta_1,$$
$$\alpha_1 > \xi > \beta_1,$$

according as $\alpha = a$ or $\alpha = b$, gives an estimation of the error committed by calculating the second approximation to the root by Newton's method.

If the limits α_1 and β_1 are still too far apart, the most practical way to proceed further is as follows: Let the smaller of these limits be a' and the larger b'. Examine the difference $b' - a'$ and suppose that the first significant decimal in it occupies the lth place. Then, retain in a' and b' only l decimals. Let a'' and b'' be the numbers obtained by thus rounding off a' and b'. Clearly a'' is less than ξ, but b'' may be greater or less than this number. Accordingly, find out first by substitution whether b'' is greater than ξ or not. Set in the first case $a_1 = a''$, $b_1 = b''$ and in the second case $a_1 = b''$, $b_1 = b'' + 10^{-l}$. Then, the interval (a_1, b_1) contains ξ. Starting with this interval, find narrower limits α_2, β_2, in the same way as α_1, β_1 were found starting with (a, b). Then, with α_2, β_2 repeat the procedure applied to α_1, β_1 and continue in this manner as long as necessary to reach the desired degree of approximation. After a few steps of this kind, the convergence becomes very rapid, but we shall not detain ourselves to examine this point.

Example 1. Let us calculate by this method the root of the equation

$$f(x) = x^3 - 3x + 1 = 0$$

that is contained between 1 and 2. The derivative

$$f'(x) = 3x^2 - 3$$

vanishes for $x = 1$ so that strictly speaking the interval $(1, 2)$ does not satisfy the assumed conditions. Since, however,

$$f'(1) = -1, \qquad f''(1) = 6$$

are of opposite signs, we have to take $\alpha = 2$, and we observe that in the interval $(\xi, 2)$ the first derivative does not vanish. With $\alpha = 2$, $\beta = 1$ we find

$$\alpha_1 = 2 - \tfrac{3}{9} = 1.66+, \qquad \beta_1 = 1.25$$

so that α_1 and β_1 are rounded off to

$$1.6 \quad \text{and} \quad 1.2.$$

On substituting 1.6 it is found that

$$f(1.6) = 0.296, \qquad f'(1.6) = 4.68,$$
$$f(1.2) = -0.872,$$
$$\alpha_2 = 1.6 - \frac{0.296}{4.68} = 1.536-, \qquad \beta_2 = 1.2 + \frac{0.872}{1.168} \times 0.4 = 1.498+.$$

These numbers are rounded off to 1.49 and 1.53, and on substitution it is found

$$f(1.53) = -0.008423$$

so that 1.53 is less than the root. Accordingly, we take $a_1 = 1.53$, $b_1 = 1.54$ and continue thus:

$$f(1.54) = 0.032264, \qquad f'(1.54) = 4.1148,$$
$$f(1.53) = -0.008423,$$
$$\alpha_3 = 1.54 - \frac{32264}{411480} = 1.53216-, \qquad \beta_3 = 1.53 + \frac{8423}{40687} = 1.53207+.$$

Without carrying out the approximations further we may conclude that by taking

$$\xi = 1.5321$$

we do not commit an error exceeding in absolute value 10^{-4}. The next step would give about eight decimals in the root. Of course, the interval $(1, 2)$ with which we started is too large and it is always advisable to find by trials, as in Horner's method, one or two decimals in the root.

Example 2. To find approximately the root of the equation

$$x = 10^{-x}.$$

If we take common logarithms of both sides, this equation is replaced by

$$f(x) = x + \log x = 0$$

and by trial it is found that its root is contained between 0.3 and 0.4. The derivatives

$$f'(x) = 1 + \frac{M}{x}, \qquad M = 0.434295$$
$$f''(x) = -\frac{M}{x^2}$$

do not vanish in the interval $(0.3, 0.4)$. With the approximation allowed by six-place tables of logarithms it is found that

$$f(0.3) = -0.222879, \qquad f'(0.3) = 2.44765,$$
$$f(0.4) = +0.002060.$$

The second derivative being negative, we take $\alpha = 0.3$, $\beta = 0.4$ and then

$$\alpha_1 = 0.3 + \frac{222879}{244765} = 0.391+, \qquad \beta_1 = 0.4 - \frac{206}{224939} = 0.3991-.$$

These numbers are rounded off to 0.391 and 0.399. On trying 0.399 it is found that

$$f(0.399) = -0.000027.$$

Consequently, we must take $a_1 = 0.399$, $b_1 = 0.400$. Next,

$$f(0.399) = -0.000027, \qquad f'(0.399) = 2.08846,$$
$$f(0.400) = 0.002060,$$

$$\alpha_2 = 0.399 + \frac{27}{208846} = 0.3990129+, \qquad \beta_2 = 0.400 - \frac{2060}{2087} \times 10^{-3} = 0.399013-.$$

Hence, with the approximation attainable with six-place tables the requested root is

$$\xi = 0.399013.$$

Problems

Apply Newton's method to the equations:

1. $x^3 - 3x + 1 = 0.$ 2. $x^3 + 4x^2 - 7 = 0.$
3. $x^3 - 7x + 7 = 0.$ 4. $x^4 - 4x + 1 = 0.$
5. $2^x = 4x.$ Write $x \log 2 - \log x - \log 4 = 0.$

6. $x - \sin x = \dfrac{\pi}{2}.$ 7. $x^2 - \log x = \frac{3}{2}.$

8. $x = \cos x.$ 9. $x^x = 1000.$

10. For what central angles is the arc of a circle twice as long as the chord subtended by the arc?

11. If the area of a sector of a circle is bisected by its chord, what is the angle at the center?

12. A chord cuts off a segment equal to one-third the area of a circle; what is the angle at the center subtended by the chord?

13. A cylindrical log of wood of specific gravity $\frac{2}{3}$ floats in the water. To what depth does it sink?

*14. Find the smallest α such that the inequality

$$e^{-\alpha x} \leqq \frac{1}{1 + x^2}$$

holds for all positive x.

*15. Writing an equation $f(x) = 0$ in the form

$$\frac{f(x)}{\sqrt{\pm f'(x)}} = 0$$

and applying Newton's method, the following formula for the second approximation is obtained:

$$x' = x - \frac{f(x) f'(x)}{f'(x)^2 - \frac{1}{2} f(x) f''(x)}.$$

Show that the derivative of the right-hand side is always positive if $f(x)$ has only real roots. Hence, show that the iterations always converge no matter how the first approximation is chosen. If the first approximation is chosen between two consecutive roots α and β of the derivative $f'(x)$, the iterations converge and very rapidly to the root of $f(x) = 0$ that lies between α and β. Examine the question of the rapidity of convergence in the neighborhood of a root. Apply this method to the examples:

$$(a)\ \ x^3 - 3x + 1 = 0, \qquad (b)\ \ x^3 - 7x + 7 = 0.$$

What iteration process results in the case of an equation $x^n - a = 0$?

CHAPTER IX

DETERMINANTS AND MATRICES

1. Determinants of Order 2. Expressions known as determinants have their natural origin in the solution of systems of linear equations with two or more unknowns. Suppose we have to solve the system of two equations

$$a_1x + b_1y = c_1$$
$$a_2x + b_2y = c_2 \tag{1}$$

with unknowns x and y. To solve this system we may seek to combine these equations so as to eliminate either y or x. To eliminate y, equations (1) are multiplied, respectively, by b_2 and $- b_1$ and the results added; using multipliers $- a_2$ and a_1 and adding, x will be eliminated. The equations, one containing only x and the other only y, obtained in this way are

$$(a_1b_2 - a_2b_1)x = c_1b_2 - c_2b_1$$
$$(a_1b_2 - a_2b_1)y = c_2a_1 - c_1a_2 \tag{2}$$

From the manner in which they are obtained equations (2) are satisfied by the values x, y that satisfy (1) so that all solutions of the system (1) are among the solutions of the system (2). Now we notice that in both equations (2) x and y have the same coefficient

$$a_1b_2 - a_2b_1$$

and, if this coefficient is not zero, the only values of x and y satisfying equations (2) are

$$x = \frac{c_1b_2 - c_2b_1}{a_1b_2 - a_2b_1}, \qquad y = \frac{c_2a_1 - c_1a_2}{a_1b_2 - a_2b_1}. \tag{3}$$

Therefore, these are the only values of the unknowns that can satisfy the original system (1), and it can be verified by direct substitution that x and y as given by (3) actually satisfy equations (1). On examining expressions (3) we notice that x and y appear as fractions with the same denominator

$$a_1b_2 - a_2b_1.$$

This expression is called the determinant of system (1) and is denoted by the sign

$$\begin{vmatrix} a_1 & b_1 \\ a_2 & b_2 \end{vmatrix} = a_1b_2 - a_2b_1.$$

On examining the numerators of formulas (3) it is seen at once that they also are determinants

$$\begin{vmatrix} c_1 & b_1 \\ c_2 & b_2 \end{vmatrix} = c_1 b_2 - c_2 b_1; \qquad \begin{vmatrix} c_1 & c_. \\ a_2 & c_2 \end{vmatrix} = a_1 c_2 - a_2 c_1.$$

Consequently, x and y appear as quotients of the determinants

$$x = \frac{\begin{vmatrix} c_1 & b_1 \\ c_2 & b_2 \end{vmatrix}}{\begin{vmatrix} a_1 & b_1 \\ a_2 & b_2 \end{vmatrix}}, \qquad y = \frac{\begin{vmatrix} a_1 & c_1 \\ a_2 & c_2 \end{vmatrix}}{\begin{vmatrix} a_1 & b_1 \\ a_2 & b_2 \end{vmatrix}}. \tag{4}$$

In the determinant

$$\begin{vmatrix} a_1 & b_1 \\ a_2 & b_2 \end{vmatrix}$$

we distinguish two *rows* a_1, b_1 and a_2, b_2, and two *columns* $\begin{matrix} a_1 \\ a_2 \end{matrix}$ and $\begin{matrix} b_1 \\ b_2 \end{matrix}$; a_1, b_1, a_2, b_2 themselves are called *elements* of the determinant. The determinants in the numerators are obtained from

$$\begin{vmatrix} a_1 & b_1 \\ a_2 & b_2 \end{vmatrix}$$

by replacing the elements of the first and second *columns* by c_1 and c_2, respectively.

By means of formulas (4), solutions of the given system can be readily computed provided the determinant of this system is not zero. Circumstances that may occur when this determinant is zero will be examined later in connection with the general treatment of systems of linear equations with any number of unknowns.

Example. If we wish to solve the system

$$7x - 3y = 1, \qquad 6x + 5y = 3$$

by application of formulas (4), we calculate the determinants:

$$\begin{vmatrix} 7 & -3 \\ 6 & 5 \end{vmatrix} = 53, \qquad \begin{vmatrix} 1 & -3 \\ 3 & 5 \end{vmatrix} = 14, \qquad \begin{vmatrix} 7 & 1 \\ 6 & 3 \end{vmatrix} = 15$$

and, dividing the last two by the first, we find at once

$$x = \tfrac{14}{53}, \qquad y = \tfrac{15}{53}.$$

Problems

Solve by determinants:

1. $6x + 2y = 25,$
   $7x + 3y = 20.$

2. $5x - 3y = 1,$
   $4x + 7y = 5.$

3. $10x - 7y = 1,$
   $15x - 11y = 2.$

4. $21x + 13y = 1,$
   $5x + 3y = 1.$

5. $ax + (a + 1)y = 1,$
$a^2x + (a^2 - 1)y = a.$

6. $\dfrac{x}{a + b} + \dfrac{y}{a - b} = 2a.$
$x - y = 4ab.$

7. $\dfrac{2}{x} - \dfrac{3}{y} = 1,$
$\dfrac{3}{x} - \dfrac{2}{y} = 1.$

8. $3x - 4y = 2xy,$
$4x - 5y = 3xy.$

2. Polynomials in Several Variables. Considering the letters a_1, b_1, a_2, b_2 as indeterminates or variables, we see that the determinant

$$\begin{vmatrix} a_1 & b_1 \\ a_2 & b_2 \end{vmatrix} = a_1b_2 - a_2b_1$$

is a rational integral function or a polynomial in four variables a_1, b_1, a_2, b_2.

Not only in discussing determinants but also later we shall have occasion to deal with rational integral functions or polynomials in several variables, and it is necessary to explain at once what meaning is given to these terms. The letters x, y, z, . . . representing variables or indeterminates, an expression of the form

$$Ax^\alpha y^\beta z^\gamma \cdots$$

where A is a numerical constant and exponents α, β, γ, . . . are non-negative integers, is called a monomial in the variables x, y, z, . . . , A being its coefficient. Similar monomials are those in which exponents of the same letters are equal. An expression consisting of several monomials connected with signs $+$ or $-$ is called an *integral rational function* of the variables x, y, z, . . . or a polynomial in x, y, z, . . . , while the monomials that compose it are its terms. Every polynomial can be written in the form where no two of the constituent monomials are similar; such a form may be called a *reduced form*. In the following examples:

$$x - 2y, \qquad x^2 + xy + y^2, \qquad x^4 - x^3y + x^2 - y + 1$$

we have polynomials in two variables presented in a reduced form. The expressions

$$x^3 + y^3 + z^3 - 6xyz, \qquad x^3y + y^3z + z^3x, \qquad 2x^2y^2z^2 + x + y + z$$

are polynomials in three variables, and similarly one can write examples of polynomials in four and more variables. Two polynomials in the same variables are considered as equal if, presented in a reduced form, they are identical term for term. Notice that by writing in terms with coefficients zero, or on the contrary deleting such terms, we do not change the polynomial. A polynomial is identically equal to zero if the coefficients of all its terms are zero. How to add, subtract, and multiply polynomials is assumed as known from elementary courses of algebra. Polynomials in several variables are classified according to their degree. The sum of the degrees of the variables in a monomial is called its *dimension*. Now the highest dimension of the constituent monomials is called the *degree* of a polynomial that is not identically equal to 0. Thus, the polynomials

$$x^3 + x^2y - 3x^2 + y + 1, \qquad x^2y^2z^2 + x^4z + x^2y - x + 4y$$

are, respectively, of degrees 3 and 6. A polynomial is called *homogeneous* of dimension m if all its constituent monomials have the same dimension m. Thus, the polynomials

$$x + 2y - 3z, \qquad x^2 - xy + y^2, \qquad x^2y^2 + x^4 + z^4 - 3x^2yz$$

are homogeneous polynomials of dimensions 1, 2, 4.

Homogeneous polynomials are oftentimes called *forms*. Forms of dimension 1 are called linear forms, those of dimensions 2, 3, . . . are called quadratic, cubic, etc., forms. The general linear form in the variables $x, y, z, \ldots , v$ is

$$ax + by + cz + \cdots + lv$$

where $a, b, c, \ldots , l$ are constants. Often in dealing with integral rational functions of several variables special attention is paid to some of these variables, say $x, y, \ldots , t$, and the function is written as a polynomial in $x, y, \ldots , t$ whose coefficients are polynomials in the remaining variables. Again, we can speak of the degree of this function with respect to the selected variables, or of its being homogeneous in them of a certain dimension. Thus, the integral rational function in four variables

$$xz + yt$$

is linear and homogeneous with respect to x, y as well as with respect to z, t. Again,

$$zxy + t^2x^2 + (z^2 + t^2)y^2$$

with respect to x, y is homogeneous of dimension 2; writing the same expression in the form

$$y^2z^2 + (x^2 + y^2)t^2 + xyz,$$

we can say that it is of the second degree in z, t but not homogeneous.

Problems

1. What are the degrees of the polynomials?

(a) $xyz + x^2y + y^2x + 1$.   (b) $x^3y + y^3z + 6x^2y^2z$.

(c) $(2x^2 + xy)(z^3 - 1) + x^2y^2$.   (d) $(z^2 - xy)^2x - 7x^4z^2$.

2. Which of the following polynomials are homogeneous and which are not?

(a) $x^3 + y^3 - 3xyz$.

(b) $x^3 + y^3 + x^2y - 3xy$.

(c) $x^4 - 6x^2y^2 + z^4 - 2x^2y^2t^2$.

(d) $x^3 + y^3 + z^3 - 3xyz$.

3. Show that a polynomial in x, y, z is divisible by $(x - y)(x - z)(y - z)$ if it vanishes for $x = y$ and $x = z$ and $y = z$.

4. Find a homogeneous polynomial of dimension 4 in the variables x, y, z that vanishes for $x = y$, $x = z$, $y = z$ and takes values 1, 2, 3, respectively, for

$$x = -1, \quad y = 1, \quad z = 0$$
$$x = 1, \quad y = -1, \quad z = 0$$
$$x = 0, \quad y = 1, \quad z = -1.$$

5. Show that $x^3 + y^3 + z^3 - 3xyz$ has a factor $x + y + z$, and find the other factor. Show that this factor cannot vanish for real values of x, y, z except when $x = y = z$.

3. Characteristic Properties of Determinants of the Second Order.
Returning to the determinant

$$\begin{vmatrix} a_1 & b_1 \\ a_2 & b_2 \end{vmatrix} = a_1b_2 - a_2b_1,$$

we notice at once that as a function of the variables a_1, b_1, a_2, b_2 it has the following properties:

1. With the variables arranged in a square array

$$\begin{pmatrix} a_1 & b_1 \\ a_2 & b_2 \end{pmatrix}$$

that is called a *matrix*, the determinant is a linear homogeneous function of the elements of each row of this matrix.

2. The determinant vanishes if two rows are identical.

3. For the special matrix

$$\begin{pmatrix} 1 & 0 \\ 0 & 1 \end{pmatrix}$$

the determinant takes the value 1.

Presently, we shall see that by these properties the determinant of the second order is characterized completely.

Let us set a more general problem of finding all the integral rational functions of four elements of a matrix

$$\begin{pmatrix} a_1 & b_1 \\ a_2 & b_2 \end{pmatrix}$$

that possess the following two properties:

1. They are linear homogeneous functions in the elements of each row.

2. They vanish if two rows of the matrix are identical.

By the property 1 the function we seek must be of the form

$$Aa_1 + Bb_1 = F(a_1, b_1, a_2, b_2)$$

and A and B in turn are linear and homogeneous expressions in a_2, b_2:

$$A = C_1a_2 + C_2b_2, \qquad B = D_1a_2 + D_2b_2$$

so that

$$F(a_1, b_1, a_2, b_2) = (C_1a_2 + C_2b_2)a_1 + (D_1a_2 + D_2b_2)b_1 \qquad (1)$$

with coefficients C_1, C_2; D_1, D_2 independent of a_1, b_1; a_2, b_2. Replacing a_1, b_1; a_2, b_2, respectively, by $a_1 + a_2$, $b_1 + b_2$; $a_2 + a_1$, $b_2 + b_1$, we must have by the property 2

$$F(a_1 + a_2, \ b_1 + b_2, \ a_2 + a_1, \ b_2 + b_1) = 0$$

because two rows of the matrix

$$\begin{pmatrix} a_1 + a_2 & b_1 + b_2 \\ a_2 + a_1 & b_2 + b_1 \end{pmatrix}$$

are identical. On the other hand, in view of the fact that F is a linear homogeneous function with respect to the two first arguments, we have

$$F(a_1 + a_2, \ b_1 + b_2, \ a_2 + a_1, \ b_2 + b_1) = F(a_1, \ b_1, \ a_2 + a_1, \ b_2 + b_1) \\ + F(a_2, \ b_2, \ a_2 + a_1, \ b_2 + b_1).$$

Taking into account that F is also linear and homogeneous with respect to the last two arguments, we have

$$F(a_1, \ b_1, \ a_2 + a_1, \ b_2 + b_1) = F(a_1, \ b_1, \ a_2, \ b_2) + F(a_1, \ b_1, \ a_1, \ b_1),$$
$$F(a_2, \ b_2, \ a_2 + a_1, \ b_2 + b_1) = F(a_2, \ b_2, \ a_2, \ b_2) + F(a_2, \ b_2, \ a_1, \ b_1).$$

But by the property 2 again

$$F(a_1, \ b_1, \ a_1, \ b_1) = 0, \qquad F(a_2, \ b_2, \ a_2, \ b_2) = 0;$$

hence,

$$F(a_1, \ b_1, \ a_2 + a_1, \ b_2 + b_1) = F(a_1, \ b_1, \ a_2, \ b_2),$$
$$F(a_2, \ b_2, \ a_2 + a_1, \ b_2 + b_1) = F(a_2, \ b_2, \ a_1, \ b_1),$$

and

$$F(a_1, \ b_1, \ a_2, \ b_2) + F(a_2, \ b_2, \ a_1, \ b_1) \\ = F(a_1 + a_2, \ b_1 + b_2, \ a_2 + a_1, \ b_2 + b_1) = 0;$$

or

$$F(a_1, \ b_1, \ a_2, \ b_2) = - F(a_2, \ b_2, \ a_1, \ b_1). \qquad (2)$$

This means that F changes its sign when two rows of the matrix on which it depends are interchanged. On account of (1)

$$F(a_1, \ b_1, \ a_2, \ b_2) = (C_1a_2 + C_2b_2)a_1 + (D_1a_2 + D_2b_2)b_1,$$
$$F(a_2, \ b_2, \ a_1, \ b_1) = (C_1a_1 + C_2b_1)a_2 + (D_1a_1 + D_2b_1)b_2,$$

which combined with (2) leads to the identity

$$(C_1a_2 + C_2b_2)a_1 + (D_1a_2 + D_2b_2)b_1 = -(C_1a_1 + C_2b_1)a_2 - (D_1a_1 + D_2b_1)b_2,$$

and from this identity it follows that

$$C_1 = 0, \qquad D_2 = 0, \qquad C_2 = -D_1.$$

On substituting $C_1 = 0$, $D_2 = 0$, $C_2 = C$, $D_1 = -C$ into the expression of F we have

$$F(a_1, b_1, a_2, b_2) = C(a_1b_2 - a_2b_1) = C\begin{vmatrix} a_1 & b_1 \\ a_2 & b_2 \end{vmatrix},$$

and consequently all functions of the matrix

$$\begin{pmatrix} a_1 & b_1 \\ a_2 & b_2 \end{pmatrix}$$

satisfying the requirements 1 and 2 differ from the determinant

$$\begin{vmatrix} a_1 & b_1 \\ a_2 & b_2 \end{vmatrix}$$

only by a factor independent of the variables a_1, b_1; a_2, b_2. In particular, if in addition to conditions 1 and 2 the function we seek takes the value 1 for the special matrix

$$\begin{pmatrix} 1 & 0 \\ 0 & 1 \end{pmatrix}$$

it will be identical with the above-mentioned determinant.

All other properties of determinants of the second order follow very easily if we consider them as functions of a matrix satisfying the characteristic properties 1, 2, and 3. Later, we shall discuss from the same point of view the properties of general determinants and will confine ourselves here to showing how easily the theorem of multiplication of determinants can be established for determinants of the second order.

Let

$$D_1 = \begin{vmatrix} a_1 & b_1 \\ a_2 & b_2 \end{vmatrix}, \qquad D_2 = \begin{vmatrix} c_1 & d_1 \\ c_2 & d_2 \end{vmatrix}$$

be two determinants. Making sums of the products of elements of the rows of D_1 by the corresponding elements of the columns of D_2, we form a new determinant

$$D = \begin{vmatrix} a_1c_1 + b_1c_2 & a_1d_1 + b_1d_2 \\ a_2c_1 + b_2c_2 & a_2d_1 + b_2d_2 \end{vmatrix}.$$

The theorem of multiplication of determinants states that

$$D = D_1D_2.$$

To prove this theorem consider a_1, b_1; a_2, b_2 as variables. Then, D is a function of the matrix

$$\begin{pmatrix} a_1 & b_1 \\ a_2 & b_2 \end{pmatrix}$$

linear and homogeneous in the elements of each row, and vanishing when two rows are identical. Consequently,

$$D = C \begin{vmatrix} a_1 & b_1 \\ a_2 & b_2 \end{vmatrix}$$

where C does not depend on the variables a_1, b_1; a_2, b_2. For the matrix

$$\begin{pmatrix} 1 & 0 \\ 0 & 1 \end{pmatrix}$$

D reduces to the determinant

$$\begin{vmatrix} c_1 & d_1 \\ c_2 & d_2 \end{vmatrix} = D_2$$

and

$$\begin{vmatrix} a_1 & b_1 \\ a_2 & b_2 \end{vmatrix} \text{ becomes } \begin{vmatrix} 1 & 0 \\ 0 & 1 \end{vmatrix} = 1.$$

Hence,

$$C = D_2$$

and

$$D = D_1 D_2$$

as we wanted to prove.

Problems

1. Verify the following properties of determinants:

(a) $\begin{vmatrix} a+e & b \\ c+f & d \end{vmatrix} = \begin{vmatrix} a & b \\ c & d \end{vmatrix} + \begin{vmatrix} e & b \\ f & d \end{vmatrix}$.

(b) $\begin{vmatrix} a+e & b+g \\ c+f & d+h \end{vmatrix} = \begin{vmatrix} a & b \\ c & d \end{vmatrix} + \begin{vmatrix} e & b \\ f & d \end{vmatrix} + \begin{vmatrix} a & g \\ c & h \end{vmatrix} + \begin{vmatrix} e & g \\ f & h \end{vmatrix}$.

2. Verify the identity:

$$\begin{vmatrix} a\alpha + b\beta + c\gamma & a\alpha' + b\beta' + c\gamma' \\ a'\alpha + b'\beta + c'\gamma & a'\alpha' + b'\beta' + c'\gamma' \end{vmatrix}$$

$$= \begin{vmatrix} a & b \\ a' & b' \end{vmatrix} \cdot \begin{vmatrix} \alpha & \beta \\ \alpha' & \beta' \end{vmatrix} + \begin{vmatrix} a & c \\ a' & c' \end{vmatrix} \cdot \begin{vmatrix} \alpha & \gamma \\ \alpha' & \gamma' \end{vmatrix} + \begin{vmatrix} b & c \\ b' & c' \end{vmatrix} \cdot \begin{vmatrix} \beta & \gamma \\ \beta' & \gamma' \end{vmatrix}$$

Make use of the identities in Prob. 1.

3. Deduce from Prob. 2 that

$$(x^2 + y^2 + z^2)(x'^2 + y'^2 + z'^2) = (xx' + yy' + zz')^2 + (xy' - x'y)^2$$
$$+ (xz' - x'z)^2 + (yz' - y'z)^2.$$

4. What identity similar to that of Prob. 2 can be found for the following determinant?

$$\begin{vmatrix} a\alpha + b\beta + c\gamma + d\delta & a\alpha' + b\beta' + c\gamma' + d\delta' \\ a'\alpha + b'\beta + c'\gamma + d'\delta & a'\alpha' + b'\beta' + c'\gamma' + d'\delta' \end{vmatrix}$$

5. Show that

$$\begin{vmatrix} x + iy & -z + it \\ z + it & x - iy \end{vmatrix} \cdot \begin{vmatrix} x' - iy' & z' - it' \\ -z' - it' & x' + iy' \end{vmatrix} = \begin{vmatrix} P + iQ & -R + iS \\ R + iS & P - iQ \end{vmatrix}$$

where

$$P = xx' + yy' + zz' + tt', \qquad Q = -xy' + yx' + zt' - tz'$$
$$R = -xz' + yt' - zx' - ty', \qquad S = -xt' + yz' - zy' + tx'.$$

Hence, deduce Euler's identity

$$(x^2 + y^2 + z^2 + t^2)(x'^2 + y'^2 + z'^2 + t'^2) = (xx' + yy' + zz' + tt')^2$$
$$+ (xy' - yx' - zt' + tz')^2 + (xz' - yt' + zx' + ty')^2 + (xt' - yz' + zy' - tx')^2.$$

4. Determinants as Functions of Matrices. Determinants of the second order as we have seen can be introduced as functions of matrices of the form

$$\begin{pmatrix} a_1 & b_1 \\ a_2 & b_2 \end{pmatrix}$$

possessing certain characteristic properties. Nothing stands in the way of generalizing these considerations. Instead of a matrix of four elements we may consider a matrix of n^2 elements

$$\begin{pmatrix} a_1 & b_1 & c_1 & \cdots & l_1 \\ a_2 & b_2 & c_2 & \cdots & l_2 \\ \cdots & \cdots & \cdots & \cdots & \cdots \\ a_n & b_n & c_n & \cdots & l_n \end{pmatrix}$$

arranged in n rows and n columns and seek all integral rational functions of the n^2 elements of this matrix considered as variables that possess the following three properties:

1. They are linear homogeneous functions in the elements of each row of the matrix.

2. They vanish identically in case two rows of the matrix are identical.

3. They take the value 1 for the special matrix

$$\begin{pmatrix} 1 & 0 & 0 & \cdots & 0 \\ 0 & 1 & 0 & \cdots & 0 \\ \cdots & \cdots & \cdots & \cdots & \cdots \\ 0 & 0 & 0 & \cdots & 1 \end{pmatrix}$$

all elements of which are zeros except diagonal elements, which are equal to 1.

We shall see that there is only one integral rational function satis-

fying these requirements, and it is this function that is called the determinant of n^2 elements or of the nth order and which is denoted by
the symbol

$$\begin{vmatrix} a_1 & b_1 & c_1 & \cdots & l_1 \\ a_2 & b_2 & c_2 & \cdots & l_2 \\ \cdots\cdots\cdots\cdots\cdots\cdots \\ a_n & b_n & c_n & \cdots & l_n \end{vmatrix}.$$

5. Determinants of Order 3. The method for the solution of the
general problem set in Sec. 4 will be better understood if we consider
first the particular case $n = 3$. It will be convenient also to adopt a
special notation for the elements of the matrix. Instead of using different letters and different indices to distinguish different elements and
different rows we shall use the same letter provided with two indices
as, for example, a_{ij}, the first index showing the number of the row (from
the top down), and the second index the number of the column (from
left to right) in which the letter occurs. Thus, in case $n = 3$ the matrix
will be

$$\begin{pmatrix} a_{11} & a_{12} & a_{13} \\ a_{21} & a_{22} & c_{23} \\ a_{31} & a_{32} & a_{33} \end{pmatrix}.$$

Rows of this matrix will be conveniently designated by the letters R_1,
R_2, R_3, and a symbol like $R_1 + R_2$ will represent a row composed of the
elements

$$a_{11} + a_{21}, \quad a_{12} + a_{22}, \quad a_{13} + a_{23}.$$

We shall denote by the sign

$$F(R_1, R_2, R_3),$$

the function of the nine elements a_{ij} $(i, j = 1, 2, 3)$ that satisfies conditions 1, 2, and 3 of Sec. 4. From properties 1 and 2, it can be shown
that this function changes its sign if two rows R are interchanged so that

$$\begin{aligned} F(R_2, R_1, R_3) &= -F(R_1, R_2, R_3), \\ F(R_3, R_2, R_1) &= -F(R_1, R_2, R_3), \\ F(R_1, R_3, R_2) &= -F(R_1, R_2, R_3). \end{aligned} \tag{1}$$

The proof being the same in all cases, it suffices to prove the first of these
identities. Consider the function

$$F(R_1 + R_2, \ R_2 + R_1, \ R_3).$$

Because F is a linear homogeneous function of the elements of the first
row of the matrix, we have

$$F(R_1 + R_2, \ R_2 + R_1, \ R_3) = F(R_1, \ R_2 + R_1, \ R_3) + F(R_2, \ R_2 + R_1, \ R_3).$$

Again, F being linear and homogeneous in the elements of the second row, we have

$$F(R_1, \ R_2 + R_1, \ R_3) = F(R_1 \ R_2, \ R_3) + F(R_1, \ R_1, \ R_3),$$
$$F(R_2, \ R_2 + R_1, \ R_3) = F(R_2, \ R_2, \ R_3) + F(R_2, \ R_1, \ R_3),$$

and

$$F(R_1 + R_2, \ R_2 + R_1, \ R_3) = F(R_1, \ R_2, \ R_3)$$
$$+ F(R_2, \ R_1, \ R_3) + F(R_1, \ R_1, \ R_3) + F(R_2, \ R_2, \ R_3).$$

By property 2, F vanishes when two rows are identical; hence,

$$F(R_1 + R_2, \ R_2 + R_1, \ R_3) = 0, \qquad F(R_1, \ R_1, \ R_3) = 0,$$
$$F(R_2, \ R_2, \ R_3) = 0.$$

Owing to this, the preceding identity reduces to

$$0 = F(R_1, \ R_2, \ R_3) + F(R_2, \ R_1, \ R_3),$$

or

$$F(R_2, \ R_1, \ R_3) = - F(R_1, \ R_2, \ R_3),$$

as we wanted to prove.

Further, F, being homogeneous of dimension 1 in the elements of each of the rows R_1, R_2, R_3, may consist only of terms of the form

$$A_{\alpha,\beta,\gamma} a_{1\alpha} a_{2\beta} a_{3\gamma},$$

where $A_{\alpha,\beta,\gamma}$ is independent of the elements a_{ij}, and where the indices α, β, γ may each take values 1, 2, 3. By using the summation sign, F may be written thus:

$$F = \Sigma A_{\alpha,\beta,\gamma} a_{1\alpha} a_{2\beta} a_{3\gamma},$$

and it remains to see what further conditions are imposed on coefficients $A_{\alpha,\beta,\gamma}$ by the requirements 2. The requirement 1, it will be noted, is satisfied independently of the values of these coefficients. Identities (1) follow from the requirement 2 and, vice versa, imply it. It suffices, therefore, to require that

$$F(R_1, \ R_2, \ R_3) = \Sigma A_{\alpha,\beta,\gamma} a_{1\alpha} a_{2\beta} a_{3\gamma}$$

must satisfy identities (1). Now

$$F(R_2, \ R_1, \ R_3) = \Sigma A_{\alpha,\beta,\gamma} a_{2\alpha} a_{1\beta} a_{3\gamma}$$

or, changing α into β and β into α, which may be done since α and β run through the same values,

$$F(R_2, \ R_1, \ R_3) = \Sigma A_{\beta,\alpha,\gamma} a_{1\alpha} a_{2\beta} a_{3\gamma}.$$

Here again α, β, γ run independently through values 1, 2, 3. Thus,

$$\Sigma A_{\beta,\alpha,\gamma} a_{1\alpha} a_{2\beta} a_{3\gamma} = - \Sigma A_{\alpha,\beta,\gamma} a_{1\alpha} a_{2\beta} a_{3\gamma},$$

and this identity implies

$$A_{\beta,\alpha,\gamma} = - A_{\alpha,\beta,\gamma}. \tag{2a}$$

Similarly,

$$A_{\gamma,\beta,\alpha} = - A_{\alpha,\beta,\gamma}, \tag{2b}$$

$$A_{\alpha,\gamma,\beta} = - A_{\alpha,\beta,\gamma}. \tag{2c}$$

If two of the indices α, β, γ are equal, the corresponding coefficient will be 0. For if, for instance, $\alpha = \beta$, we have by (2a)

$$A_{\alpha,\alpha,\gamma} = - A_{\alpha,\alpha,\gamma},$$

that is,

$$A_{\alpha,\alpha,\gamma} = 0.$$

Hence, in the expression of F, only terms corresponding to distinct values α, β, γ occur so that $\alpha\beta\gamma$ is one of the six permutations of the numbers 1, 2, 3. These six permutations are

$$\begin{matrix} 123 & 213 \\ 231 & 321 \\ 312 & 132 \end{matrix}$$

and, on account of (2a), (2b), and (2c), between the corresponding coefficients there are the following relations:

$$A_{2,1,3} = - A_{1,2,3};$$
$$A_{3,2,1} = - A_{2,3,1} = - A_{1,2,3};$$
$$A_{2,3,1} = - A_{2,1,3} = A_{1,2,3};$$
$$A_{3,1,2} = - A_{3,2,1} = A_{1,2,3};$$
$$A_{1,3,2} = - A_{1,2,3},$$

so that the five coefficients are expressed through the sixth. Writing for brevity

$$A_{1,2,3} = C,$$

we shall have

$$A_{1,2,3} = A_{2,3,1} = A_{3,1,2} = C,$$
$$A_{2,1,3} = A_{3,2,1} = A_{1,3,2} = - C,$$

and so

$$F = C(a_{11}a_{22}a_{33} + a_{12}a_{23}a_{31} + a_{13}a_{21}a_{32} - a_{12}a_{21}a_{33} - a_{13}a_{22}a_{31} - a_{11}a_{23}a_{32}).$$

Such is the general expression of functions satisfying requirements 1 and 2. If, in addition, F takes the value 1 for the particular matrix

$$\begin{pmatrix} 1 & 0 & 0 \\ 0 & 1 & 0 \\ 0 & 0 & 1 \end{pmatrix},$$

considering that the expression by which C is multiplied takes the value 1 for this matrix, it is clear that $C = 1$. Consequently, there exists only one rational integral function of nine elements a_{ij} that satisfies conditions 1, 2, and 3, namely,

$$a_{11}a_{22}a_{33} + a_{12}a_{23}a_{31} + a_{13}a_{21}a_{32} - a_{12}a_{21}a_{33} - a_{13}a_{22}a_{31} - a_{11}a_{23}a_{32}.$$

This expression is called the determinant of nine elements or of the third order and is represented by the symbol

$$\begin{vmatrix} a_{11} & a_{12} & a_{13} \\ a_{21} & a_{22} & a_{23} \\ a_{31} & a_{32} & a_{33} \end{vmatrix}$$

Here we shall not develop properties of determinants of the third order, for it is just as easy to study these properties in the general case of determinants of any order. We shall confine ourselves to a mnemonical rule for the formation of terms of the determinant of the third order that is good only in this particular case. In order to write the six terms of the determinant

$$\begin{vmatrix} a_1 & b_1 & c_1 \\ a_2 & b_2 & c_2 \\ a_3 & b_3 & c_3 \end{vmatrix}$$

form a table

and take the products of elements on descending diagonal lines with the sign $+$ and of those on ascending diagonal lines with the sign $-$. The aggregate of the six terms thus obtained

$$a_1b_2c_3 + b_1c_2a_3 + c_1a_2b_3 - a_3b_2c_1 - b_3c_2a_1 - c_3a_2b_1$$

will be the determinant

$$\begin{vmatrix} a_1 & b_1 & c_1 \\ a_2 & b_2 & c_2 \\ a_3 & b_3 & c_3 \end{vmatrix}.$$

For example, the value of the determinant

$$\begin{vmatrix} 1 & 2 & 3 \\ 3 & 1 & 2 \\ 6 & 4 & 5 \end{vmatrix}$$

found by this rule is

$$5 + 24 + 36 - 18 - 8 - 30 = 9.$$

Problems

The following properties hold for determinants of any order and will be proved later in a general manner. For determinants of the third order they can be verified easily.

1. Prove that

$$\begin{vmatrix} a_1 & b_1 & c_1 \\ a_2 & b_2 & c_2 \\ a_3 & b_3 & c_3 \end{vmatrix} = \begin{vmatrix} a_1 & a_2 & a_3 \\ b_1 & b_2 & b_3 \\ c_1 & c_2 & c_3 \end{vmatrix}.$$

How can one state this property in words?

2. Show that

$$\begin{vmatrix} ma_1 & mb_1 & mc_1 \\ na_2 & nb_2 & nc_2 \\ pa_3 & pb_3 & pc_3 \end{vmatrix} = mnp \begin{vmatrix} a_1 & b_1 & c_1 \\ a_2 & b_2 & c_2 \\ a_3 & b_3 & c_3 \end{vmatrix}.$$

Also

$$\begin{vmatrix} ma_1 & nb_1 & pc_1 \\ ma_2 & nb_2 & pc_2 \\ ma_3 & nb_3 & pc_3 \end{vmatrix} = mnp \begin{vmatrix} a_1 & b_1 & c_1 \\ a_2 & b_2 & c_2 \\ a_3 & b_3 & c_3 \end{vmatrix}.$$

3. Referring to Probs. 1 and 2, show that

$$\begin{vmatrix} 0 & a & b \\ -a & 0 & c \\ -b & -c & 0 \end{vmatrix} = 0.$$

4. Prove that

$$\begin{vmatrix} a_1 + d_1 & b_1 & c_1 \\ a_2 + d_2 & b_2 & c_2 \\ a_3 + d_3 & b_3 & c_3 \end{vmatrix} = \begin{vmatrix} a_1 & b_1 & c_1 \\ a_2 & b_2 & c_2 \\ a_3 & b_3 & c_3 \end{vmatrix} + \begin{vmatrix} d_1 & b_1 & c_1 \\ d_2 & b_2 & c_2 \\ d_3 & b_3 & c_3 \end{vmatrix}.$$

What are the corresponding results if the first column is left unchanged but b_1, b_2, b_3 (or c_1, c_2, c_3) are replaced by $b_1 + d_1$, $b_2 + d_2$, $b_3 + d_3$ (or $c_1 + d_1$, $c_2 + d_2$, $c_3 + d_3$)? Is there a similar property with respect to rows?

5. Show that a determinant of the third order is not changed if to elements of any row (or column) are added elements of another row (or column) multiplied by an arbitrary factor.

6. Show that

$$(a)\ \begin{vmatrix} 1 & 2 & 3 \\ 2 & 3 & 4 \\ 3 & 4 & 5 \end{vmatrix} = 0. \qquad (b)\ \begin{vmatrix} 1 & 2 & 3 \\ 3 & 4 & 5 \\ 5 & 6 & 7 \end{vmatrix} = 0.$$

7. Show that

$$(a)\ \begin{vmatrix} 1 & 1 & 1 \\ 3 & 4 & 2 \\ 5 & 1 & 6 \end{vmatrix} = \begin{vmatrix} 1 & 0 & 0 \\ 3 & 1 & -1 \\ 5 & -4 & 1 \end{vmatrix} = \begin{vmatrix} 1 & 0 & 0 \\ 8 & -3 & 0 \\ 5 & -4 & 1 \end{vmatrix}.$$

$$(b)\ \begin{vmatrix} 1 & 6 & 7 \\ 2 & 3 & 5 \\ 4 & 1 & 7 \end{vmatrix} = 3 \begin{vmatrix} 1 & 6 & 2 \\ 2 & 3 & 1 \\ 4 & 1 & 1 \end{vmatrix} = 3 \begin{vmatrix} -3 & 0 & 0 \\ -2 & 2 & 0 \\ 4 & 1 & 1 \end{vmatrix} = 18 \begin{vmatrix} 1 & 0 & 0 \\ 1 & -1 & 0 \\ 4 & 1 & 1 \end{vmatrix}.$$

8. Show that

$$\begin{vmatrix} 1 & 1 + ac & 1 + bc \\ 1 & 1 + ad & 1 + bd \\ 1 & 1 + ae & 1 + be \end{vmatrix} = 0.$$

9. Show that

$$\begin{vmatrix} 1 & a & a^2 \\ 1 & b & b^2 \\ 1 & c & c^2 \end{vmatrix} = (b-a)(c-a)\begin{vmatrix} 1 & a & 0 \\ 0 & 1 & b \\ 0 & 1 & c \end{vmatrix}.$$

10. A determinant can be expanded by elements of any row or any column. For instance,

$$\begin{vmatrix} a_1 & b_1 & c_1 \\ a_2 & b_2 & c_2 \\ a_3 & b_3 & c_3 \end{vmatrix} = a_1 \begin{vmatrix} b_2 & c_2 \\ b_3 & c_3 \end{vmatrix} - b_1 \begin{vmatrix} a_2 & c_2 \\ a_3 & c_3 \end{vmatrix} + c_1 \begin{vmatrix} a_2 & b_2 \\ a_3 & b_3 \end{vmatrix},$$

$$\begin{vmatrix} a_1 & b_1 & c_1 \\ a_2 & b_2 & c_2 \\ a_3 & b_3 & c_3 \end{vmatrix} = - b_1 \begin{vmatrix} a_2 & c_2 \\ a_3 & c_3 \end{vmatrix} + b_2 \begin{vmatrix} a_1 & c_1 \\ a_3 & c_3 \end{vmatrix} - b_3 \begin{vmatrix} a_1 & c_1 \\ a_2 & c_2 \end{vmatrix}.$$

In these expansions elements of a row or column are multiplied by determinants of order 2 taken with sign $\pm$ and which are obtained by crossing out the row and the column intersecting at the element under consideration. Find a rule for determining whether the $+$ or $-$ sign is used. These expansions immediately lower the order of a determinant in which all elements but one in a row or a column are zeros.

11. Use the method given in Prob. 10 to evaluate determinants (a) and (b) in Prob. 7.

12. Show that

(a)
$$\begin{vmatrix} 1 & a & a^2 \\ 1 & b & b^2 \\ 1 & c & c^2 \end{vmatrix} = (c-a)(c-b)(b-a).$$

(b)
$$\begin{vmatrix} 1 & a & a^3 \\ 1 & b & b^3 \\ 1 & c & c^3 \end{vmatrix} = (c-a)(c-b)(b-a)(a+b+c).$$

13. Evaluate the numerical determinants

(a)
$$\begin{vmatrix} 2 & 10 & 7 \\ 5 & 3 & 1 \\ 7 & 2 & 4 \end{vmatrix}.$$
(b)
$$\begin{vmatrix} 10 & 5 & 4 \\ 8 & 11 & 7 \\ 6 & 12 & 14 \end{vmatrix}.$$

14. Show that

$$\begin{vmatrix} x & a & b \\ -a & x & c \\ -b & -c & x \end{vmatrix} = x(x^2 + a^2 + b^2 + c^2).$$

6. Even and Odd Permutations. To introduce a determinant of higher order, the procedure to be used is quite analogous to that employed in the case of determinants of the third order except that it will be necessary to refer to some properties of permutations which, in that particular case, did not impose themselves with such necessity as they do in the general case. Let numbers 1, 2, 3, . . . , n be written in increasing order

$$1 \quad 2 \quad 3 \quad \cdots \quad n$$

This will be called the natural order. On placing these integers so that the first place is occupied by i_1, the second by i_2, the third by i_3, etc., we have a permutation

$$i_1 \quad i_2 \quad i_3 \quad \cdots \quad i_n.$$

Thus,

$$6 \quad 5 \quad 1 \quad 3 \quad 4 \quad 2$$

is a permutation of six numbers 1, 2, 3, 4, 5, 6. The number of permutations of n numbers is

$$1 \cdot 2 \cdot 3 \cdots n = n\,!.$$

Thus, the number of permutations of 2, 3, 4, 5, 6, 7 numbers is, respectively,

$$2\,! = 2, \qquad 3\,! = 6, \qquad 4\,! = 24,$$
$$5\,! = 120, \qquad 6\,! = 720, \qquad 7\,! = 5040.$$

Exchanging in a permutation

$$i_1 i_2 \cdots i_\alpha i_{\alpha+1} \cdots i_\beta \cdots i_n$$

two elements i_α and i_β and keeping the other elements in their places, we pass to another permutation

$$i_1 \cdots i_\beta \cdots i_\alpha \cdots i_n,$$

which is said to result from the former by a transposition of i_α and i_β. Every permutation can be obtained from the permutation

$$1 \quad 2 \quad 3 \quad \cdots \quad n,$$

in which the elements are placed in natural order, by a series of successive transpositions. For instance, the permutation

$$7 \quad 6 \quad 8 \quad 3 \quad 5 \quad 1 \quad 4 \quad 2$$

can be obtained from

$$1 \quad 2 \quad 3 \quad 4 \quad 5 \quad 6 \quad 7 \quad 8$$

by the succession of the following transpositions: (a) 1 and 6; (b) 2 and 8; (c) 3 and 4; (d) 4 and 7; (e) 6 and 8; (f) 7 and 8. The same can be achieved by another series of transpositions: (a) 1 and 4; (b) 1 and 7; (c) 2 and 8; (d) 3 and 8; (e) 6 and 1; (f) 6 and 4; (g) 6 and 7; (h) 6 and 3.

In general there is an infinite variety of ways to pass from one permutation to another by a series of successive transpositions. But it is a remarkable fact that the number of transpositions used to do that is always either even or odd no matter what transpositions are employed. The proof of this fact may be based on the notion of *inversion*. If in a permutation

$$i_1 \quad i_2 \quad \cdots \quad i_n$$

an element i_α is followed by a smaller element, we say that there is an inversion relative to i_α and that element. The number of elements following i_α and smaller than it gives the total number of inversions relative to i_α. The number of inversions relative to all elements of a permutation may be called the *index I* of that permutation. The index is a number completely determined by the permutation. Considering, for example, the permutation

$$7 \quad 6 \quad 8 \quad 3 \quad 5 \quad 1 \quad 4 \quad 2,$$

we count

Number of inversions	relative to
6	7
5	6
5	8
2	3
3	5
0	1
1	4

The index of the permutation is therefore

$$I = 6 + 5 + 5 + 2 + 3 + 0 + 1 = 22.$$

LEMMA. *If in a permutation*

$$i_1 \quad i_2 \quad \cdots \quad i_n$$

two elements i_α and i_β are transposed the index is changed by an odd number.

PROOF. Suppose at first that i_β immediately follows i_α so that $\beta = \alpha + 1$. In counting the number of inversions it is convenient to divide the permutation into three sections:

$$i_1 \quad i_2 \quad \cdots \quad i_{\alpha-1}; \quad i_\alpha \quad i_{\alpha+1}; \quad i_{\alpha+2} \quad \cdots \quad i_n$$

Let A be the number of inversions relative to the first section, and B that relative to the third section, while P and Q are the numbers of inversions relative to i_α and $i_{\alpha+1}$. Then,

$$I = A + B + P + Q.$$

After transposing i_α and $i_{\alpha+1}$ we have another permutation, which is again divided in three sections:

$$i_1 \quad i_2 \quad \cdots \quad i_{\alpha-1}; \quad i_{\alpha+1} \quad i_\alpha; \quad i_{\alpha+2} \quad \cdots \quad i_n.$$

Its index I', denoting by P' and Q' the numbers of inversions relative to $i_{\alpha+1}$ and i_α, is

$$I' = A + B + P' + Q'.$$

Let M and N be the numbers of elements in the section

$$i_{\alpha+2} \quad \ldots \quad i_n$$

respectively less than i_α and $i_{\alpha+1}$. Then,

$$P = M, \qquad Q = N$$

in case $i_\alpha < i_{\alpha+1}$, and

$$P = M + 1, \qquad Q = N$$

in case $i_\alpha > i_{\alpha+1}$. Similarly,

$$P' = N + 1, \qquad Q' = M$$

in case $i_\alpha < i_{\alpha+1}$, and

$$P' = N, \qquad Q' = M$$

in case $i_\alpha > i_{\alpha+1}$. Hence,

$$P' + Q' - (P + Q) = 1 \quad \text{or} \quad -1$$

according as $i_\alpha < i_{\alpha+1}$ or $i_\alpha > i_{\alpha+1}$, and consequently

$$I' = I \pm 1;$$

that is, the index is either increased or decreased by 1 after transposing two adjoining elements.

Now suppose that the transposed elements, i_α and i_β, are not adjoining, and let l be the number of elements between them. Then, from the permutation

$$i_1 \quad \cdots \quad i_\alpha \quad \cdots \quad i_\beta \quad \cdots \quad i_n$$

we pass to the permutation

$$i_1 \quad \cdots \quad i_\beta \quad \cdots \quad i_\alpha \quad \cdots \quad i_n$$

by the following transpositions of adjoining elements: (a) l transpositions of i_α with adjoining elements place it before i_β; (b) one transposition places i_α in the position occupied by i_β, and i_β precedes i_α; (c) l transpositions of i_β with adjoining elements bring it to the place formerly occupied by i_α. The passage from one permutation to the other is thus accomplished by $2l + 1$ successive transpositions of adjoining elements, and since each such transposition decreases or increases the index by 1, the index of the permutation

$$i_1 \quad \cdots \quad i_\beta \quad \cdots \quad i_\alpha \quad \cdots \quad i_n$$

differs from that of the permutation

$$i_1 \quad \cdots \quad i_\alpha \quad \cdots \quad i_\beta \quad \cdots \quad i_n$$

by an odd number, and so the lemma is proved.

Suppose now that the permutation

$$i_1 \quad i_2 \quad \cdots \quad i_n$$

is obtained from

$$1 \quad 2 \quad \cdots \quad n$$

by r transpositions. Since for the last permutation the index is 0, and since the r transpositions cause odd increments $2h_1 + 1$, $2h_2 + 1$, ... , $2h_r + 1$, the index I of the permutation

$$i_1 \quad i_2 \quad \cdots \quad i_n$$

will be

$$I = r + 2(h_1 + h_2 + \cdots + h_r) = r + 2h.$$

Hence, r will be even or odd according as I is even or odd. Permutations resulting from $1\ 2\ \cdots\ n$ by an even or odd number of transpositions are called, respectively, even or odd permutations. The permutation resulting from an even permutation by one transposition will be odd, and vice versa. It is easy to prove that among the $n\,!$ permutations of $1\ 2\ 3\ \cdots\ n$ the even and odd permutations are in equal numbers. Let R and R' represent the numbers of even and odd permutations. Transposing the two first elements of each of the R even permutations, R distinct odd permutations result; hence, $R' \geqq R$. But by the same reasoning $R \geqq R'$ and so $R' = R$. Since the total number of permutations is $n\,!$, there are

$$\tfrac{1}{2}n\,!$$

even and odd permutations.

Suppose that

$$A_{i_1, i_2, \ldots, i_n}$$

is a quantity depending on n distinct indices i_1, i_2, ... , i_n, each of which must be one of the numbers $1, 2, \ldots, n$ so that

$$i_1 \quad i_2 \quad \cdots \quad i_n$$

is some permutation of these numbers. Suppose further that each of the $n\,!$ such quantities possesses the property of changing its sign if any two indices are transposed while preserving its absolute value; in other words, suppose that

$$A_{i_1, \ldots, i_\alpha, \ldots, i_\beta, \ldots, i_n} = - A_{i_1, \ldots, i_\beta, \ldots, i_\alpha, \ldots, i_n}. \tag{1}$$

Let $I(i_1, i_2, \ldots, i_n)$ be the index of the permutation $i_1, i_2, \ldots, i_n$. By the lemma just proved

$$(-1)^{I(i_1, i_2, \ldots, i_n)}$$

also satisfies condition (1) and consequently

$$A_{i_1, i_2, \ldots, i_n}/(-1)^{I(i_1, i_2, \ldots, i_n)}$$

is not changed by transposition of any two indices. Since every permutation results from $1\ 2\ \cdots\ n$ by successive transpositions, it follows that the above quotient has the same value for all permutations. Calling this quotient C, we have therefore

$$A_{i_1, i_2, \ldots, i_n} = \pm C$$

taking the sign $+$ or $-$ according as the permutation $i_1\ i_2 \cdots i_n$ is even or odd.

Problems

1. Find the number of inversions in each of the permutations:

     (a) 6 8 1 7 5 3 2 4,   (b) 2 1 3 9 8 7 6 5 4,   (c) 3 7 6 1 2 5 8 4 9.

2. Which of the permutations are even and which are odd?

     (a) 6 3 1 2 5 4,   (b) 1 9 2 8 3 7 4 6 5,   (c) 7 1 5 6 3 2 4.

3. Show that the permutation 1 6 5 3 2 4 can be obtained from 1 4 5 3 6 2 by an even number of transpositions.

4. Write down all even and all odd permutations of elements 1, 2, 3, 4.

7. General Determinants. It will be easy now to give a general definition of a determinant of the order n. Let

$$\begin{pmatrix} a_{11} & a_{12} & \cdots & a_{1n} \\ a_{21} & a_{22} & \cdots & a_{2n} \\ \cdot\cdot\cdot\cdot\cdot\cdot\cdot\cdot\cdot\cdot\cdot \\ a_{n1} & a_{n2} & \cdots & a_{nn} \end{pmatrix}$$

be a matrix of n^2 elements a_{ij} distributed in n rows $R_1, R_2, \ldots, R_n$ and n columns $C_1, C_2, \ldots, C_n$. For convenience in reasoning we choose the notation of elements with two indices that clearly indicate the row and the column to which the element belongs. Considering these n^2 elements as variables, we set the problem of finding all rational integral functions of them satisfying the following three requirements:

1. They must be linear and homogeneous in the elements of each row of the matrix.

2. They must vanish identically whenever two rows are identical.

3. They take the value 1 for the special matrix

$$\begin{pmatrix} 1 & 0 & \cdots & 0 \\ 0 & 1 & \cdots & 0 \\ 0 & 0 & \cdots & 1 \end{pmatrix}$$

in which $a_{ii} = 1$ and $a_{ij} = 0$ for $j \neq i$.

The examination of the problem will show that there is just one function satisfying these requirements. The function we seek may be conveniently denoted by

$$F(R_1, R_2, \ldots, R_n).$$

In the first place, we are going to show that

$$F(R_1, \ldots, R_\alpha, \ldots, R_\beta, \ldots, R_n)$$
$$= - F(R_1, \ldots, R_\beta, \ldots, R_\alpha, \ldots, R_n) \quad (1)$$

so that F changes sign when two rows R_α and R_β are transposed. Consider the function

$$F(R_1, \ldots, R_\alpha + R_\beta, \ldots, R_\beta + R_\alpha, \ldots, R_n).$$

Since rows $R_\alpha + R_\beta$ and $R_\beta + R_\alpha$ are identical, we have

$$F(R_1, \ldots, R_\alpha + R_\beta, \ldots, R_\beta + R_\alpha, \ldots, R_n) = 0. \quad (2)$$

On the other hand, because F is linear and homogeneous in the elements of each row,

$$F(R_1, \ldots, R_\alpha + R_\beta, \ldots, R_\beta + R_\alpha, \ldots, R_n)$$
$$= F(R_1, \ldots, R_\alpha, \ldots, R_\beta + R_\alpha, \ldots, R_n)$$
$$+ F(R_1, \ldots, R_\beta, \ldots, R_\beta + R_\alpha, \ldots, R_n),$$

and for the same reason

$$F(R_1, \ldots, R_\alpha, \ldots, R_\beta + R_\alpha, \ldots, R_n)$$
$$= F(R_1, \ldots, R_\alpha, \ldots, R_\beta, \ldots, R_n)$$
$$+ F(R_1, \ldots, R_\alpha, \ldots, R_\alpha, \ldots, R_n),$$
$$F(R_1, \ldots, R_\beta, \ldots, R_\beta + R_\alpha, \ldots, R_n)$$
$$= F(R_1, \ldots, R_\beta, \ldots, R_\beta, \ldots, R_n)$$
$$+ F(R_1, \ldots, R_\beta, \ldots, R_\alpha, \ldots, R_n).$$

Taking into consideration equation (2) and similar identities like

$$F(R_1, \ldots, R_\alpha, \ldots, R_\alpha, \ldots, R_n)$$
$$= F(R_1, \ldots, R_\beta, \ldots, R_\beta, \ldots, R_n) = 0,$$

we find

$$0 = F(R_1, \ldots, R_\alpha, \ldots, R_\beta, \ldots, R_n)$$
$$+ F(R_1, \ldots, R_\beta, \ldots, R_\alpha, \ldots, R_n),$$

which is equivalent to equation (1). Conversely, supposing equation (1) is satisfied, it follows that

$$F(R_1, \ldots, R_\alpha, \ldots, R_\alpha, \ldots, R_n) = 0.$$

That is, requirements 1 and 2 together are equivalent to requirement 1 and the fact that F changes into $- F$ when any two rows are transposed.

By virtue of requirement 1 the requested function will consist of terms of the form

$$A_{i_1, i_2, \ldots, i_n} a_{1i_1} a_{2i_2} \cdots a_{ni_n}$$

where indices $i_1, i_2, \ldots, i_n$ run independently through the values, $1, 2, \ldots, n$ and where

$$A_{i_1, i_2, \ldots, i_n}$$

is a coefficient independent of the variables a_{ij}. Thus, we can set

$$F(R_1, R_2, \ldots, R_n) = \Sigma A_{i_1, i_2, \ldots, i_n} a_{1i_1} a_{2i_2} \cdots a_{ni_n}. \qquad (3)$$

Transposing rows R_α and R_β, we have

$$
\begin{aligned}
F(R_1, &\ldots, R_\beta, \ldots, R_\alpha, \ldots, R_n) \\
&= \Sigma A_{i_1, i_2, \ldots, i_n} a_{1i_1} \cdots a_{\beta i_\alpha} \cdots a_{\alpha i_\beta} \cdots a_{ni_n}
\end{aligned}
$$

or, replacing i_α by i_β and vice versa, which is allowable since both i_α and i_β run through the same set of values,

$$
\begin{aligned}
F(R_1, &\ldots, R_\beta, \ldots, R_\alpha, \ldots, R_n) \\
&= \Sigma A_{i_1, \ldots, i_\beta, \ldots, i_\alpha, \ldots, i_n} a_{1i_1} a_{2i_2} \cdots a_{ni_n} \qquad (4)
\end{aligned}
$$

Comparing equations (3) and (4) and taking into account the identity (1) we conclude that

$$A_{i_1, \ldots, i_\alpha, \ldots, i_\beta, \ldots, i_n} = - A_{i_1, \ldots, i_\beta, \ldots, i_\alpha, \ldots, i_n} \qquad (5)$$

for any two distinct indices α and β taken among the numbers $1, 2, \ldots, n$. In case $i_\alpha = i_\beta$ it follows from (5) that

$$A_{i_1, \ldots, i_\alpha, \ldots, i_\alpha, \ldots, i_n} = 0$$

and this means that only those terms are actually present in (3) that correspond to different values $i_1, i_2, \ldots, i_n$, so that $i_1 i_2 \cdots i_n$ is some permutation of numbers $1, 2, 3, \ldots, n$. Again, from (5) and the last remark in Sec. 6 it follows that

$$A_{i_1, i_2, \ldots, i_n} = \pm C$$

where the sign is $+$ or $-$ according as $i_1 i_2 \cdots i_n$ is an even or odd permutation.

Thus, all functions F satisfying requirements 1 and 2 are of the form

$$F = C \Sigma \pm a_{1i_1} a_{1i_2} \cdots a_{1i_n}$$

where the summation extends over all permutations of the indices, sign $\pm$ is chosen as explained, and C is a quantity independent of the elements of the matrix. Now for the special matrix in which $a_{ii} = 1$, and $a_{ij} = 0$ if $j \neq i$, the sum reduces to one term

$$a_{11} a_{22} \cdots a_{nn} = 1,$$

and, on the other hand, if requirement 3 is satisfied, F takes the value **1**, so that in this case $C = 1$. Thus, if all three requirements are fulfilled,

$$F = \Sigma \pm a_{1i_1}a_{2i_2} \cdot \cdot \cdot a_{ni_n},$$

the sum being extended over all permutations of 1, 2, . . . , n and for each permutation the product

$$a_{1i_1}a_{2i_2} \cdot \cdot \cdot a_{ni_n}$$

is taken with the sign $+$ or $-$ according as the permutation is even or odd.

The sum representing F is called the determinant of the nth order or the determinant of the n^2 elements a_{ij}, and it is denoted by the symbol

$$\begin{vmatrix} a_{11} & a_{12} & \cdot \cdot \cdot & a_{1n} \\ a_{21} & a_{22} & \cdot \cdot \cdot & a_{2n} \\ \cdot & \cdot & \cdot \cdot \cdot \cdot \cdot \cdot & \cdot \\ a_{n1} & a_{n2} & \cdot \cdot \cdot & a_{nn} \end{vmatrix}.$$

Sometimes a shorter notation

$$|a_{ij}|$$

is used, which should not be confused, of course, with the absolute value of the quantity a_{ij}.

The number of terms in the determinant of order n is the same as the number of permutations of n things, that is, $n!$, and just one half of them is preceded by the sign $+$ and the other half by the sign $-$.

8. Properties of Determinants. Since the number of terms of a determinant increases very rapidly with its order, the direct evaluation of determinants based on their definition is impracticable but in many cases can be achieved and often without much labor by resorting to certain of their properties that we shall now examine. Take the general term

$$\pm a_{1i_1}a_{2i_2} \cdot \cdot \cdot a_{ni_n}.$$

Here the second indices $i_1, i_2, \ldots, i_n$ form a certain permutation of numbers 1, 2, . . . , n. Therefore, the factors can be arranged so that these second indices follow in natural order while the first form a permutation $j_1 j_2 \cdot \cdot \cdot j_n$ of the numbers $1, 2, \ldots, n$. Thus, the same term can be written

$$\pm a_{j_1 1}a_{j_2 2} \cdot \cdot \cdot a_{j_n n}.$$

Now, if m transpositions are used to pass from $1\ 2 \cdot \cdot \cdot n$ to $i_1\ i_2 \cdot \cdot \cdot i_n$, the same m transpositions performed in the reverse order will restore the second indices to their natural order and will place the first indices, originally in the natural order, in the order $j_1 j_2 \cdots j_n$. Therefore, the permutations $i_1 i_2 \cdots i_n$ and $j_1 j_2 \cdots j_n$ are both even or both odd, and the sign $\pm$ may be determined with reference to the second instead of the first permutation. Consequently, the determinant can be presented as the sum

$$\Sigma \pm a_{j_1 1} a_{j_2 2} \cdots a_{j_n n}$$

extended over all permutations of the first indices, the sign $+$ or $-$ being chosen according as this permutation is even or odd.

Suppose now that in the determinant

$$D = \begin{vmatrix} a_{11} & a_{12} & \cdots & a_{1n} \\ a_{21} & a_{22} & \cdots & a_{2n} \\ \cdots & \cdots & \cdots & \cdots \\ a_{n1} & a_{n2} & \cdots & a_{nn} \end{vmatrix}$$

rows are replaced by columns, and vice versa; this gives another determinant

$$D' = \begin{vmatrix} a_{11} & a_{21} & \cdots & a_{n1} \\ a_{12} & a_{22} & \cdots & a_{n2} \\ \cdots & \cdots & \cdots & \cdots \\ a_{1n} & a_{2n} & \cdots & a_{nn} \end{vmatrix}$$

whose element b_{ij} belonging to the ith row and jth column is a_{ji}. By definition

$$D' = \Sigma \pm b_{1 j_1} b_{1 j_2} \cdots b_{n j_n}$$

where terms are preceded by the sign $+$ or $-$ according as $j_1 j_2 \cdots j_n$ is an even or odd permutation. Since $b_{ij} = a_{ji}$, we can write also

$$D' = \Sigma \pm a_{j_1 1} a_{j_2 2} \cdots a_{j_n n};$$

but by the remark made above the same sum represents also the determinant D; hence,

$$D' = D.$$

This important result can be stated briefly as follows: *A determinant does not change by changing rows into columns, and vice versa.* A determinant, being a *homogeneous linear function in the elements of each row*, therefore is also a *homogeneous linear function in the elements of each column*. A determinant *vanishes when two of its rows are identical*. Hence, also it *vanishes when two of its columns are identical*. In general, any property that determinants may have with respect to their rows has as a counterpart a similar property with respect to their columns.

As was shown in Sec. 7, a determinant *merely changes its sign by inter-change of two rows.* Hence, also it *merely changes its sign by interchange of two columns.*

Let

$$f(x_1, x_2, \ldots, x_n) = a_1x_1 + a_2x_2 + \cdots + a_nx_n$$

be a linear homogeneous function of the variables $x_1, x_2, \ldots, x_n$. It is clear that

$$f(mx_1, mx_2, \ldots, mx_n) = mf(x_1, x_2, \ldots, x_n)$$

and

$$f(y_1 + z_1, y_2 + z_2, \ldots, y_n + z_n) = f(y_1, y_2, \ldots, y_n) \\ + f(z_1, z_2, \ldots, z_n).$$

Determinants being linear and homogeneous in the elements of each row (and of each column), it follows that replacing in a determinant D a row $a_{i1}, a_{i2}, \ldots, a_{in}$ by $ma_{i1}, ma_{i2}, \ldots, ma_{in}$, or a column $a_{1i}, a_{2i}, \ldots, a_{ni}$ by $ma_{1i}, ma_{2i}, \ldots, ma_{ni}$, the resulting determinant D' is equal to D multiplied by m:

$$D' = mD,$$

or, putting it in another way, if *elements of a row (or column) are multiplied by some factor m, the determinant is multiplied by m.* Thus,

$$\begin{vmatrix} a_1 & b_1 & c_1 \\ ma_2 & mb_2 & mc_2 \\ a_3 & b_3 & c_3 \end{vmatrix} = m \begin{vmatrix} a_1 & b_1 & c_1 \\ a_2 & b_2 & c_2 \\ a_3 & b_3 & c_3 \end{vmatrix}$$

and

$$\begin{vmatrix} ma_1 & b_1 & c_1 \\ ma_2 & b_2 & c_2 \\ ma_3 & b_3 & c_3 \end{vmatrix} = m \begin{vmatrix} a_1 & b_1 & c_1 \\ a_2 & b_2 & c_2 \\ a_3 & b_3 & c_3 \end{vmatrix}.$$

A determinant vanishes if the elements of two rows or two columns are proportional. Suppose, for instance, that in a determinant of the fourth order

$$D = \begin{vmatrix} a_1 & b_1 & c_1 & d_1 \\ a_2 & b_2 & c_2 & d_2 \\ a_3 & b_3 & c_3 & d_3 \\ a_4 & b_4 & c_4 & d_4 \end{vmatrix}$$

the elements of the first and third columns are proportional so that

$$c_1 = ma_1, \qquad c_2 = ma_2, \qquad c_3 = ma_3, \qquad c_4 = ma_4.$$

Then,

$$D = m \begin{vmatrix} a_1 & b_1 & a_1 & c_1 \\ a_2 & b_2 & a_2 & c_2 \\ a_3 & b_3 & a_3 & c_3 \\ a_4 & b_4 & a_4 & c_4 \end{vmatrix} = 0.$$

since the determinant on the right has two identical columns.

The second of the above-mentioned properties of linear homogeneous functions implies the following proposition: *If in a determinant D the elements of some row are sums*

$$b_{i1} + c_{i1}, \quad b_{i2} + c_{i2}, \ldots, b_{in} + c_{in},$$

then this determinant will be the sum of two determinants D′ and D″ in which the corresponding rows are, respectively, $b_{i1}, b_{i2}, \ldots, b_{in}$ and $c_{i1}, c_{i2}, \ldots, c_{in}$, the other rows being the same as in D. Clearly, a similar property holds with respect to columns. Thus,

$$\begin{vmatrix} a_1 + d_1 & b_1 & c_1 \\ a_2 + d_2 & b_2 & c_2 \\ a_3 + d_3 & b_3 & c_3 \end{vmatrix} = \begin{vmatrix} a_1 & b_1 & c_1 \\ a_2 & b_2 & c_2 \\ a_3 & b_3 & c_3 \end{vmatrix} + \begin{vmatrix} d_1 & b_1 & c_1 \\ d_2 & b_2 & c_2 \\ d_3 & b_3 & c_3 \end{vmatrix}.$$

A particularly important consequence of this proposition is the following: *A determinant D does not change if to each element of a row (or column) is added the corresponding element of another row (or column) multiplied by the same factor.* In fact, the new determinant D' is equal to D plus another determinant that is equal to 0 since elements of two rows (or columns) are proportional. Thus,

$$\begin{vmatrix} a_1 & b_1 & c_1 \\ a_2 & b_2 & c_2 \\ a_3 & b_3 & c_3 \end{vmatrix} = \begin{vmatrix} a_1 & b_1 + mc_1 & c_1 \\ a_2 & b_2 + mc_2 & c_2 \\ a_3 & b_3 + mc_3 & c_3 \end{vmatrix}.$$

9. Examples. The properties of determinants just derived may be used to transform determinants and sometimes facilitate their evaluation. The following examples will show how such transformations can be done:

Example 1. Consider the following determinant:

$$D = \begin{vmatrix} 1 & 2 & 3 & 4 \\ 2 & 3 & 4 & 5 \\ 3 & 4 & 5 & 6 \\ 4 & 5 & 6 & 7 \end{vmatrix}.$$

Subtracting the first row from the second (which means that to elements of the second row are added those of the first row multiplied by -1) and afterward again the first row from the third, we get the determinant

$$D' = \begin{vmatrix} 1 & 2 & 3 & 4 \\ 1 & 1 & 1 & 1 \\ 2 & 2 & 2 & 2 \\ 4 & 5 & 6 & 7 \end{vmatrix}$$

having the same value as D. But in D' the elements of the third and second rows are proportional; hence, $D' = 0$ and also $D = 0$.

Example 2. Consider the determinant

$$D = \begin{vmatrix} 3 & 1 & 2 & 3 \\ 4 & -1 & 2 & 4 \\ 1 & -1 & 1 & 1 \\ 4 & -1 & 2 & 5 \end{vmatrix}.$$

Adding the second column to the first and again to the third and fourth, we have another determinant equal to D:

$$D = \begin{vmatrix} 4 & 1 & 3 & 4 \\ 3 & -1 & 1 & 3 \\ 0 & -1 & 0 & 0 \\ 3 & -1 & 1 & 4 \end{vmatrix}.$$

Add the first row of this determinant to the second, third, and fourth. This gives

$$D = \begin{vmatrix} 4 & 1 & 3 & 4 \\ 7 & 0 & 4 & 7 \\ 4 & 0 & 3 & 4 \\ 7 & 0 & 4 & 8 \end{vmatrix}.$$

Now subtract the third row from the first and the fourth from the second; then,

$$D = \begin{vmatrix} 0 & 1 & 0 & 0 \\ 0 & 0 & 0 & -1 \\ 4 & 0 & 3 & 4 \\ 7 & 0 & 4 & 8 \end{vmatrix}$$

and after exchanging first and second columns:

$$D = - \begin{vmatrix} 1 & 0 & 0 & 0 \\ 0 & 0 & 0 & -1 \\ 0 & 4 & 3 & 4 \\ 0 & 7 & 4 & 8 \end{vmatrix}.$$

We will discuss this determinant further in Sec. 11.

Example 3. Consider the determinant

$$D = \begin{vmatrix} 1 & a & b & c+d \\ 1 & b & c & a+d \\ 1 & c & d & a+b \\ 1 & d & a & b+c \end{vmatrix}.$$

Adding the second and third column to the last, which does not change the value of the determinant, we have

$$D = \begin{vmatrix} 1 & a & b & a+b+c+d \\ 1 & b & c & a+b+c+d \\ 1 & c & d & a+b+c+d \\ 1 & d & a & a+b+c+d \end{vmatrix} = (a+b+c+d) \begin{vmatrix} 1 & a & b & 1 \\ 1 & b & c & 1 \\ 1 & c & d & 1 \\ 1 & d & a & 1 \end{vmatrix} = 0.$$

Example 4. Let

$$D = \begin{vmatrix} 1 & a & bc \\ 1 & b & ac \\ 1 & c & ab \end{vmatrix}.$$

Multiplying the first, second, and third rows by a, b, and c, respectively, the determinant will be multiplied by abc; thus,

$$abcD = \begin{vmatrix} a & a^2 & abc \\ b & b^2 & abc \\ c & c^2 & abc \end{vmatrix} = abc \begin{vmatrix} a & a^2 & 1 \\ b & b^2 & 1 \\ c & c^2 & 1 \end{vmatrix}$$

whence

$$D = \begin{vmatrix} a & a^2 & 1 \\ b & b^2 & 1 \\ c & c^2 & 1 \end{vmatrix},$$

and by interchange of columns

$$D = - \begin{vmatrix} 1 & a^2 & a \\ 1 & b^2 & b \\ 1 & c^2 & c \end{vmatrix} = \begin{vmatrix} 1 & a & a^2 \\ 1 & b & b^2 \\ 1 & c & c^2 \end{vmatrix} = \begin{vmatrix} 1 & 1 & 1 \\ a & b & c \\ a^2 & b^2 & c^2 \end{vmatrix}.$$

The last operation consists in changing rows into columns and does not affect the value of the determinant.

Example 5. Let

$$D = \begin{vmatrix} 1 & a & a^2 \\ 1 & b & b^2 \\ 1 & c & c^2 \end{vmatrix}.$$

Subtract the second row from the first and then the third from the second and factor out $a - b$ and $b - c$. Then,

$$D = (a - b)(b - c) \begin{vmatrix} 0 & 1 & a + b \\ 0 & 1 & b + c \\ 1 & c & c^2 \end{vmatrix}.$$

Now subtract the second row from the first and remove the factor $a - c$; then,

$$\begin{vmatrix} 0 & 1 & a + b \\ 0 & 1 & a + c \\ 1 & c & c^2 \end{vmatrix} = (a - c) \begin{vmatrix} 0 & 0 & 1 \\ 0 & 1 & b + c \\ 1 & c & c^2 \end{vmatrix}$$

and

$$D = (a - b)(a - c)(b - c) \begin{vmatrix} 0 & 0 & 1 \\ 0 & 1 & b + c \\ 1 & c & c^2 \end{vmatrix}.$$

Considering a, b, c as variables, the determinant D is of the second degree in each of them; but so is the product

$$(a - b)(a - c)(b - c).$$

Consequently,

$$\Delta = \begin{vmatrix} 0 & 0 & 1 \\ 0 & 1 & b + c \\ 1 & c & c^2 \end{vmatrix}$$

does not actually contain b and c, and its value can be found by attributing to b and c special values, for instance, $b = c = 0$. Thus,

$$\Delta = \begin{vmatrix} 0 & 0 & 1 \\ 0 & 1 & 0 \\ 1 & 0 & 0 \end{vmatrix} = - \begin{vmatrix} 1 & 0 & 0 \\ 0 & 1 & 0 \\ 0 & 0 & 1 \end{vmatrix} = -1,$$

and, finally,

$$D = -(a-b)(a-c)(b-c) = (a-b)(a-c)(c-b).$$

Problems

Evaluate the determinants:

1. $\begin{vmatrix} 1 & 3 & 5 & 7 \\ 2 & 4 & 6 & 8 \\ 5 & 7 & 9 & 11 \\ a & b & c & d \end{vmatrix}.$

2. $\begin{vmatrix} 1 & 4 & 9 & 16 \\ 4 & 9 & 16 & 25 \\ 9 & 16 & 25 & 36 \\ 16 & 25 & 36 & 49 \end{vmatrix}.$

3. $\begin{vmatrix} a & b & c \\ 3a+2b & 3b+2c & 3c+2a \\ b & c & a \end{vmatrix}.$

4. $\begin{vmatrix} 1 & a & b+c \\ 1 & b & a+c \\ 1 & c & a+b \end{vmatrix}.$

5. $\begin{vmatrix} 1+x_1y_1 & 1+x_2y_1 & 1+x_3y_1 & 1+x_4y_1 \\ 1+x_1y_2 & 1+x_2y_2 & 1+x_3y_2 & 1+x_4y_2 \\ 1+x_1y_3 & 1+x_2y_3 & 1+x_3y_3 & 1+x_4y_3 \\ 1+x_1y_4 & 1+x_2y_4 & 1+x_3y_4 & 1+x_4y_4 \end{vmatrix}.$

6. $\begin{vmatrix} 0 & a & b & c & d \\ -a & 0 & e & f & g \\ -b & -e & 0 & h & k \\ -c & -f & -h & 0 & l \\ -d & -g & -k & -l & 0 \end{vmatrix}.$

Show that

7. $\begin{vmatrix} a+b & b+c & c+a \\ a_1+b_1 & b_1+c_1 & c_1+a_1 \\ a_2+b_2 & b_2+c_2 & c_2+a_2 \end{vmatrix} = 2 \begin{vmatrix} a & b & c \\ a_1 & b_1 & c_1 \\ a_2 & b_2 & c_2 \end{vmatrix}.$

8. $\begin{vmatrix} a+b & a+c & b+c \\ a+c & a+b & b+c \\ b+c & b+a & a+c \end{vmatrix} = 2(a+b+c) \begin{vmatrix} 1 & c & b \\ 1 & b & b \\ 1 & b & a \end{vmatrix}.$

9. $\begin{vmatrix} 1 & 2x & x^2 & 0 \\ 0 & 1 & 2x & x^2 \\ 1 & x & x^2 & 0 \\ 0 & 1 & x & x^2 \end{vmatrix} = x^4 \begin{vmatrix} 1 & 2 & 1 & 0 \\ 0 & 1 & 2 & 1 \\ 1 & 1 & 1 & 0 \\ 0 & 1 & 1 & 1 \end{vmatrix}.$

10. $\begin{vmatrix} 0 & a^2 & b^2 \\ a^2 & 0 & c^2 \\ b^2 & c^2 & 0 \end{vmatrix} = a^2b^2c^2 \begin{vmatrix} 0 & 1 & 1 \\ 1 & 0 & 1 \\ 1 & 1 & 0 \end{vmatrix}.$

11. $\begin{vmatrix} 0 & a & b & c \\ a & 0 & c & b \\ b & c & 0 & a \\ c & b & a & 0 \end{vmatrix} = \begin{vmatrix} 0 & 1 & 1 & 1 \\ 1 & 0 & c^2 & b^2 \\ 1 & c^2 & 0 & a^2 \\ 1 & b^2 & a^2 & 0 \end{vmatrix}.$

12. $\begin{vmatrix} x^4 & x^9 & x^{16} \\ x^9 & x^{16} & x^{25} \\ x^{16} & x^{25} & x^{36} \end{vmatrix} = x^{36} \begin{vmatrix} 1 & x & x^4 \\ x & x^4 & x^9 \\ x^4 & x^9 & x^{16} \end{vmatrix}.$

10. Expansions by Rows and Columns. Minors and Cofactors. The determinant as a linear homogeneous function of elements of any row, say of the ith row, can be expanded in the elements of this row thus:

$$D = A_{i1}a_{i1} + \cdots + A_{ij}a_{ij} + \cdots + A_{in}a_{in}$$

where the coefficients A_{ij} $(j = 1, 2, \ldots, n)$ do not contain a_{i1}, a_{i2} $\ldots$, a_{in}. The coefficient A_{ij} of a_{ij} in this expansion is called the *complement* or *cofactor* of the element a_{ij}. This complement stands in a close relation to the determinant of order $n - 1$ obtained from D by striking out the ith row and the jth column without changing the order of the other rows and columns. This determinant D_{ij} of order $n - 1$ is called the *minor* corresponding to the element a_{ij}. For example, the minors corresponding to the elements a_1, b_2, and c_1 in the determinant

$$\begin{vmatrix} a_1 & b_1 & c_1 \\ a_2 & b_2 & c_2 \\ a_3 & b_3 & c_3 \end{vmatrix}$$

are

$$\begin{vmatrix} b_2 & c_2 \\ b_3 & c_3 \end{vmatrix}, \quad \begin{vmatrix} a_1 & c_1 \\ a_3 & c_3 \end{vmatrix}, \quad \begin{vmatrix} a_2 & b_2 \\ a_3 & b_3 \end{vmatrix}.$$

The complement A_{ij} and the minor D_{ij} are closely related; in fact, we are going to prove that

$$A_{ij} = (-1)^{i+j}D_{ij},$$

that is,

$$A_{ij} = D_{ij}$$

if the row and the column occupied by a_{ij} have the same parity, and

$$A_{ij} = -D_{ij}$$

in the contrary case. We shall prove this important relation first for the left-hand top element a_{11}. In the expansion

$$D = A_{11}a_{11} + A_{12}a_{12} + \cdots + A_{1n}a_{1n}$$

the complement A_{11} does not depend on a_{11}, a_{12}, $\ldots$, a_{1n} and consequently does not change by taking particular values

$$a_{11} = 1, \qquad a_{12} = a_{13} = \cdots = a_{1n} = 0.$$

Hence,

$$A_{11} = \begin{vmatrix} 1 & 0 & 0 & \cdots & 0 \\ a_{21} & a_{22} & a_{23} & \cdots & a_{2n} \\ \cdots & \cdots & \cdots & \cdots & \cdots \\ a_{n1} & a_{n2} & a_{n3} & \cdots & a_{nn} \end{vmatrix}.$$

When in this determinant from the second, third, . . . , nth row is subtracted the first row multiplied by $a_{21}, a_{31}, \ldots, a_{n1}$, respectively, the complement A_{11} is represented by the determinant

$$A_{11} = \begin{vmatrix} 1 & 0 & 0 & \cdots & 0 \\ 0 & a_{22} & a_{23} & \cdots & a_{2n} \\ 0 & a_{32} & a_{33} & \cdots & a_{3n} \\ \cdots & \cdots & \cdots & \cdots & \cdots \\ 0 & a_{n2} & a_{n3} & \cdots & a_{nn} \end{vmatrix}$$

depending only on the matrix

$$B = \begin{pmatrix} a_{22} & a_{23} & \cdots & a_{2n} \\ a_{32} & a_{33} & \cdots & a_{3n} \\ \cdots & \cdots & \cdots & \cdots \\ a_{n2} & a_{n3} & \cdots & a_{nn} \end{pmatrix}.$$

Clearly, A_{11} is an integral rational function of the elements of B, is linear and homogeneous in elements of each row of B, and vanishes when two of these rows are identical, taking the value 1 in case

$$B = \begin{pmatrix} 1 & 0 & \cdots & 0 \\ 0 & 1 & \cdots & 0 \\ 0 & 0 & \cdots & 1 \end{pmatrix}.$$

By these properties A_{11} is characterized as the determinant of matrix B, that is,

$$A_{11} = \begin{vmatrix} a_{22} & a_{23} & \cdots & a_{2n} \\ a_{32} & a_{33} & \cdots & a_{3n} \\ \cdots & \cdots & \cdots & \cdots \\ a_{n2} & a_{n3} & \cdots & a_{nn} \end{vmatrix} = D_{11}.$$

To establish the relationship between the complement A_{ij} and the minor D_{ij} of an arbitrary element a_{ij}, the column to which this element belongs is transposed j times with neighboring columns until it occupies the place of the first, after which the ith row is transposed i times with adjacent rows until it occupies the place of the first. After these $i + j$ transpositions of rows and columns, in the new determinant D' the element a_{ij} is in the left top corner and its complement in D', as can be seen easily, is D_{ij}; but

$$D = D'(-1)^{i+j};$$

consequently, the complement of a_{ij} in D is

$$A_{ij} = (-1)^{i+j}D_{ij}.$$

A determinant can also be expanded by elements of any column. The expansion by elements of the jth column is

$$D = A_{1j}a_{1j} + A_{2j}a_{2j} + \cdots + A_{nj}a_{nj}.$$

Notice the following important property of complements: When in the expansion

$$D = A_{i1}a_{i1} + A_{i2}a_{i2} + \cdots + A_{in}a_{in}$$

the elements of the ith row are replaced by the elements of some other row $a_{k1}, a_{k2}, \ldots, a_{kn}$ $(k \neq i)$, the sum

$$A_{i1}a_{k1} + A_{i2}a_{k2} + \cdots + A_{in}a_{kn}$$

is the expansion of a determinant in which ith and kth rows are the same; hence,

$$A_{i1}a_{k1} + A_{i2}a_{k2} + \cdots + A_{in}a_{kn} = 0$$

provided k and i are different. Similarly,

$$A_{1j}a_{1k} + A_{2j}a_{2k} + \cdots + A_{nj}a_{nk} = 0$$

if k and j are not equal.

11. Examples. The expansion of a determinant by rows or columns together with the transformations to which it may be subjected provides a practical way for evaluating determinants. The following examples will show how to do this:

Example 1. In Example 2, Sec. 9, it was shown that

$$D = \begin{vmatrix} 3 & 1 & 2 & 3 \\ 4 & -1 & 2 & 4 \\ 1 & -1 & 1 & 1 \\ 4 & -1 & 2 & 5 \end{vmatrix} = - \begin{vmatrix} 1 & 0 & 0 & 0 \\ 0 & 0 & 0 & -1 \\ 0 & 4 & 3 & 4 \\ 0 & 7 & 4 & 8 \end{vmatrix}.$$

Expanding the last determinant by the elements of the first row, all of which are zeros except the first, we see that the expansion reduces to one term so that

$$\begin{vmatrix} 1 & 0 & 0 & 0 \\ 0 & 0 & 0 & -1 \\ 0 & 4 & 3 & 4 \\ 0 & 7 & 4 & 8 \end{vmatrix} = \begin{vmatrix} 0 & 0 & -1 \\ 4 & 3 & 4 \\ 7 & 4 & 8 \end{vmatrix}.$$

In this determinant of the third order the complement of -1 being

$$\begin{vmatrix} 4 & 3 \\ 7 & 4 \end{vmatrix} = 16 - 21 = -5$$

we see at once that

$$\begin{vmatrix} 0 & 0 & -1 \\ 4 & 3 & 4 \\ 7 & 4 & 8 \end{vmatrix} = 5$$

and, consequently,

$$D = -5.$$

Example 2. Evaluate

$$D = \begin{vmatrix} 3 & 7 & 1 & 2 & 5 \\ 6 & 4 & 3 & 0 & 2 \\ 0 & 3 & 0 & 1 & 2 \\ 1 & 0 & 6 & 5 & 3 \\ 2 & 1 & 0 & 2 & 0 \end{vmatrix}.$$

Subtract the second column multiplied by 2 from the first and the fourth. This gives

$$D = \begin{vmatrix} -11 & 7 & 1 & -12 & 5 \\ -2 & 4 & 3 & -8 & 2 \\ -6 & 3 & 0 & -5 & 2 \\ 1 & 0 & 6 & 5 & 3 \\ 0 & 1 & 0 & 0 & 0 \end{vmatrix}.$$

Expand by elements of the fifth row, noticing that the complement of 1 in this row is

$$-\begin{vmatrix} -11 & 1 & -12 & 5 \\ -2 & 3 & -8 & 2 \\ -6 & 0 & -5 & 2 \\ 1 & 6 & 5 & 3 \end{vmatrix} = -D'.$$

This gives

$$D = -D'$$

so that everything is reduced to the evaluation of a determinant of the fourth order. To simplify D' add to columns 1 and 3, column 4 multiplied by 2; then,

$$D' = \begin{vmatrix} -1 & 1 & -2 & 5 \\ 2 & 3 & -4 & 2 \\ -2 & 0 & -1 & 2 \\ 7 & 6 & 11 & 3 \end{vmatrix}.$$

Now add column 1 to column 4, and to column 1 add column 3 multiplied by -2; this gives

$$D' = \begin{vmatrix} 3 & 1 & -2 & 4 \\ 10 & 3 & -4 & 4 \\ 0 & 0 & -1 & 0 \\ -15 & 6 & 11 & 10 \end{vmatrix}.$$

Expanding by the elements of row 3, we have

$$D' = -\begin{vmatrix} 3 & 1 & 4 \\ 10 & 3 & 4 \\ -15 & 6 & 10 \end{vmatrix} = -2\begin{vmatrix} 3 & 1 & 2 \\ 10 & 3 & 2 \\ -15 & 6 & 5 \end{vmatrix}.$$

Now

$$\begin{vmatrix} 3 & 1 & 2 \\ 10 & 3 & 2 \\ -15 & 6 & 5 \end{vmatrix} = \begin{vmatrix} 3 & 1 & 2 \\ 1 & 0 & -4 \\ -15 & 6 & 5 \end{vmatrix} = \begin{vmatrix} 3 & 1 & 14 \\ 1 & 0 & 0 \\ -15 & 6 & -55 \end{vmatrix},$$

and, expanding by the elements of the second row,

$$\begin{vmatrix} 3 & 1 & 14 \\ 1 & 0 & 0 \\ -15 & 6 & -55 \end{vmatrix} = -\begin{vmatrix} 1 & 14 \\ 6 & -55 \end{vmatrix} = 55 + 84 = 139.$$

Hence, $D' = -278$ and finally $D = 278$.

Example 3. To evaluate the determinant

$$D = \begin{vmatrix} 1+a & 1 & 1 & 1 \\ 1 & 1+b & 1 & 1 \\ 1 & 1 & 1+c & 1 \\ 1 & 1 & 1 & 1+d \end{vmatrix}$$

subtract the second column from the first, and expand by the elements of the first column; this gives

$$D = a \begin{vmatrix} 1+b & 1 & 1 \\ 1 & 1+c & 1 \\ 1 & 1 & 1+d \end{vmatrix} + b \begin{vmatrix} 1 & 1 & 1 \\ 1 & 1+c & 1 \\ 1 & 1 & 1+d \end{vmatrix}.$$

The first determinant in the right-hand side is similar to the proposed one; we shall write

$$D' = \begin{vmatrix} 1+b & 1 & 1 \\ 1 & 1+c & 1 \\ 1 & 1 & 1+d \end{vmatrix}.$$

The other determinant becomes, after subtracting the first column from the second and third,

$$\begin{vmatrix} 1 & 0 & 0 \\ 1 & c & 0 \\ 1 & 0 & d \end{vmatrix} = \begin{vmatrix} c & 0 \\ 0 & d \end{vmatrix} = cd.$$

Hence,

$$D = aD' + bcd.$$

Applying the same transformations to D', we find

$$D' = bD'' + cd,$$

where

$$D'' = \begin{vmatrix} 1+c & 1 \\ 1 & 1+d \end{vmatrix} = cd + c + d.$$

Hence,

$$D' = bcd + bc + bd + cd$$

and

$$D = abcd + abc + abd + acd + bcd$$
$$= abcd \left(1 + \frac{1}{a} + \frac{1}{b} + \frac{1}{c} + \frac{1}{d} \right).$$

Example 4. The following determinant of order n

$$\begin{vmatrix} 1 & x_1 & x_1^2 & \cdots & x_1^{n-1} \\ 1 & x_2 & x_2^2 & \cdots & x_2^{n-1} \\ \cdots & \cdots & \cdots & \cdots & \cdots \\ 1 & x_n & x_n^2 & \cdots & x_n^{n-1} \end{vmatrix}$$

occurs often and is called Vandermonde's determinant. The simplest way to evaluate it is to replace x_n by a variable x. Then, the determinant becomes a polynomial $D_n(x)$ of degree $n - 1$ in x, as can be seen by expanding it by elements of the last row. For $x = x_1, x_2, \ldots, x_{n-1}$ this polynomial vanishes since $D(x_\alpha)$ for $\alpha = 1, 2, \ldots, n - 1$ appears as a determinant with two identical rows; hence,

$$D_n(x) = C(x - x_1)(x - x_2) \cdots (x - x_{n-1}),$$

where C is the leading coefficient in $D_n(x)$. This coefficient is the minor

$$D_{n-1} = \begin{vmatrix} 1 & x_1 & x_1^2 & \cdots & x_1^{n-2} \\ 1 & x_2 & x_2^2 & \cdots & x_2^{n-2} \\ \cdots & \cdots & \cdots & \cdots & \cdots \\ 1 & x_{n-1} & x_{n-1}^2 & \cdots & x_{n-1}^{n-2} \end{vmatrix}$$

corresponding to x_n^{n-1}, and so we have

$$D_n(x_n) = D_n = D_{n-1}(x_n - x_1)(x_n - x_2) \cdots (x_n - x_{n-1}). \tag{1}$$

The determinant D_{n-1} is of the same type as D_n and can be treated in a similar manner
Now

$$D_2 = \begin{vmatrix} 1 & x_1 \\ 1 & x_2 \end{vmatrix} = x_2 - x_1;$$

hence, as follows from (1) for $n = 3$,

$$D_3 = (x_3 - x_1)(x_3 - x_2)(x_2 - x_1).$$

Further,

$$D_4 = (x_4 - x_1)(x_4 - x_2)(x_4 - x_3)(x_3 - x_1)(x_3 - x_2)(x_2 - x_1),$$

etc. The general expression of Vandermonde's determinant is

$$\begin{aligned} D_n = &(x_n - x_1)(x_n - x_2) \cdots (x_n - x_1) \\ &(x_{n-1} - x_1)(x_{n-1} - x_2) \cdots (x_{n-1} - x_{n-2}) \\ &\cdots\cdots\cdots\cdots\cdots\cdots\cdots\cdots\cdots\cdots . \\ &(x_3 - x_2)(x_3 - x_1) \\ &(x_2 - x_1). \end{aligned}$$

It is a rational integral function of $x_1, x_2, \ldots, x_n$ that merely changes its sign when two of the variables are transposed and for this reason is called an alternating function. For the exchange of two variables like x_1 and x_2 corresponds to the exchange of the first and second rows, and this causes the change of sign of Vandermonde's determinant.

Problems

Evaluate the numerical determinants:

1. $\begin{vmatrix} 0 & 1 & 1 \\ 1 & 0 & 1 \\ 1 & 1 & 0 \end{vmatrix}$.

2. $\begin{vmatrix} 1 & 1 & 1 \\ 1 & 2 & 2 \\ 1 & 3 & 6 \end{vmatrix}$.

3. $\begin{vmatrix} 5 & 2 & 2 \\ 7 & 3 & 2 \\ 9 & 6 & -5 \end{vmatrix}$.

4. $\begin{vmatrix} 5 & -2 & -3 \\ 10 & -4 & -3 \\ -5 & 3 & 4 \end{vmatrix}$.

5. $\begin{vmatrix} 1 & 1 & 1 & 1 \\ 1 & 2 & 3 & 4 \\ 1 & 3 & 6 & 10 \\ 1 & 4 & 9 & 16 \end{vmatrix}$.

6. $\begin{vmatrix} 1 & 2 & 3 & 4 \\ 2 & 3 & 4 & 1 \\ 3 & 4 & 1 & 2 \\ 4 & 1 & 2 & 3 \end{vmatrix}$.

7. $\begin{vmatrix} 1 & 3 & 5 & 7 \\ 2 & 4 & 6 & 7 \\ 3 & 1 & 2 & 2 \\ 5 & 1 & 3 & 1 \end{vmatrix}$.

8. $\begin{vmatrix} 3 & 1 & 2 & 3 \\ 4 & -1 & 2 & 4 \\ 1 & -1 & 1 & 1 \\ 4 & -1 & 2 & 5 \end{vmatrix}$.

9. $\begin{vmatrix} 9 & 13 & 17 & 4 \\ 18 & 28 & 33 & 8 \\ 30 & 40 & 54 & 13 \\ 24 & 37 & 46 & 11 \end{vmatrix}$.

Evaluate the literal determinants:

10. $\begin{vmatrix} 2a & a+b & a+c \\ b+a & 2b & b+c \\ c+a & c+b & 2c \end{vmatrix}.$

11. $\begin{vmatrix} a+b & a+c & b+c \\ a+c & a+b & b+c \\ b+c & b+a & a+c \end{vmatrix}.$

12. $\begin{vmatrix} 0 & c & b \\ b & a & 0 \\ c & 0 & a \end{vmatrix}.$

13. $\begin{vmatrix} b+c & a & a \\ b & c+a & b \\ c & c & a+b \end{vmatrix}.$

14. $\begin{vmatrix} a & b & a+b \\ b & a+b & a \\ a+b & a & b \end{vmatrix}.$

15. $\begin{vmatrix} 1 & 1 & 1 \\ a & b & c \\ bc & ac & ab \end{vmatrix}.$

16. $\begin{vmatrix} 1 & bc+ad & b^2c^2+a^2d^2 \\ 1 & ac+bd & a^2c^2+b^2d^2 \\ 1 & ab+cd & a^2b^2+c^2d^2 \end{vmatrix}.$

17. $\begin{vmatrix} 1 & a & a^3 \\ 1 & b & b^3 \\ 1 & c & c^3 \end{vmatrix}.$ When c is considered as a variable, the determinant is of degree 3 in c, vanishing for $c = a$, $c = b$, and lacking a term in c^2.

18. $\begin{vmatrix} 1 & a^2 & a^3 \\ 1 & b^2 & b^3 \\ 1 & c^2 & c^3 \end{vmatrix}.$

19. $\begin{vmatrix} a & b & c \\ a^2 & b^2 & c^2 \\ bc & ac & ab \end{vmatrix}.$

20. $\begin{vmatrix} 1 & a & a^4 \\ 1 & b & b^4 \\ 1 & c & c^4 \end{vmatrix}.$ The determinant has the form $(c-a)(c-b)(Ac^2+Bc+C)$, where A, B, C do not depend on c, and lacks terms in c^2 and c^3.

21. $\begin{vmatrix} a^2 & a^2-(b-c)^2 & bc \\ b^2 & b^2-(c-a)^2 & ac \\ c^2 & c^2-(a-b)^2 & ab \end{vmatrix}.$

22. $\begin{vmatrix} (x-a)^2 & (y-a)^2 & (z-a)^2 \\ (x-b)^2 & (y-b)^2 & (z-b)^2 \\ (x-c)^2 & (y-c)^2 & (z-c)^2 \end{vmatrix}.$

23. $\begin{vmatrix} -2a & a+b & a+c \\ b+a & -2b & b+c \\ c+a & c+b & -2c \end{vmatrix}.$

24. $\begin{vmatrix} (b+c)^2 & a^2 & a^2 \\ b^2 & (c+a)^2 & b^2 \\ c^2 & c^2 & (a+b)^2 \end{vmatrix}.$

25. $\begin{vmatrix} a & a & a & a \\ a & b & b & b \\ a & b & c & c \\ a & b & c & d \end{vmatrix}.$

26. $\begin{vmatrix} a & b & c & d \\ b & a & d & c \\ c & d & a & b \\ d & c & b & a \end{vmatrix}.$

27. $\begin{vmatrix} x & a_1 & a_2 & a_3 & 1 \\ a_1 & x & a_2 & a_3 & 1 \\ a_1 & a_2 & x & a_3 & 1 \\ a_1 & a_2 & a_3 & x & 1 \\ a_1 & a_2 & a_3 & a_4 & 1 \end{vmatrix}.$

28. $\begin{vmatrix} 1 & 1 & 1 & 1 & 1 \\ 1 & 0 & 0 & 0 & x_1 \\ x_2 & 1 & 0 & 0 & x_2 \\ x_3 & x_3 & 1 & 0 & x_3 \\ x_4 & x_4 & x_4 & 1 & x_4 \end{vmatrix}.$ Generalize.

29. $\begin{vmatrix} 1 & 1 & 1 & 1 \\ 1 & 1+b & 1 & 1 \\ 1 & 1 & 1+c & 1 \\ 1 & 1 & 1 & 1+d \end{vmatrix}.$ Generalize.

30. $\begin{vmatrix} 1+a & 1 & 1 & 1 & 1 \\ 1 & 1+b & 1 & 1 & 1 \\ 1 & 1 & 1+c & 1 & 1 \\ 1 & 1 & 1 & 1+d & 1+e \end{vmatrix}.$ Generalize.

31.
$$\begin{vmatrix} x & 1 & 1 & \cdots & 1 & 1 \\ 1 & x & 1 & \cdots & 1 & 1 \\ 1 & 1 & x & \cdots & 1 & 1 \\ \multicolumn{6}{c}{\cdots\cdots\cdots\cdots} \\ 1 & 1 & 1 & \cdots & 1 & x \end{vmatrix}.$$

32.
$$\begin{vmatrix} a_1 & x & x & \cdots & x \\ x & a_2 & x & \cdots & x \\ x & x & a_3 & \cdots & x \\ \multicolumn{5}{c}{\cdots\cdots\cdots\cdots} \\ x & x & x & \cdots & a_n \end{vmatrix}.$$

33.
$$\begin{vmatrix} 1 & a & a^2 & a^4 \\ 1 & b & b^2 & b^4 \\ 1 & c & c^2 & c^4 \\ 1 & d & d^2 & d^4 \end{vmatrix}.$$ Generalize.

HINT: Replace a by a variable x; then the determinant is of the fourth degree in x, has roots b, c, d, and lacks term in x^3.

34.
$$\begin{vmatrix} 1 & a & a^2 & a^5 \\ 1 & b & b^2 & b^5 \\ 1 & c & c^2 & c^5 \\ 1 & d & d^2 & d^5 \end{vmatrix}.$$ Generalize.

35. If

$$D_n = \begin{vmatrix} a_1 & 1 & 0 & 0 & \cdots & 0 \\ -1 & a_2 & 1 & 0 & \cdots & 0 \\ 0 & -1 & a_3 & 1 & \cdots & 0 \\ \multicolumn{6}{c}{\cdots\cdots\cdots\cdots\cdots\cdots} \\ 0 & 0 & 0 & \cdots & \cdots & a_n \end{vmatrix}$$

show that $D_n = a_n D_{n-1} + D_{n-2}$.

*36. If

$$(m,n) = \begin{vmatrix} x^{m^2} & x^{(m+1)^2} & \cdots & x^{(m+n-1)^2} \\ x^{(m+1)^2} & x^{(m+2)^2} & \cdots & x^{(m+n)^2} \\ \multicolumn{4}{c}{\cdots\cdots\cdots\cdots\cdots\cdots} \\ x^{(m+n-1)^2} & x^{(m+n)^2} & \cdots & x^{(m+2n-2)^2} \end{vmatrix},$$

show that

$$(m,n) = x^{mn(m+2n-2)}(0,n).$$

*37. Using the notation of Prob. *36, show that

$$(0,n+1) = (1 - x^{-2})(1 - x^{-4}) \cdots (1 - x^{-2n})(2,n).$$

*38. Hence, deduce

$$(0,n+1) = x^{4n^2}(1 - x^{-2})(1 - x^{-4}) \cdots (1 - x^{-2n})(0,n),$$

and

$$(0,n) = x^{n(n-1)^2}(x^2 - 1)^{n-1}(x^4 - 1)^{n-2}(x^6 - 1)^{n-3} \cdots (x^{2n-2} - 1)^1.$$

MATRICES

★12. Equality and Addition of Matrices. Thus far the term matrix has been used only to designate a square array of n^2 elements a_{ij}. Such arrays of elements can be introduced, however, as new mathematical quantities once we define the notion of equality of matrices and the two direct operations usually called addition and multiplication. In this manner quite an extensive algebra of matrices can be developed, of which, in this book, only a brief introduction will be given. A matrix of n^2 elements is called a matrix of order n. Thus, for $n = 2, 3, 4, \ldots$ we can speak of matrices of orders 2, 3, 4, etc. Confining ourselves to the consideration of matrices of the same order, we define *two matrices as equal if corresponding elements, that is, elements belonging to the same rows and columns in both, are equal.* If two matrices are denoted by letters A and B and their elements by a_{ij} and b_{ij}, the equality

$$A = B$$

stands for the n^2 equalities

$$a_{ij} = b_{ij}$$

where i and j run independently through 1, 2, . . . , n. Thus, we have equality of the matrices

$$\begin{pmatrix} 1 & 2 & 0 \\ 0 & 1 & 2 \\ 1 & 0 & 0 \end{pmatrix} = \begin{pmatrix} 1 & \tfrac{6}{3} & 0 \\ 0 & 1 & \sqrt{4} \\ 1 & 0 & 0 \end{pmatrix};$$

but the matrices

$$\begin{pmatrix} 0 & 1 & 0 \\ 0 & 1 & 0 \\ 0 & 0 & 1 \end{pmatrix} \quad \text{and} \quad \begin{pmatrix} 1 & 0 & 0 \\ 0 & 1 & 0 \\ 0 & 0 & 1 \end{pmatrix}$$

are not equal.

On two matrices A and B (of the same order) with elements a_{ij} and b_{ij} we can perform an operation called addition and which is indicated by means of the usual sign +. *To add a matrix B to a matrix A means to form a new matrix C whose elements c_{ij} are the sums $a_{ij} + b_{ij}$ of the corresponding elements of A and B.* Thus, if

$$A = \begin{pmatrix} 1 & 2 & -1 \\ 3 & 1 & 4 \\ -5 & 6 & 1 \end{pmatrix} \quad \text{and} \quad B = \begin{pmatrix} -1 & -2 & 0 \\ 3 & -2 & -4 \\ 5 & -6 & 2 \end{pmatrix},$$

then

$$C = A + B = \begin{pmatrix} 0 & 0 & -1 \\ 6 & -1 & 0 \\ 0 & 0 & 3 \end{pmatrix}.$$

From the definition it follows immediately that the associative and the commutative laws hold for addition of matrices, that is,

$$(A + B) + C = A + (B + C)$$
$$A + B = B + A$$

with all their consequences. The matrix 0 all of whose elements are zeros is the only matrix such that

$$A + 0 = A$$

for every matrix A, and therefore 0 may be called the zero matrix.★

★13. Multiplication of Matrices. The definition of multiplication of matrices can be based on the notion of the *scalar* product of two ordered n-tuples of numbers or *vectors*. By a scalar product of two vectors (ordered n-tuples)

$$R = (a_1, a_2, \ldots, a_n), \qquad S = (b_1, b_2, \ldots, b_n)$$

we mean the following quantity

$$R \cdot S = a_1b_1 + a_2b_2 + \cdots + a_nb_n.$$

Now let A and B be two matrices of order n. The vector

$$R_i = (a_{i1}, a_{i2}, \ldots, a_{in})$$

is associated with the ith row of A, and similarly the vector

$$C_j = (b_{1j}, b_{2j}, \ldots, b_{nj})$$

is associated with the jth column of B. The matrix

$$C = \begin{pmatrix} R_1{\cdot}C_1 & R_1{\cdot}C_2 & \cdots & R_1{\cdot}C_n \\ R_2{\cdot}C_1 & R_2{\cdot}C_2 & \cdots & R_2{\cdot}C_n \\ \cdots\cdots\cdots\cdots\cdots\cdots\cdots \\ R_n{\cdot}C_1 & R_n{\cdot}C_2 & \cdots & R_n{\cdot}C_n \end{pmatrix}$$

by definition is the product of the matrix A by the matrix B, and we write

$$AB = C.$$

In other words,

$$\begin{pmatrix} a_{11} & a_{12} & \cdots & a_{1n} \\ a_{21} & a_{22} & \cdots & a_{2n} \\ \cdots\cdots\cdots\cdots\cdots \\ a_{n1} & a_{n2} & \cdots & a_{nn} \end{pmatrix} \begin{pmatrix} b_{11} & b_{12} & \cdots & b_{1n} \\ b_{21} & b_{22} & \cdots & b_{2n} \\ \cdots\cdots\cdots\cdots\cdots \\ b_{n1} & b_{n2} & \cdots & b_{nn} \end{pmatrix} = \begin{pmatrix} c_{11} & c_{12} & \cdots & c_{1n} \\ c_{21} & c_{22} & \cdots & c_{2n} \\ \cdots\cdots\cdots\cdots\cdots \\ c_{n1} & c_{n2} & \cdots & c_{nn} \end{pmatrix}$$

where

$$c_{ij} = a_{i1}b_{1j} + a_{i2}b_{2j} + \cdots + a_{in}b_{nj} = \sum_{k=1}^{n} a_{ik}b_{kj}$$

is the scalar product $R_i \cdot C_j$. For example, according to this definition,

$$\begin{pmatrix} 1 & -1 & 1 \\ 0 & 1 & -1 \\ 1 & 0 & 0 \end{pmatrix} \begin{pmatrix} -1 & 1 & 1 \\ 1 & 0 & 0 \\ -1 & -1 & -1 \end{pmatrix} = \begin{pmatrix} -3 & 0 & 0 \\ 2 & 1 & 1 \\ -1 & 1 & 1 \end{pmatrix},$$

but

$$\begin{pmatrix} -1 & 1 & 1 \\ 1 & 0 & 0 \\ -1 & -1 & -1 \end{pmatrix} \begin{pmatrix} 1 & -1 & 1 \\ 0 & 1 & -1 \\ 1 & 0 & 0 \end{pmatrix} = \begin{pmatrix} 0 & 2 & -2 \\ 1 & -1 & 1 \\ -2 & 0 & 0 \end{pmatrix}.$$

This example shows that, in general, multiplication of matrices is not commutative so that we have to distinguish BA from AB. On the contrary, the associative law of multiplication holds for matrices, that is,

$$(AB)C = A(BC).$$

This can be verified by computing the elements of the matrices in the left- and the right-hand members in accordance with the definition of a product. Also, direct verification shows the validity of the two distributive laws:

$$(A + B)C = AC + BC$$

and

$$C(A + B) = CA + CB.$$

The matrix

$$E = \begin{pmatrix} 1 & 0 & \cdots & 0 \\ 0 & 1 & \cdots & 0 \\ \cdot & \cdot & \cdots & \cdot \\ 0 & 0 & \cdots & 1 \end{pmatrix},$$

in which diagonal elements are 1 and all others 0, possesses the property that for any matrix A

$$AE = EA = A,$$

and E is the only matrix with this property. For let E' be a second matrix such that

$$AE' = E'A = A$$

for all matrices A. Taking in particular $A = E$, we must have

$$EE' = E$$

But, on the other hand, taking $A = E'$ in

$$EA = A,$$

we have

$$EE' = E',$$

and so $E' = E$. The matrix E plays the role of a unit with respect to multiplication of matrices, and for this reason it is called the *unit matrix*. It is a particular case of the so-called *scalar matrices*

$$(a) = \begin{pmatrix} a & 0 & \cdots & 0 \\ 0 & a & \cdots & 0 \\ 0 & 0 & \cdots & a \end{pmatrix}$$

in which all the diagonal elements are equal to the same number a, the other elements being 0. On multiplying any matrix

$$A = \begin{pmatrix} a_{11} & a_{12} & \cdots & a_{1n} \\ a_{21} & a_{22} & \cdots & a_{2n} \\ \cdots\cdots\cdots\cdots\cdots \\ a_{n1} & a_{n2} & \cdots & a_{nn} \end{pmatrix}$$

by (a), or (a) by A, in both cases the resulting matrix is the same:

$$A(a) = (a)A = \begin{pmatrix} aa_{11} & aa_{12} & \cdots & aa_{1n} \\ aa_{21} & aa_{22} & \cdots & aa_{2n} \\ \cdots\cdots\cdots\cdots\cdots \\ aa_{n1} & aa_{n2} & \cdots & aa_{nn} \end{pmatrix}.$$

The product of two scalar matrices (a) and (b) is a scalar matrix (ab) and, similarly, $(a) + (b)$ is a scalar matrix $(a + b)$. In view of this, scalar matrices (a) are represented simply by a, and the symbol

$$aA = Aa$$

means the product of a scalar matrix (a) by another matrix A. A matrix in which all elements except those of the first column are 0 is called a *column matrix*. Thus, column matrices are of the type

$$\begin{pmatrix} x_1 & 0 & \cdots & 0 \\ x_2 & 0 & \cdots & 0 \\ \cdots\cdots\cdots\cdots \\ x_n & 0 & \cdots & 0 \end{pmatrix}$$

and may be conveniently denoted by

$$(x_1, x_2, \ldots, x_n).$$

The product of any matrix A by a column matrix $(x_1, x_2, \ldots, x_n)$ is another column matrix $(y_1, y_2, \ldots, y_n)$ where

$$\begin{aligned} y_1 &= a_{11}x_1 + a_{12}x_2 + \cdots + a_{1n}x_n, \\ y_2 &= a_{21}x_1 + a_{22}x_2 + \cdots + a_{2n}x_n, \\ &\cdots\cdots\cdots\cdots\cdots\cdots\cdots\cdots \\ y_n &= a_{n1}x_1 + a_{n2}x_2 + \cdots + a_{nn}x_n. \end{aligned}$$

These relations are therefore equivalent to a single matrix equation

$$A(x_1, x_2, \ldots, x_n) = (y_1, y_2, \ldots, y_n).\star$$

Problems

Prove that the matrices given below satisfy the following equations:

1. $S = \begin{pmatrix} 3 & 6 \\ -1 & -2 \end{pmatrix}$; $S^2 = S$.

2. $S = \begin{pmatrix} 1 & 3 \\ -2 & 2 \end{pmatrix}$; $S^2 - 3S + 8E = 0$.

3. $S = \begin{pmatrix} a & b \\ c & d \end{pmatrix}$; $\quad S^2 - (a + d)S + (ad - bc)E = 0$.

4. $S = \begin{pmatrix} 3 & -2 & 1 \\ 2 & -1 & 1 \\ -2 & 2 & 0 \end{pmatrix}$; $\quad S^2 = S$.

5. $S = \begin{pmatrix} -1 & 2 & -2 \\ 4 & -3 & 4 \\ 4 & -4 & 5 \end{pmatrix}$; $\quad S^2 = E$.

6. $S = \begin{pmatrix} -2 & -2 - \omega & -3 \\ -1 - 2\omega & -1 - \omega & -1 - 2\omega \\ -\omega & -\omega & 1 - \omega \end{pmatrix}$; $\quad \omega^2 + \omega + 1 = 0$; $\quad S^3 = E$.

7. If

$$A = \begin{pmatrix} 1 & 0 & 1 & -1 \\ 2 & -1 & 0 & -2 \\ -3 & 1 & -2 & 4 \\ 1 & -1 & 0 & -1 \end{pmatrix}, \quad B = \begin{pmatrix} 2 & 1 & 1 & 0 \\ 0 & 1 & 0 & -2 \\ 1 & -1 & 0 & 1 \\ 2 & 0 & 1 & 1 \end{pmatrix},$$

show that $AB = E$. What is BA?

8. Show that

$$\begin{pmatrix} 1 & 1 + i & -i \\ 0 & i & 1 - 2i \\ 1 & 1 & i \end{pmatrix} \begin{pmatrix} -2 + 2i & 1 - 2i & 2 - i \\ 1 - 2i & 2i & -1 + 2i \\ -i & i & i \end{pmatrix} = \begin{pmatrix} i & 0 & 0 \\ 0 & i & 0 \\ 0 & 0 & i \end{pmatrix}.$$

★14. Multiplication of Determinants. With the matrix

$$A = \begin{pmatrix} a_{11} & a_{12} & \cdots & a_{1n} \\ a_{21} & a_{22} & \cdots & a_{2n} \\ \cdots\cdots\cdots\cdots\cdots\cdots \\ a_{n1} & a_{n2} & \cdots & a_{nn} \end{pmatrix}$$

is associated the determinant

$$\begin{vmatrix} a_{11} & a_{12} & \cdots & a_{1n} \\ a_{21} & a_{22} & \cdots & a_{2n} \\ \cdots\cdots\cdots\cdots\cdots\cdots \\ a_{n1} & a_{n2} & \cdots & a_{nn} \end{vmatrix}$$

which is called the determinant of A and may be denoted by the sign det A. Between the determinants of the product of matrices and the determinants of the factors there exists a simple relation that is expressed in the following important theorem called the theorem of multiplication of determinants:

THEOREM. *The determinant of the product of the matrices A and B, or det (AB), is equal to the product of the determinants of the factors, or*

$$\det(AB) = \det A \cdot \det B.$$

Proof. Let

$$A = \begin{pmatrix} a_{11} & a_{12} & \cdots & a_{1n} \\ a_{21} & a_{22} & \cdots & a_{2n} \\ \cdots\cdots\cdots\cdots\cdots \\ a_{n1} & a_{n2} & \cdots & a_{nn} \end{pmatrix}, \quad B = \begin{pmatrix} b_{11} & b_{12} & \cdots & b_{1n} \\ b_{21} & b_{22} & \cdots & b_{2n} \\ \cdots\cdots\cdots\cdots\cdots \\ b_{n1} & b_{n2} & \cdots & b_{nn} \end{pmatrix},$$

and

$$C = AB = \begin{pmatrix} c_{11} & c_{12} & \cdots & c_{1n} \\ c_{21} & c_{22} & \cdots & c_{2n} \\ \cdots\cdots\cdots\cdots\cdots \\ c_{n1} & c_{n2} & \cdots & c_{nn} \end{pmatrix},$$

where

$$c_{ij} = a_{i1}b_{1j} + a_{i2}b_{2j} + \cdots + a_{in}b_{nj}.$$

Considering the elements a_{ij} as variables, the determinant of C will be a rational integral function of the n^2 variables a_{ij}, linear and homogeneous in the elements of each row of A, and will vanish when two of these rows are identical, since then C will have two identical rows. From the discussion in Sec. 7 it follows that det C differs from det A by a factor Γ independent of the elements a_{ij} so that

$$\det C = \Gamma \det A.$$

To determine Γ we choose $A = E$, then $C = B$, and since det $E = 1$, we shall have

$$\Gamma = \det B,$$

and so

$$\det C = \det AB = \det A \cdot \det B.$$

In other words, the product of two determinants, one with elements a_{ij} and the other with elements b_{ij}, can be expressed as a determinant with elements c_{ij} obtained by combining them according to the rule of multiplication of matrices in which rows of the first matrix are multiplied with columns of the second or, speaking shortly, by row-column multiplication. Since a determinant does not change if rows are replaced by columns, we can use also row-row, column-row, or column-column multiplication, obtaining by any of these procedures a determinant equal to the product of given determinants. Which way of multiplication we choose depends chiefly on convenience.

Example 1. Let us multiply the determinants

$$d = \begin{vmatrix} 1 & 1 & 1 \\ 1 & \omega & \omega^2 \\ 1 & \omega^2 & \omega \end{vmatrix} \quad \text{and} \quad D = \begin{vmatrix} a & b & c \\ b & c & a \\ c & a & b \end{vmatrix},$$

ω being an imaginary cube root of unity. Multiplying row by row, we have

$$dD = \begin{vmatrix} a+b+c & b+c+a & c+a+b \\ a+b\omega+c\omega^2 & b+c\omega+a\omega^2 & c+a\omega+b\omega^2 \\ a+b\omega^2+c\omega & b+c\omega^2+a\omega & c+a\omega^2+b\omega \end{vmatrix}.$$

But

$$b+c\omega+a\omega^2 = \omega^2(a+b\omega+c\omega^2), \qquad c+a\omega+b\omega^2 = \omega(a+b\omega+c\omega^2),$$
$$b+c\omega^2+a\omega = \omega(a+b\omega^2+c\omega), \qquad c+a\omega^2+b\omega = \omega^2(a+b\omega^2+c\omega),$$

and, consequently,

$$dD = (a+b+c)(a+b\omega+c\omega^2)(a+b\omega^2+c\omega) \begin{vmatrix} 1 & 1 & 1 \\ 1 & \omega^2 & \omega \\ 1 & \omega & \omega^2 \end{vmatrix},$$

or

$$dD = -d(a+b+c)(a+b\omega+c\omega^2)(a+b\omega^2+c\omega).$$

Since d is Vandermonde's determinant and is consequently not zero, it can be canceled out and thus

$$D = \begin{vmatrix} a & b & c \\ b & c & a \\ c & a & b \end{vmatrix} = -(a+b+c)(a+b\omega+c\omega^2)(a+b\omega^2+c\omega).$$

This determinant is called a cyclic determinant for the reason that its second and third rows are obtained by permuting cyclically elements of the first row. By a similar procedure cyclic determinants of any order can be presented in factorized form.

Example 2. Let, as usual. A_{ij} be the complement of a_{ij} in the determinant

$$D = \begin{vmatrix} a_{11} & a_{12} & \cdots & a_{1n} \\ a_{21} & a_{22} & \cdots & a_{2n} \\ \cdots\cdots\cdots\cdots\cdots \\ a_{n1} & a_{n2} & \cdots & a_{nn} \end{vmatrix},$$

and consider the *adjoint* determinant

$$\Delta = \begin{vmatrix} A_{11} & A_{12} & \cdots & A_{1n} \\ A_{21} & A_{22} & \cdots & A_{2n} \\ \cdots\cdots\cdots\cdots\cdots \\ A_{n1} & A_{n2} & \cdots & A_{nn} \end{vmatrix}.$$

When D is multiplied by Δ, row by row, the product is presented as a determinant

$$D\Delta = \begin{vmatrix} c_{11} & c_{12} & \cdots & c_{1n} \\ c_{21} & c_{22} & \cdots & c_{2n} \\ \cdots\cdots\cdots\cdots\cdots \\ c_{n1} & c_{n2} & \cdots & c_{nn} \end{vmatrix}$$

in which

$$c_{ij} = a_{i1}A_{j1} + a_{i2}A_{j2} + \cdots + a_{in}A_{jn}.$$

But according to the identities established in Sec. 10,

$$c_{ij} = 0 \quad \text{if} \quad j \neq i, \quad \text{and} \quad c_{ii} = D.$$

Hence,

$$DΔ = \begin{vmatrix} D & 0 & \cdots & 0 \\ 0 & D & \cdots & 0 \\ \multicolumn{4}{c}{\cdots\cdots\cdots} \\ 0 & 0 & \cdots & D \end{vmatrix} = D^n,$$

or

$$D(Δ - D^{n-1}) = 0.$$

Now, if the a_{ij} are considered as variables, D and $Δ$ are rational integral functions of them, and D does not vanish identically. Since the product

$$D(Δ - D^{n-1})$$

vanishes identically, the second factor must vanish identically so that

$$Δ = D^{n-1}$$

or

$$\begin{vmatrix} A_{11} & A_{12} & \cdots & A_{1n} \\ A_{21} & A_{22} & \cdots & A_{2n} \\ \multicolumn{4}{c}{\cdots\cdots\cdots\cdots} \\ A_{n1} & A_{n2} & \cdots & A_{nn} \end{vmatrix} = D^{n-1}$$

identically in the elements a_{ij}. Naturally, this equality remains true for particular values of a_{ij} even though for such particular values it may happen that $D = 0$. The reasoning employed is based upon the following theorem: *If the product of two polynomials F and ϕ in the variables $x_1, x_2, \ldots, x_n$ vanishes identically, at least one of the factors must be an identically vanishing polynomial.*

The theorem is true in case of polynomials in one variable, and it suffices to show that it is true for polynomials in n variables supposing its validity for polynomials in $n - 1$ variables. If neither F nor ϕ vanishes identically, we can arrange them in powers of one of the variables, say x_1, and write

$$F = F_0 x_1^n + F_1 x_1^{n-1} + \cdots,$$
$$\phi = \phi_0 x_1^m + \phi_1 x_1^{m-1} + \cdots,$$

where $F_0, F_1, \ldots; \phi_0, \phi_1, \ldots$ are polynomials in the $n - 1$ variables $x_2, \ldots, x_n$, and neither F_0 nor ϕ_0 vanishes identically. But

$$F\phi = F_0\phi_0 x_1^{n+m} + \cdots = 0$$

identically by hypothesis; hence, it also vanishes identically in the variables $x_2, \ldots, x_n$, therefore,

$$F_0\phi_0 = 0.$$

By hypothesis this implies that either F_0 or ϕ_0 vanishes identically, and this is a contradiction.★

Problems

1. Show that

$$\begin{vmatrix} a+b & c & c \\ a & b+c & a \\ b & b & c+a \end{vmatrix} \cdot \begin{vmatrix} a+b+\tfrac{1}{2}c & -\tfrac{1}{2}a & -\tfrac{1}{2}b \\ -\tfrac{1}{2}c & b+c+\tfrac{1}{2}a & -\tfrac{1}{2}b \\ -\tfrac{1}{2}c & -\tfrac{1}{2}a & c+a+\tfrac{1}{2} \end{vmatrix}$$
$$= \begin{vmatrix} (a+b)^2 & c^2 & c^2 \\ a^2 & (b+c)^2 & a^2 \\ b^2 & b^2 & (a+c)^2 \end{vmatrix},$$

and write in factorized form the determinant on the right.

2. Show that

$$\begin{vmatrix} a & a & a & a \\ a & b & b & b \\ a & b & c & c \\ a & b & c & d \end{vmatrix} \cdot \begin{vmatrix} -1 & 1 & 0 & 0 \\ 0 & -1 & 1 & 0 \\ 0 & 0 & -1 & 1 \\ 1 & 1 & 1 & -1 \end{vmatrix} = 2a(b-a)(c-b)(d-c).$$

3. Show that

$$\begin{vmatrix} 1 & \omega & \omega^2 & \omega^3 \\ 1 & \omega^2 & \omega^4 & \omega \\ 1 & \omega^3 & \omega & \omega^4 \\ 1 & \omega^4 & \omega^3 & \omega^2 \end{vmatrix}^2 = 125,$$

if ω is an imaginary fifth root of unity.

4. A determinant

$$\Delta = \begin{vmatrix} c_{11} & c_{12} & \cdots & c_{1n} \\ c_{21} & c_{22} & \cdots & c_{2n} \\ \cdots & \cdots & \cdots & \cdots \\ c_{n1} & c_{n2} & \cdots & c_{nn} \end{vmatrix}$$

in which

$$\sum_{k=1}^{n} c_{ik}c_{jk} = 0 \quad \text{if} \quad j \neq i, \quad \text{and} \quad \sum_{k=1}^{n} c_{ik}^2 = 1,$$

is called an "orthogonal determinant." Show that $\Delta = \pm 1$.

★15. Reciprocal Matrices.   A matrix

$$A = \begin{pmatrix} a_{11} & a_{12} & \cdots & a_{1n} \\ a_{21} & a_{22} & \cdots & a_{2n} \\ \cdots & \cdots & \cdots & \cdots \\ a_{n1} & a_{n2} & \cdots & a_{nn} \end{pmatrix}$$

is called *singular* or *nonsingular* according as its determinant is zero or not. Let A be a nonsingular matrix and $D \neq 0$ its determinant. Denoting as before the complement of the element a_{ij} in D by A_{ij} let us consider the matrix

$$X = \begin{pmatrix} \dfrac{A_{11}}{D} & \dfrac{A_{21}}{D} & \cdots & \dfrac{A_{n1}}{D} \\[2mm] \dfrac{A_{12}}{D} & \dfrac{A_{22}}{D} & \cdots & \dfrac{A_{n?}}{D} \\[2mm] \cdots & \cdots & \cdots & \cdots \\[2mm] \dfrac{A_{1n}}{D} & \dfrac{A_{2n}}{D} & \cdots & \dfrac{A_{nn}}{D} \end{pmatrix}.$$

The products AX and XA are

$$AX = \begin{pmatrix} c_{11} & c_{12} & \cdots & c_{1n} \\ c_{21} & c_{22} & \cdots & c_{2n} \\ \cdots & \cdots & \cdots & \cdots \\ c_{n1} & c_{n2} & \cdots & c_{nn} \end{pmatrix}; \quad XA = \begin{pmatrix} c'_{11} & c'_{12} & \cdots & c'_{1n} \\ c'_{12} & c'_{22} & \cdots & c'_{2n} \\ \cdots & \cdots & \cdots & \cdots \\ c'_{n1} & c'_{n2} & \cdots & c'_{nn} \end{pmatrix},$$

where

$$c_{ij} = \frac{a_{i1}A_{j1} + a_{i2}A_{j2} + \cdots + a_{in}A_{jn}}{D},$$

$$c'_{ij} = \frac{A_{1i}a_{1j} + A_{2i}a_{2j} + \cdots + A_{ni}a_{nj}}{D}.$$

By the identities of Sec. 10

$$c_{ij} = 0 \quad \text{if} \quad j \neq i, \quad \text{and} \quad c_{ii} = 1;$$
$$c'_{ij} = 0 \quad \text{if} \quad j \neq i, \quad \text{and} \quad c'_{ii} = 1.$$

Hence,

$$XA = AX = E.$$

This matrix X is called the *reciprocal* of A and is denoted by A^{-1}. The matrix A^{-1} is unique, that is, there is no other matrix Y such that

$$YA = AY = E.$$

If there is such a matrix, then by multiplying both members of the relation

$$YA = E$$

by A^{-1} we have

$$(YA)A^{-1} = Y(AA^{-1}) = EA^{-1}.$$

But

$$AA^{-1} = E, \qquad EA^{-1} = A^{-1},$$

and so

$$YE = Y = A^{-1}.$$

On the other hand, multiplying A^{-1} by $AY = E$, we have

$$A^{-1}(AY) = (A^{-1}A)Y = A^{-1}E;$$

but

$$A^{-1}A = E, \qquad A^{-1}E = A^{-1},$$

and again

$$EY = Y = A^{-1}.$$

Thus, for any nonsingular matrix A there exists a unique reciprocal matrix

$$A^{-1} = \begin{pmatrix} \dfrac{A_{11}}{D} & \dfrac{A_{21}}{D} & \cdots & \dfrac{A_{n1}}{D} \\ \dfrac{A_{12}}{D} & \dfrac{A_{22}}{D} & \cdots & \dfrac{A_{n2}}{D} \\ \cdots & \cdots & \cdots & \cdots \\ \dfrac{A_{1n}}{D} & \dfrac{A_{2n}}{D} & \cdots & \dfrac{A_{nn}}{D} \end{pmatrix}$$

such that

$$A^{-1}A = AA^{-1} = E.$$

The product of several matrices, say four, A, B, C, D, taken in this order is defined as

$$ABCD = ((AB)C)D,$$

and it follows from the associative law that, in performing multiplication, factors can be combined in arbitrary groups if the order of the factors is undisturbed.

$$ABCD = A(BCD) = (AB)(CD).$$

If all factors are nonsingular matrices, their product is a nonsingular matrix. For by the theorem of multiplication of determinants

$$\det(ABCD) = \det A \cdot \det B \cdot \det C \cdot \det D \neq 0.$$

The reciprocal matrix of the product $ABCD$ is

$$D^{-1}C^{-1}B^{-1}A^{-1},$$

for

$$\begin{aligned}
(ABCD)(D^{-1}C^{-1}B^{-1}A^{-1}) &= ABC(DD^{-1})C^{-1}B^{-1}A^{-1} = AB(CE)C^{-1}B^{-1}A^{-1} \\
&= AB(CC^{-1})B^{-1}A^{-1} = A(BE)B^{-1}A^{-1} \\
&= A(BB^{-1})A^{-1} = (AE)A^{-1} = AA^{-1} = E.
\end{aligned}$$

The product of n equal matrices

$$AA \cdots A$$

is, by definition, the nth power of A:

$$A^n$$

whose exponent is a positive integer n, and it follows from the associative law for multiplication that

$$A^mA^n = A^{m+n}$$

for any two positive integers m and n. The symbol A^0 is defined by

$$A^0 = E;$$

and a power A^{-n}, with negative integral exponent, is by definition

$$A^{-n} = (A^{-1})^n$$

provided that A is nonsingular. With these definitions the laws of exponents

$$A^mA^n = A^{m+n}; \qquad (A^m)^n = A^{mn}$$

hold for any integral exponents, but negative exponents can be admitted only if A is a nonsingular matrix.

If A is a nonsingular matrix, the matrix equation

$$AX = B$$

has a unique solution

$$X = A^{-1}B.$$

For

$$A(A^{-1}B) = (AA^{-1})B = EB = B$$

so that $A^{-1}B$ is a solution. On the other hand, from the equation

$$AX = B$$

it follows that

$$A^{-1}(AX) = (A^{-1}A)X = EX = X = A^{-1}B$$

so that the solution

$$X = A^{-1}B$$

is unique. The equation

$$XA = B$$

has also a unique solution

$$X = BA^{-1},$$

which can be verified in the same way. Of course, all this holds only for nonsingular matrices A. If A is a singular matrix, neither of the equations

$$AX = B, \qquad XA = B$$

has a solution unless det $B = 0$. For

$$\det B = \det A \cdot \det X = 0.$$

In particular, a singular matrix has no reciprocal.★

Problems

1. If rows and columns are exchanged in a matrix A, another matrix called the *conjugate* or the *transposed* of A is obtained. Denoting in general by A_0 the matrix conjugate to A, show that

$$(AB)_0 = B_0 A_0,$$

and, more generally, that

$$(ABC \cdots L)_0 = L_0 \cdots C_0 B_0 A_0.$$

2. Show that $(A_0)^{-1} = (A^{-1})_0$ if A is nonsingular.

3. A matrix whose elements A_{ij} are the complements of the elements a_{ij} of a matrix A is called the *adjoint* of A. Denoting the adjoint matrix by A^*, show that

$$(AB)^* = A^* \cdot B^*.$$

4. The relations

$$x_1 = a_{11}y_1 + a_{12}y_2 + \cdots + a_{1n}y_n,$$
$$x_2 = a_{21}y_1 + a_{22}y_2 + \cdots + a_{2n}y_n,$$
$$\cdots\cdots\cdots\cdots\cdots\cdots\cdots$$
$$x_n = a_{n1}y_1 + a_{n2}y_2 + \cdots + a_{nn}y_n,$$

define a linear transformation of the variables x_1, x_2, . . . , x_n into the variables y_1, y_2, . . . , y_n. A linear transformation is characterized by the matrix $A = (a_{ij})$ of its coefficients. If y_1, y_2, . . . , y_n are transformed into z_1, z_2, . . . , z_n by a linear transformation with the matrix B, show that x_1, x_2, . . . , x_n are tansformed into z_1, z_2, . . . , z_n by the linear transfcrmation with the matrix AB. If A is nonsingular, show also that y_1, y_2, . . . , y_n are transformed into x_1, x_2, . . . , x_n by the linear transformation whose matrix is A^{-1}.

5. Variables x_1, x_2, . . . , x_n are transformed into y_1, y_2, . . . , y_n by a linear transformation with matrix A. Variables x_1, x_2, . . . , x_n and y_1, y_2, . . . , y_n are transformed, respectively, into x_1', x_2', . . . , x_n' and y_1', y_2', . . . , y_n' by the same nonsingular transformation whose matrix is T. Show that the transformation of x_1', x_2', . . . , x_n' into y_1', y_2', . . . , y_n' has the matrix

$$T^{-1}AT.$$

Matrices of this type are called *similar* to A.

CHAPTER X

SOLUTION OF LINEAR EQUATIONS BY DETERMINANTS. SOME APPLICATIONS OF DETERMINANTS TO GEOMETRY

1. Cramer's Rule. Determinants have many applications. One of the most important is the application to the solution of systems of linear equations with several unknowns, and, in fact, determinants were introduced for this purpose by Leibnitz and Cramer although without the convenient symbolical notation which is of later origin. Let

$$
\begin{aligned}
a_{11}x_1 + a_{12}x_2 + \cdots + a_{1n}x_n &= b_1, \\
a_{21}x_1 + a_{22}x_2 + \cdots + a_{2n}x_n &= b_2, \\
&\cdots\cdots\cdots\cdots\cdots\cdots\cdots \\
a_{n1}x_1 + a_{n2}x_2 + \cdots + a_{nn}x_n &= b_n,
\end{aligned} \tag{1}
$$

be a system of n linear equations with n unknowns. Out of the coefficients of this system we form the determinant

$$
D = \begin{vmatrix}
a_{11} & a_{12} & \cdots & a_{1n} \\
a_{21} & a_{22} & \cdots & a_{2n} \\
\cdots & \cdots & \cdots & \cdots \\
a_{n1} & a_{n2} & \cdots & a_{nn}
\end{vmatrix}
$$

called determinant of system (1). On multiplying equations (1), respectively, by the complements of the elements of the first column and adding them, in the resulting equation all unknowns except x_1 will be eliminated. In fact, the coefficient of x_i for $i > 1$ will be

$$
A_{11}a_{1i} + A_{21}a_{2i} + \cdots + A_{n1}a_{ni} = 0,
$$

and that of x_1

$$
A_{11}a_{11} + A_{21}a_{21} + \cdots + A_{n1}a_{n1} = D.
$$

Thus,

$$
Dx_1 = A_{11}b_1 + A_{21}b_2 + \cdots + A_{n1}b_n, \tag{2a}
$$

and similarly

$$
Dx_i = A_{1i}b_1 + A_{2i}b_2 + \cdots + A_{ni}b_n \tag{2b}
$$

for $i > 1$. Equations (2) are necessary consequences of equations (1); and in case $D \neq 0$ the converse is also true, that is, equations (1) follow from equations (2). To show this, multiply equations (2), respectively, by $a_{11}, a_{12}, \ldots, a_{1n}$ and add the results. Observing that

$$
A_{11}a_{11} + A_{12}a_{12} + \cdots + A_{1n}a_{1n} = D
$$

whereas

$$A_{i1}a_{11} + A_{i2}a_{12} + \cdots + A_{in}a_{1n} = 0$$

for $i > 1$, it follows that

$$D(a_{11}x_1 + a_{12}x_2 + \cdots + a_{1n}x_n) = Db_1,$$

whence, canceling $D \neq 0$,

$$a_{11}x_1 + a_{12}x_2 + \cdots + a_{1n}x_n = b_1,$$

and it can be shown similarly that

$$a_{i1}x_1 + a_{i2}x_2 + \cdots + a_{in}x_n = b_i$$

for $i > 1$. Thus, in case $D \neq 0$, the systems (1) and (2) are equivalent. The solution of the latter system is immediate and leads to the expression of x_i in the form

$$x_i = \frac{D_i}{D},$$

where

$$D_i = A_{1i}b_1 + A_{2i}b_2 + \cdots + A_{ni}b_n$$

is the expansion by elements of the ith column of the determinant resulting from D by replacing its ith column by $b_1, b_2, \ldots, b_n$—the second members of equations (1). Thus, we come to the following rule for solving systems of linear equations:

Cramer's Rule. If the determinant D of a system of linear equations is different from 0, the system has a solution and the solution is unique. The values of the unknowns appear as quotients of determinants:

$$x_i = \frac{D_i}{D}; \qquad i = 1, 2, \ldots, n$$

where D_i results from D by replacing the ith column of D by $b_1, b_2, \ldots, b_n$.

Example. To solve by determinants the system

$$\begin{aligned}
5x +   4z + 2t &= 3, \\
x - y + 2z + t &= 1, \\
4x + y + 2z &= 1, \\
x + y + z + t &= 0.
\end{aligned}$$

The determinant of this system is

$$D = \begin{vmatrix} 5 & 0 & 4 & 2 \\ 1 & -1 & 2 & 1 \\ 4 & 1 & 2 & 0 \\ 1 & 1 & 1 & 1 \end{vmatrix} = \begin{vmatrix} 1 & -1 & 2 & 2 \\ 1 & -1 & 2 & 1 \\ 4 & 1 & 2 & 0 \\ 1 & 1 & 1 & 1 \end{vmatrix} = \begin{vmatrix} 0 & 0 & 0 & 1 \\ 1 & -1 & 2 & 1 \\ 4 & 1 & 2 & 0 \\ 1 & 1 & 1 & 1 \end{vmatrix} = - \begin{vmatrix} 1 & -1 & 2 \\ 4 & 1 & 2 \\ 1 & 1 & 1 \end{vmatrix}$$

$$= - \begin{vmatrix} 1 & 0 & 0 \\ 4 & 5 & -6 \\ 1 & 2 & -1 \end{vmatrix} = \begin{vmatrix} 5 & 6 \\ 2 & 1 \end{vmatrix} = -7.$$

Moreover,

$$D_1 = \begin{vmatrix} 3 & 0 & 4 & 2 \\ 1 & -1 & 2 & 1 \\ 1 & 1 & 2 & 0 \\ 0 & 1 & 1 & 1 \end{vmatrix} = \begin{vmatrix} 3 & 0 & 4 & 2 \\ 0 & -2 & 0 & 1 \\ 1 & 0 & 1 & -1 \\ 0 & 1 & 1 & 1 \end{vmatrix} = \begin{vmatrix} 3 & 4 & 4 & 2 \\ 0 & 0 & 0 & 1 \\ 1 & -2 & 1 & -1 \\ 0 & 3 & 1 & 1 \end{vmatrix} = \begin{vmatrix} 3 & 4 & 4 \\ 1 & -2 & 1 \\ 0 & 3 & 1 \end{vmatrix}$$

$$= \begin{vmatrix} 3 & 10 & 1 \\ 1 & 0 & 0 \\ 0 & 3 & 1 \end{vmatrix} = -\begin{vmatrix} 10 & 1 \\ 3 & 1 \end{vmatrix} = -7,$$

and similarly

$$D_2 = \begin{vmatrix} 5 & 3 & 4 & 2 \\ 1 & 1 & 2 & 1 \\ 4 & 1 & 2 & 0 \\ 1 & 0 & 1 & 1 \end{vmatrix} = 7, \qquad D_3 = \begin{vmatrix} 5 & 0 & 3 & 2 \\ 1 & -1 & 1 & 1 \\ 4 & 1 & 1 & 0 \\ 1 & 1 & 0 & 1 \end{vmatrix} = 7,$$

$$D_4 = \begin{vmatrix} 5 \cdot & 0 & 4 & 3 \\ 1 & -1 & 2 & 1 \\ 4 & 1 & 2 & 1 \\ 1 & 1 & 1 & 0 \end{vmatrix} = -7.$$

Hence, finally,

$$x = 1, \qquad y = -1, \qquad z = -1, \qquad t = 1.$$

Problems

Solve by determinants:

1. $2x - z = 1,$
$2x + 4y - z = 1,$
$x - 8y - 3z = -2.$

2. $x + y + z = a,$
$x + (1 + a)y + z = 2a,$
$x + y + (1 + a)z = 0.$

3. $5x + 4z + 2t = 3,$
$x - y + 2z + t = 1,$
$4x + y + 2z = 1,$
$x + y + z + t = 0.$

4. $3x + 2y - z = 5,$
$x - y - t = 0,$
$3x - 2y - z - t = 4,$
$y - t = 1.$

5. $x + 2y + z + t = -1,$
$2x + y + z + 2t = 0,$
$x + 2y + 2z + t = 0,$
$x + y + z + 2t = 2.$

6. $x_1 + x_2 + x_3 + x_4 = a,$
$x_1 + x_2 + x_3 - x_4 = b,$
$x_1 + x_2 - x_3 - x_4 = c,$
$x_1 - x_2 - x_3 - x_4 = d.$

7. $2x_1 + x_2 + x_3 + x_4 = 1,$
$x_1 + 2x_2 + x_3 + x_4 = 0,$
$x_1 + x_2 + 2x_3 + x_4 = 1,$
$x_1 + x_2 + x_3 + 2x_4 = 0.$

8. $x_1 + 2x_2 + 2x_3 + 2x_4 = 1,$
$x_1 + x_2 + 2x_3 + 2x_4 = 2,$
$x_1 + x_2 + x_3 + 2x_4 = 3,$
$x_1 + x_2 + x_3 + x_4 = 4.$

9. $x_1 + x_2 - x_3 = 1,$
$x_2 + x_3 - x_4 = 1,$
$x_3 + x_4 - x_5 = 1,$
$x_5 + x_4 - x_3 = 1,$
$x_4 + x_3 - x_2 = 1.$

10. $x_1 + x_2 + x_3 = 0,$
$x_2 + x_3 + x_4 = 0,$
$x_3 + x_4 + x_5 = 1,$
$-x_5 + x_4 + x_3 = 1,$
$-x_4 + x_3 + x_2 = 2.$

2. Linear Homogeneous Equations.

When the determinant of a system of linear equations is zero, the system may be inconsistent or impossible to satisfy by any values of the unknowns, or indeterminate, that is, admitting of infinitely many solutions. Before discussing this matter in a general manner we shall examine the case of n homogeneous equations with n unknowns:

$$\begin{aligned}
f_1 &= a_{11}x_1 + a_{12}x_2 + \cdots + a_{1n}x_n = 0, \\
f_2 &= a_{21}x_1 + a_{22}x_2 + \cdots + a_{2n}x_n = 0, \\
&\;\,\cdots\cdots\cdots\cdots\cdots\cdots\cdots\cdots\cdots\cdots\cdots \\
f_n &= a_{n1}x_1 + a_{n2}x_2 + \cdots + a_{nn}x_n = 0.
\end{aligned}$$

This system, when its determinant is different from zero, has only the trivial solution

$$x_1 = 0, \qquad x_2 = 0, \qquad \ldots, \qquad x_n = 0$$

as follows from Cramer's rule. But if the determinant is zero, we are going to show that besides the trivial solution it is possible to satisfy the system by values of the unknowns not all zero simultaneously. The proof will be by induction. First, the statement is true for two equations

$$a_{11}x_1 + a_{12}x_2 = 0, \qquad a_{21}x_1 + a_{22}x_2 = 0.$$

For if all coefficients a_{11}, a_{12}, a_{21}, a_{22} are zeros, the system is satisfied by arbitrary values x_1, x_2. In the contrary case let, for instance, $a_{11} \neq 0$. Then,

$$x_1 = -\frac{a_{12}}{a_{11}} x_2,$$

and on substituting this expression into the second equation it becomes

$$\frac{a_{11}a_{22} - a_{12}a_{21}}{a_{11}} x_2 = 0,$$

or

$$0 \cdot x_2 = 0,$$

supposing

$$a_{11}a_{22} - a_{12}a_{21} = 0.$$

Hence, x_2 can be chosen arbitrarily; for instance, let $x_2 = -a_{11} \neq 0$ and then both equations will be satisfied if $x_1 = a_{12}$. Now to complete the proof by induction it suffices to show that a homogeneous system of n equations with n unknowns has nontrivial solutions if its determinant is zero provided this property holds for $n - 1$ equations with $n - 1$ unknowns. There is nothing to prove if all the coefficients a_{ij} are zeros. In the contrary case, by a suitable change in the notation of the unknowns and in the order of the equations we can suppose that $a_{11} \neq 0$. The system

$$f_1 = 0, \qquad f_2 = 0, \qquad \ldots, \qquad f_n = 0 \tag{1}$$

is equivalent to the system

$$f_1 = 0, \qquad f_2 - \frac{a_{21}}{a_{11}} f_1 = 0, \qquad \cdots, \qquad f_n - \frac{a_{n1}}{a_{11}} f_1 = 0 \qquad (2)$$

whose determinant is equal to

$$D = \begin{vmatrix} a_{11} & a_{12} & \cdots & a_{1n} \\ a_{21} & a_{22} & \cdots & a_{2n} \\ \cdots & \cdots & \cdots & \cdots \\ a_{n1} & a_{n2} & \cdots & a_{nn} \end{vmatrix}.$$

In fact, the determinant of system (2) is obtained from D by multiplying its first row, respectively, by $\frac{a_{21}}{a_{11}}, \frac{a_{31}}{a_{11}}, \cdots, \frac{a_{n1}}{a_{11}}$ and subtracting it from the second, third, $\ldots$, nth rows; and by these operations the value of the determinant is not changed. Expanding the determinant thus obtained by the elements of the first column, we have

$$D = a_{11}\Delta,$$

where Δ is the determinant of the system of equations

$$f_2 - \frac{a_{21}}{a_{11}} f_1 = 0, \qquad \cdots, \qquad f_n - \frac{a_{n1}}{a_{11}} f_1 = 0 \qquad (3)$$

not containing x_1. Since by hypothesis $D = 0$ and $a_{11} \neq 0$, the determinant of the system (3) of the $n - 1$ equations with the $n - 1$ unknowns $x_2, \ldots, x_n$ is zero. Consequently, supposing the theorem true for $n - 1$ equations, it is possible to satisfy (3) by values $x_2, \ldots, x_n$ not all equal to 0. Determining then x_1 from the equation

$$f_1 = 0,$$

the system (2) and the equivalent system (1) are satisfied in a way that not all the unknowns are zeros.

Thus, we have proved the following theorem, which, despite its simplicity, is among the most useful and frequently applied tools in mathematical investigations:

THEOREM. *A system of n homogeneous linear equations with n unknowns has nontrivial solutions if, and only if, its determinant is zero.*

As a corollary it follows that *a homogeneous system in which the number of equations is less than the number of unknowns has always a nontrivial solution.* In fact, it is possible to complete the system by adding a number of equations of the form

$$0 \cdot x_1 + 0 \cdot x_2 + \cdots + 0 \cdot x_n = 0,$$

which do not restrict $x_1, x_2, \ldots, x_n$ in any way whatsoever, so as to make the number of equations equal to the number of unknowns. But such a system has its determinant equal to 0.

★3. Rank of a Matrix. Linear Independence. Let

$$f_1 = a_{11}x_1 + a_{12}x_2 + \cdots + a_{1n}x_n,$$
$$f_2 = a_{21}x_1 + a_{22}x_2 + \cdots + a_{2n}x_n,$$
$$\cdots \cdots \cdots \cdots \cdots \cdots \cdots \cdots$$
$$f_m = a_{m1}x_1 + a_{m2}x_2 + \cdots + a_{mn}x_n,$$

be a system of linear homogeneous functions or forms in n variables. With these forms we associate the matrix of their coefficients

$$A = \begin{pmatrix} a_{11} & a_{12} & \cdots & a_{1n} \\ a_{21} & a_{22} & \cdots & a_{2n} \\ \cdots & \cdots & \cdots & \cdots \\ a_{m1} & a_{m2} & \cdots & a_{mn} \end{pmatrix}$$

consisting of m rows (number of forms) and n columns (number of variables). Selecting in this matrix certain r rows and certain r columns ($r \leqq m$, $r \leqq n$), we can form out of these r^2 elements a determinant of order r. There are several determinants of order r that can be formed in this manner. They are called the *minors* of order r of the matrix A. In the case $r = 1$, the elements of the matrix are considered as determinants of order 1. For instance, with elements of the matrix

$$X = \begin{pmatrix} 0 & 1 & 1 & 0 & 1 & 0 \\ 1 & 2 & -1 & 3 & 2 & 4 \\ 4 & 1 & 3 & 0 & 1 & 2 \end{pmatrix}$$

we can form (a) 20 minors of order 3; (b) $15 \times 3 = 45$ minors of order 2; (c) $6 \times 3 = 18$ minors of order 1. Corresponding to the matrix

$$Y = \begin{pmatrix} 0 & 1 & 0 \\ 1 & 2 & 3 \\ -1 & -1 & -3 \\ 1 & 4 & 3 \end{pmatrix}$$

there are (a) 4 minors of order 3; (b) $6 \times 3 = 18$ minors of order 2; (c) $4 \times 3 = 12$ minors of order 1. By definition, a matrix is of *rank* ρ if it contains minors of order ρ different from 0, while all minors of order $\rho + 1$ (if there are such) are zero. The fact that all the minors of order $\rho + 1$ are equal to 0 implies that the minors of order $> \rho + 1$, if there are any, are also equal to 0. For example, the matrix X is of rank 3 since it contains the minor

$$\begin{vmatrix} 0 & 1 & 1 \\ 1 & 2 & -1 \\ 4 & 1 & 3 \end{vmatrix} = -14$$

of order 3, different from 0, while there are no minors of order 4. Matrix Y is of rank 2 since it contains the minor of order 2

$$\begin{vmatrix} 0 & 1 \\ 1 & 2 \end{vmatrix} = -1,$$

different from 0, while all 4 minors of order 3 are found to be equal to 0.

Linear forms $f_1, f_2, \ldots, f_m$ corresponding to the matrix A are said to be *linearly independent* if the identical relation

$$\lambda_1 f_1 + \lambda_2 f_2 + \cdots + \lambda_m f_m = 0$$

cannot be satisfied by any choice of constants $\lambda_1, \lambda_2, \ldots, \lambda_m$ except $\lambda_1 = \lambda_2 = \cdots = \lambda_m = 0$. On the contrary, if constants $\lambda_1, \lambda_2, \ldots, \lambda_m$ not all zeros can be found so that identically in the variables $x_1, x_2, \ldots, x_n$

$$\lambda_1 f_1 + \lambda_2 f_2 + \cdots + \lambda_m f_m = 0,$$

then the forms $f_1, f_2, \ldots, f_m$ are said to be linearly dependent. Let σ be the maximum number of linearly independent forms among $f_1, f_2, \ldots, f_m$. Between this number σ and the rank ρ of the matrix corresponding to the forms $f_1, f_2, \ldots, f_m$, there exists a simple relation expressed in the following theorem:

THEOREM. *If the matrix corresponding to the forms $f_1, f_2, \ldots, f_m$ is of rank ρ, then among these forms there are ρ linearly independent forms, and any $\rho + 1$ forms of the system $f_1, f_2, \ldots, f_m$ are linearly dependent, so that $\sigma = \rho$.*

PROOF. By an eventual change in the notation of the variables and numbering of the forms we can assume that the minor of order ρ, different from 0, is

$$D = \begin{vmatrix} a_{11} & a_{12} & \cdots & a_{1\rho} \\ a_{21} & a_{22} & \cdots & a_{2\rho} \\ \cdots \cdots \cdots \cdots \cdots \\ a_{\rho 1} & a_{\rho 2} & \cdots & a_{\rho\rho} \end{vmatrix}.$$

Then, the forms $f_1, f_2, \ldots, f_\rho$ are linearly independent. In fact, let us try to find constants $\lambda_1, \lambda_2, \ldots, \lambda_n$ such that identically

$$\lambda_1 f_1 + \lambda_2 f_2 + \cdots + \lambda_\rho f_\rho = 0.$$

In this equation equate the coefficients of $x_1, x_2, \ldots, x_\rho$ to zero, getting the system of linear homogeneous equations

$$\lambda_1 a_{11} + \lambda_2 a_{21} + \cdots + \lambda_\rho a_{\rho 1} = 0,$$
$$\lambda_1 a_{12} + \lambda_2 a_{22} + \cdots + \lambda_\rho a_{\rho 2} = 0,$$
$$\cdots \cdots \cdots \cdots \cdots \cdots \cdots$$
$$\lambda_1 a_{1\rho} + \lambda_2 a_{2\rho} + \cdots + \lambda_\rho a_{\rho\rho} = 0,$$

the determinant D' of which differs from D only in that the rows of D' are columns of D, and vice versa. Hence, $D' \neq 0$ and therefore $\lambda_1 = \lambda_2 = \cdots = \lambda_\rho = 0$, which proves that the ρ forms $f_1, f_2, \ldots, f_\rho$ are independent. Now let $k > \rho$; we want to show that the forms $f_1, f_2, \ldots, f_\rho$ and f_k are linearly dependent. To this end consider the determinants

$$\Delta = \begin{vmatrix} a_{11} & a_{12} & \cdots & a_{1\rho} & f_1 \\ a_{21} & a_{22} & \cdots & a_{2\rho} & f_2 \\ \cdots & \cdots & \cdots & \cdots & \cdots \\ a_{\rho 1} & a_{\rho 2} & \cdots & a_{\rho\rho} & f_\rho \\ a_{k1} & a_{k2} & \cdots & a_{k\rho} & f_k \end{vmatrix}.$$

On replacing in the last column $f_1, f_2, \ldots, f_k$ by their expressions in the variables $x_1, x_2, \ldots, x_n$, the determinant Δ splits into a sum of determinants of the type

$$\begin{vmatrix} a_{11} & a_{12} & \cdots & a_{1\rho} & a_{1\sigma}x_\sigma \\ a_{21} & a_{22} & \cdots & a_{2\rho} & a_{2\sigma}x_\sigma \\ \cdots & \cdots & \cdots & \cdots & \cdots \\ a_{\rho 1} & a_{\rho 2} & \cdots & a_{\rho\rho} & a_{\rho\sigma}x_\sigma \\ a_{k1} & a_{k2} & \cdots & a_{k\rho} & a_{k\sigma}x_\sigma \end{vmatrix} = \begin{vmatrix} a_{11} & a_{12} & \cdots & a_{1\rho} & a_{1\sigma} \\ a_{21} & a_{22} & \cdots & a_{2\rho} & a_{2\sigma} \\ \cdots & \cdots & \cdots & \cdots & \cdots \\ a_{\rho 1} & a_{\rho 2} & \cdots & a_{\rho\rho} & a_{\rho\sigma} \\ a_{k1} & a_{k2} & \cdots & a_{k\rho} & a_{k\sigma} \end{vmatrix} x_\sigma,$$

ρ of which, corresponding to $\sigma \leq \rho$, are equal to 0 as having two identical columns, and the remaining $n - \rho$ are equal to 0 as being proportional to minors of order $\rho + 1$ derived from the matrix A. Hence,

$$\Delta = \begin{vmatrix} a_{11} & a_{12} & \cdots & a_{1\rho} & f_1 \\ a_{21} & a_{22} & \cdots & a_{2\rho} & f_2 \\ \cdots & \cdots & \cdots & \cdots & \cdots \\ a_{\rho 1} & a_{\rho 2} & \cdots & a_{\rho\rho} & f_\rho \\ a_{k1} & a_{k2} & \cdots & a_{k\rho} & f_k \end{vmatrix} = 0$$

identically in the variables $x_1, x_2, \ldots, x_n$. Expanding by elements of the last column, we have an identity of the form

$$D f_k + D_1 f_1 + D_2 f_2 + \cdots + D_\rho f_\rho = 0$$

in which $D \neq 0$. Hence, the forms $f_1, f_2, \ldots, f_\rho, f_k$ are linearly dependent, and f_k for $k > \rho$ can be expressed as follows through $f_1, f_2, \ldots, f_\rho$:

$$f_k = l_1 f_1 + l_2 f_2 + \cdots + l_\rho f_\rho.$$

Take now any $\rho + 1$ forms

$$f\alpha, f\beta, \ldots, f\lambda$$

and express them through $f_1, f_2, \ldots, f_\rho$:

$$f_\alpha = A_1 f_1 + A_2 f_2 + \cdots + A_\rho f_\rho,$$
$$f_\beta = B_1 f_1 + B_2 f_2 + \cdots + B_\rho f_\rho,$$
$$\cdots \cdots \cdots \cdots \cdots \cdots \cdots \cdots$$
$$f_\lambda = L_1 f_1 + L_2 f_2 + \cdots + L_\rho f_\rho.$$

Choose the numbers $a, b, \ldots, l$, not all equal to zero, so as to satisfy the ρ equations

$$A_1 a + B_1 b + \cdots + L_1 l = 0,$$
$$A_2 a + B_2 b + \cdots + L_2 l = 0,$$
$$\cdots \cdots \cdots \cdots \cdots \cdots \cdots$$
$$A_\rho a + B_\rho b + \cdots + L_\rho l = 0,$$

with $\rho + 1$ unknowns. Then,

$$a f_\alpha + b f_\beta + \cdots + l f_\lambda = 0$$

so that any $\rho + 1$ forms are linearly dependent.

COROLLARY. *Any $n + 1$ linear forms in n variables are linearly dependent. For the corresponding matrix has no minors of order higher than n and its rank is $\leq n$.*

Example. Let us consider the linear forms

$$f_1 = y,$$
$$f_2 = x + 2y + 3z,$$
$$f_3 = -x - y - 3z,$$
$$f_4 = x + 4y + 3z,$$

in the three variables x, y, z. Their matrix is the above matrix Y of rank 2. Hence, among these forms there are only two that are linearly independent. For these we can take f_1 and f_2. To express f_3 and f_4 through f_1 and f_2 consider the determinants

$$\begin{vmatrix} 0 & 1 & f_1 \\ 1 & 2 & f_2 \\ -1 & -1 & f_3 \end{vmatrix} = 0, \qquad \begin{vmatrix} 0 & 1 & f_1 \\ 1 & 2 & f_2 \\ 1 & 4 & f_4 \end{vmatrix} = 0.$$

Expanding, we find

$$-f_3 - f_2 + f_1 = 0, \qquad -f_4 + f_2 + 2f_1 = 0$$

whence

$$f_3 = f_1 - f_2, \qquad f_4 = 2f_1 + f_2. \star$$

★4. How to Find the Rank of a Matrix. To find the rank of a matrix, this matrix by a series of transformations that preserve rank can be reduced to a certain *normal* form from which the requested rank can be found by inspection. These transformations are the following:

1. Interchange of two rows.
2. Interchange of two columns.
3. Addition to elements of some row of elements of another row multiplied by an arbitrary factor.
4. Addition to elements of some column of elements of another column multiplied by an arbitrary factor.

To show that these operations leave the rank of a matrix unaltered, introduce linear forms corresponding to that matrix:

$$f_1, f_2, \ldots, f_m. \tag{1}$$

The number of independent forms among these is the rank of the matrix. Now an interchange of two rows amounts to altering the order of two of these forms, which evidently does not change the number of independent forms in system (1). To show that the addition of elements of some column multiplied by a factor λ to the elements of another column does not change the rank of the matrix assume, for example, that the elements of the second column multiplied by λ are added to the elements of the first column. The new matrix corresponds to forms $f_1', f_2', \ldots, f_m'$ obtained from $f_1, f_2, \ldots, f_m$ by introducing the new variables $x_1', x_2', \ldots, x_n'$ so that

$$x_1 = x_1' + \lambda x_2', \qquad x_2 = x_2', \qquad \ldots, \qquad x_n = x_n';$$

whence, conversely,

$$x_1' = x_1 - \lambda x_2, \qquad x_2' = x_2, \qquad \ldots, \qquad x_n' = x_n.$$

But forms independent when expressed through variables $x_1, x_2, \ldots, x_n$ will evidently be independent when expressed through new variables $x_1', x_2', \ldots, x_n'$ and conversely; hence, the number of independent forms among $f_1', f_2', \ldots, f_m'$ is the same as among $f_1, f_2, \ldots, f_m$. Finally, to show that the third operation does not alter the rank, assume that to the elements of the first row are added the elements of the second row multiplied by λ. The new matrix corresponds to the forms

$$\phi_1 = f_1 + \lambda f_2, \qquad \phi_2 = f_2, \qquad \ldots, \qquad \phi_m = f_m,$$

whence, conversely,

$$f_1 = \phi_1 - \lambda \phi_2, \qquad f_2 = \phi_2, \qquad \ldots, \qquad f_m = \phi_m.$$

Now we can apply the following more general proposition: If forms $\phi_1, \phi_2, \ldots, \phi_m$ can be expressed linearly through $f_1, f_2, \ldots, f_m$ among which there are ρ independent forms, the number ρ' of independent forms among $\phi_1, \phi_2, \ldots, \phi_m$ is not greater than ρ. For let $f_1, f_2, \ldots, f_\rho$ be ρ independent forms among $f_1, f_2, \ldots, f_m$ through which all other forms of this system can be linearly expressed. Then, taking any $\sigma \geqq \rho + 1$ among the forms $\phi_1, \phi_2, \ldots, \phi_m$ and denoting them for simplicity by $\psi_1, \psi_2, \ldots, \psi_\sigma$, we can express them through $f_1, f_2, \ldots, f_\rho$ thus:

$$\psi_1 = c_{11}f_1 + c_{12}f_2 + \cdots + c_{1\rho}f_\rho,$$
$$\psi_2 = c_{21}f_1 + c_{22}f_2 + \cdots + c_{2\rho}f_\rho,$$
$$\cdots \cdots \cdots \cdots \cdots \cdots \cdots \cdots$$
$$\psi_\sigma = c_{\sigma1}f_1 + c_{\sigma2}f_2 + \cdots + c_{\sigma\rho}f_\rho.$$

But ρ equations with σ unknowns $\lambda_1, \lambda_2, \ldots, \lambda_\sigma$:

$$c_{11}\lambda_1 + c_{21}\lambda_2 + \cdots + c_{\sigma1}\lambda_\sigma = 0,$$
$$c_{12}\lambda_1 + c_{22}\lambda_2 + \cdots + c_{\sigma2}\lambda_\sigma = 0,$$
$$\cdots \cdots \cdots \cdots \cdots \cdots \cdots \cdots$$
$$c_{1\rho}\lambda_1 + c_{2\rho}\lambda_2 + \cdots + c_{\sigma\rho}\lambda_\sigma = 0,$$

can be satisfied by values $\lambda_1, \lambda_2, \ldots, \lambda_\sigma$ not all zeros since there are more unknowns than equations. But then

$$\lambda_1\psi_1 + \lambda_2\psi_2 + \cdots + \lambda_\sigma\psi_\sigma = 0,$$

which shows that the forms $\psi_1, \psi_2, \ldots, \psi_\sigma$ are linearly dependent. Hence, it follows that $\rho' \leqq \rho$.

Returning to the forms

$$\rho_1 = f_1 + \lambda f_2, \qquad \phi_2 = f_2, \qquad \ldots, \qquad \phi_m = f_m,$$

and calling by ρ' the greatest number of independent forms among them, we conclude that $\rho' \leqq \rho$. Since $f_1, f_2, \ldots, f_m$ can be expressed through $\phi_1, \phi_2, \ldots, \phi_m$, we have, for the same reason, $\rho \leqq \rho'$ and so $\rho' = \rho$, which shows that the rank of a matrix is not changed by the operation 3.

To reduce a matrix to normal form we assume that it contains elements not equal to 0. Then, by a number of interchanges of rows and columns an element not 0 can be placed at the intersection of the first row and the first column. Multiplying now the first column by properly chosen factors and adding to the remaining columns, we can make all elements of columns 2, 3, . . . belonging to row 1 equal to 0. Similarly, we can reduce to 0 all elements of the first column belonging to rows 2, 3, ... Now the matrix has in the first row and in the first columns only one element different from 0. If besides this there are some other elements different from 0, bring one of them to the intersection of the second row and second column. Next, by operations 4 and 3, reduce to 0 all elements of the second row and second column except the di-

agonal element. Continuing in the same manner we shall be able to
reduce the matrix to the form

$$
\begin{pmatrix}
a & 0 & \cdots & \cdots & 0 & 0 \\
0 & b & \cdots & \cdots & 0 & 0 \\
\cdot & & \cdot\, l & \cdots & 0 & 0 \\
\cdot & & & 0 & & \\
0 & 0 & \cdots & \cdots & 0 & 0
\end{pmatrix}
$$

where all elements not on the diagonal are zeros and on the diagonal
the first ρ elements are different from 0. The rank of a matrix of this
type is evidently ρ and it is the same as the rank of the original matrix.
It should be noted that in case elements of the original matrix are
integers the reduction can be made without introducing fractions, as
will be seen in the following example. Also, by watching carefully the
interchanges of rows it can be found which of the ρ forms corresponding to the original matrix are independent.

Example. To find the rank of the matrix

$$
\begin{pmatrix}
2 & 3 & 4 & 5 & 6 \\
3 & 4 & 5 & 6 & 7 \\
4 & 5 & 6 & 7 & 8 \\
9 & 10 & 11 & 12 & 13 \\
14 & 15 & 16 & 17 & 18
\end{pmatrix}.
$$

If column 1 is subtracted from the remaining columns, the matrix is reduced to

$$
\begin{pmatrix}
2 & 1 & 2 & 3 & 4 \\
3 & 1 & 2 & 3 & 4 \\
4 & 1 & 2 & 3 & 4 \\
9 & 1 & 2 & 3 & 4 \\
14 & 1 & 2 & 3 & 4
\end{pmatrix}.
$$

Next, from columns 1, 3, 4, 5 subtract column 2 multiplied by 2, 2, 3, 4, respectively;
this reduces the preceding matrix to

$$
\begin{pmatrix}
0 & 1 & 0 & 0 & 0 \\
1 & 1 & 0 & 0 & 0 \\
2 & 1 & 0 & 0 & 0 \\
7 & 1 & 0 & 0 & 0 \\
12 & 1 & 0 & 0 & 0
\end{pmatrix}.
$$

Interchange now columns 1 and 2:

$$
\begin{pmatrix}
1 & 0 & 0 & 0 & 0 \\
1 & 1 & 0 & 0 & 0 \\
1 & 2 & 0 & 0 & 0 \\
1 & 7 & 0 & 0 & 0 \\
1 & 12 & 0 & 0 & 0
\end{pmatrix}
$$

and subtract row 1 from rows 2, 3, 4, 5. This gives

$$\begin{pmatrix} 1 & 0 & 0 & 0 & 0 \\ 0 & 1 & 0 & 0 & 0 \\ 0 & 2 & 0 & 0 & 0 \\ 0 & 7 & 0 & 0 & 0 \\ 0 & 12 & 0 & 0 & 0 \end{pmatrix}.$$

Now subtract row 2 multiplied by 2, 7, 12 from rows 3, 4, 5, after which the matrix appears in the reduced form

$$\begin{pmatrix} 1 & 0 & 0 & 0 & 0 \\ 0 & 1 & 0 & 0 & 0 \\ 0 & 0 & 0 & 0 & 0 \\ 0 & 0 & 0 & 0 & 0 \\ 0 & 0 & 0 & 0 & 0 \end{pmatrix}$$

and its rank is 2. This is the same, therefore, as the rank of the original matrix. Since rows were not interchanged among the forms

$$\begin{aligned} f_1 &= 2x + 3y + 4z + 5t + 6u, \\ f_2 &= 3x + 4y + 5z + 6t + 7u, \\ f_3 &= 4x + 5y + 6z + 7t + 8u, \\ f_4 &= 9x + 10y + 11z + 12t + 13u, \\ f_5 &= 14x + 15y + 16z + 17t + 18u, \end{aligned}$$

forms f_1 and f_2 are independent and the remaining forms can be expressed through them.★

Problems

Find the rank of matrices:

1. $\begin{pmatrix} 2 & -1 & -1 \\ -1 & 2 & -1 \\ -1 & -1 & 2 \end{pmatrix}.$

2. $\begin{pmatrix} 2 & 3 & 0 \\ 3 & 4 & 1 \\ 0 & 1 & -2 \end{pmatrix}.$

3. $\begin{pmatrix} 2 & 1 & 2 & 1 \\ 1 & 2 & 1 & 2 \\ 1 & -1 & -1 & -1 \\ 1 & 1 & 1 & 1 \end{pmatrix}.$

4. $\begin{pmatrix} 2 & 3 & 0 & -2 \\ 3 & 4 & -1 & -3 \\ 0 & -1 & -2 & 0 \\ -2 & -3 & 0 & 2 \end{pmatrix}.$

Find the number of independent forms among the following and express the rest through them:

5. $\begin{aligned} f_1 &= x + y - z, \\ f_2 &= 3x - y + 4z, \\ f_3 &= -3x + 5y - 11z. \end{aligned}$

6. $\begin{aligned} f_1 &= 2x - y - z, \\ f_2 &= -x + 2y - z, \\ f_3 &= -x - y + 2z. \end{aligned}$

7. $\begin{aligned} f_1 &= 2x + 3y - 2t, \\ f_2 &= 3x + 4y - z - 3t, \\ f_3 &= -y - 2z, \\ f_4 &= -2x - 3y + 2t. \end{aligned}$

8. $\begin{aligned} f_1 &= x + 7z + t, \\ f_2 &= 3x + 2y + 15z + t, \\ f_3 &= x + 2y + z - t, \\ f_4 &= y - 3z - t. \end{aligned}$

★5. General Discussion of Linear Systems. We can return now to systems of linear equations and examine the question of their solution.

in a more general manner. Let the system to be solved consist of m equations with n unknowns:

$$
\begin{aligned}
f_1 &= a_{11}x_1 + a_{12}x_2 + \cdots + a_{1n}x_n = b_1, \\
f_2 &= a_{21}x_1 + a_{22}x_2 + \cdots + a_{2n}x_n = b_2, \\
&\cdots\cdots\cdots\cdots\cdots\cdots\cdots\cdots\cdots\cdots \\
f_m &= a_{m1}x_1 + a_{m2}x_2 + \cdots + a_{mn}x_n = b_m.
\end{aligned}
\tag{1}
$$

Suppose that the rank of the matrix

$$
\begin{pmatrix}
a_{11} & a_{12} & \cdots & a_{1n} \\
a_{21} & a_{22} & \cdots & a_{2n} \\
\cdots & \cdots & \cdots & \cdots \\
a_{m1} & a_{m2} & \cdots & a_{mn}
\end{pmatrix}
$$

corresponding to the linear forms $f_1, f_2, \ldots, f_m$ is ρ. Then, the maximum number of independent forms among these will be ρ. We can suppose that these independent forms are $f_1, f_2, \ldots, f_\rho$. Also, by an eventual change in the notation of the unknowns we can assume that

$$
d =
\begin{vmatrix}
a_{11} & a_{12} & \cdots & a_{1\rho} \\
a_{21} & a_{22} & \cdots & a_{2\rho} \\
\cdots & \cdots & \cdots & \cdots \\
a_{\rho 1} & a_{\rho 2} & \cdots & a_{\rho\rho}
\end{vmatrix}
\neq 0.
$$

Denoting by σ any one of the numbers $\rho + 1$, $\rho + 2$, $\ldots$, m, consider the determinant

$$
D_\sigma =
\begin{vmatrix}
a_{11} & a_{12} & \cdots & a_{1\rho} & f_1 - b_1 \\
a_{21} & a_{22} & \cdots & a_{2\rho} & f_2 - b_2 \\
\cdots & \cdots & \cdots & \cdots & \cdots \\
a_{\rho 1} & a_{\rho 2} & \cdots & a_{\rho\rho} & f_\rho - b_\rho \\
a_{\sigma 1} & a_{\sigma 2} & \cdots & a_{\sigma\rho} & f_\sigma - b_\sigma
\end{vmatrix}
=
\begin{vmatrix}
a_{11} & a_{12} & \cdots & a_{1\rho} & f_1 \\
a_{21} & a_{22} & \cdots & a_{2\rho} & f_2 \\
\cdots & \cdots & \cdots & \cdots & \cdots \\
a_{\rho 1} & a_{\rho 2} & \cdots & a_{\rho\rho} & f_\rho \\
a_{\sigma 1} & a_{\sigma 2} & \cdots & a_{\sigma\rho} & f_\sigma
\end{vmatrix}
-
\begin{vmatrix}
a_{11} & a_{12} & \cdots & a_{1\rho} & b_1 \\
a_{21} & a_{22} & \cdots & a_{2\rho} & b_2 \\
\cdots & \cdots & \cdots & \cdots & \cdots \\
a_{\rho 1} & a_{\rho 2} & \cdots & a_{\rho\rho} & b_\rho \\
a_{\sigma 1} & a_{\sigma 2} & \cdots & a_{\sigma\rho} & b_\sigma
\end{vmatrix}.
$$

The first determinant in the right-hand side reduces identically to zero as was shown in Sec. 3; hence, for variable $x_1, x_2, \ldots, x_n$ we have the identity

$$
D_\sigma = -
\begin{vmatrix}
a_{11} & a_{12} & \cdots & a_{1\rho} & b_1 \\
a_{21} & a_{22} & \cdots & a_{2\rho} & b_2 \\
\cdots & \cdots & \cdots & \cdots & \cdots \\
a_{\rho 1} & a_{\rho 2} & \cdots & a_{\rho\rho} & b_\rho \\
a_{\sigma 1} & a_{\sigma 2} & \cdots & a_{\sigma\rho} & b_\sigma
\end{vmatrix}.
$$

Suppose now that system (1) is solvable and that $x_1, x_2, \ldots, x_n$ are numbers satisfying it. Then, $f_1 - b_1 = 0$, $f_2 - b_2 = 0$, $\ldots$,

$f_m - b_m = 0$ and all the determinants $D_\sigma = 0$ for $\sigma = \rho + 1, \ldots, m$, which lead us to the conclusion that system (1) has no solution unless the following conditions of compatibility

$$\Delta_\sigma = \begin{vmatrix} a_{11} & a_{12} & \cdots & a_{1\rho} & b_1 \\ a_{21} & a_{22} & \cdots & a_{2\rho} & b_2 \\ \cdots & \cdots & \cdots & \cdots & \cdots \\ a_{\rho 1} & a_{\rho 2} & \cdots & a_{\rho\rho} & b_\rho \\ a_{\sigma 1} & a_{\sigma 2} & \cdots & a_{\sigma\rho} & b_\sigma \end{vmatrix} = 0$$

are satisfied for $\sigma = \rho + 1, \ldots, m$. Conversely, if these conditions are satisfied, we have identically

$$D_\sigma = 0,$$

for $\sigma = \rho + 1, \ldots, m$. But expanding D_σ by the elements of the last column, we have

$$d(f_\sigma - b_\sigma) + d_1(f_1 - b_1) + \cdots + d_\rho(f_\rho - b_\rho) = 0$$

identically in $x_1, x_2, \ldots, x_n$, whence, since $d \neq 0$, it follows that

$$f_\sigma - b_\sigma = c_{1\sigma}(f_1 - b_1) + c_{2\sigma}(f_2 - b_2) + \cdots + c_{\rho\sigma}(f_\rho - b_\rho)$$

for $\sigma = \rho + 1, \ldots, m$, which shows that all the equations (1) will be satisfied once the first ρ of them

$$f_1 - b_1 = 0, \qquad f_2 - b_2 = 0, \qquad \ldots, \qquad f_\rho - b_\rho = 0 \qquad (2)$$

are satisfied. Now these are satisfied in the most general way by attributing to $x_{\rho+1}, \ldots, x_n$ arbitrary values (provided $\rho < n$) and solving (2) for $x_1, x_2, \ldots, x_\rho$ by Cramer's rule, which is possible since

$$d = \begin{vmatrix} a_{11} & a_{12} & \cdots & a_{1\rho} \\ a_{21} & a_{22} & \cdots & a_{2\rho} \\ \cdots & \cdots & \cdots & \cdots \\ a_{\rho 1} & a_{\rho 2} & \cdots & a_{\rho\rho} \end{vmatrix} \neq 0.$$

The conclusion to which we are led by this discussion is the following:

If the matrix of a system of m equations with n unknowns is of rank ρ and the conditions of compatibility

$$\Delta_\sigma = 0$$

for $\sigma = \rho + 1, \ldots, m$ are not satisfied, the system has no solution or is incompatible. On the contrary, if conditions of compatibility are satisfied, the system is solvable and $n - \rho$ unknowns may have arbitrary values, the remaining ρ being then determined.

This criterion may be expressed in an elegant form by taking into consideration, besides the matrix of the system

$$A = \begin{pmatrix} a_{11} & a_{12} & \cdots & a_{1n} \\ a_{21} & a_{22} & \cdots & a_{2n} \\ \cdots & \cdots & \cdots & \cdots \\ a_{m1} & a_{m2} & \cdots & a_{mn} \end{pmatrix},$$

the so-called *augmented matrix*

$$B = \begin{pmatrix} a_{11} & a_{12} & \cdots & a_{1n} & b_1 \\ a_{21} & a_{22} & \cdots & a_{2n} & b_2 \\ \cdots & \cdots & \cdots & \cdots & \cdots \\ a_{m1} & a_{m2} & \cdots & a_{mn} & b_m \end{pmatrix}.$$

All minors of A occur among minors of B so that the rank of B cannot be less than that of A, which was denoted by ρ. The only determinants of order $\rho + 1$ of the augmented matrix that do not occur in A are the determinants Δ_σ. Hence, if all these determinants vanish, and only in this case, the rank of B is ρ. Now the vanishing of the determinants Δ_σ is the necessary and sufficient condition for the solvability of the system. Therefore, *the condition necessary and sufficient for the solvability of a system of linear equations is that the rank of the augmented matrix be the same as the rank of the matrix of the system.*

Example. Let us examine the system

$$\begin{aligned} f_1 &= 3x + 2y + 5z + 4t = 0, \\ f_2 &= 5x + 3y + 2z + t = 1, \\ f_3 &= 11x + 7y + 12z + 9t = k, \\ f_4 &= 4x + 3y + 13z + 11t = l. \end{aligned}$$

The rank of this system is 2, and forms f_1 and f_2 are independent. Moreover,

$$d = \begin{vmatrix} 3 & 2 \\ 5 & 3 \end{vmatrix} \neq 0,$$

and the two conditions of compatibility are

$$\begin{vmatrix} 3 & 2 & 0 \\ 5 & 3 & 1 \\ 11 & 7 & k \end{vmatrix} = 0, \qquad \begin{vmatrix} 3 & 2 & 0 \\ 5 & 3 & 1 \\ 4 & 3 & l \end{vmatrix} = 0.$$

whence $k = 1, l = -1$. Unless therefore $k = 1, l = -1$, the proposed system has no solution. In case $k = 1, l = -1$, we take the first two equations, write them in the form

$$\begin{aligned} 3x + 2y &= -5z - 4t \\ 5x + 3y &= -2z - t + 1, \end{aligned}$$

and solve for x and y. This gives

$$\begin{aligned} x &= 11z + 10t + 2, \\ y &= -19z - 17t - 3, \end{aligned}$$

and with x, y so determined all the equations will be satisfied, while z and t remain absolutely arbitrary.★

Problems

Examine for compatibility the following systems:

1. $2x - y - z = 1$,
$x - 2y + z = 1$,
$x + y - 2z = 1$.

2. $2x + 3y = 1$,
$3x + 4y + z = 0$,
$y - 2z = 3$.

3. $2v + y + 2z + t = 0$,
$x + 2y + z + 2t = 0$,
$x - y - z - t = 2$,
$x + y + z + t = l$.

4. $2x + 3y - 2t = k$,
$3x + 4y - z - 3t = l$,
$y + 2z = 1$,
$2x + 3y - 2t = 0$.

5. $x + y - z + t = 1$,
$x + 2y - z + 3t = 1$,
$x - z - t = 1$,
$x - y + z - 3t = 3$.

6. $x + y + z + t = 0$,
$2x + 3y - 2z + t = 1$,
$-y + 4z + t = 2$,
$x + 5z + 2t = -1$.

GEOMETRIC APPLICATIONS OF DETERMINANTS

★6. Equations of Lines, Planes, and Circles in Determinant Form.
It is easy to write in determinant form the equation of a line determined
by two distinct points, or that of a plane determined by three noncollinear points. Let $A_1(x_1, y_1)$ and $A_2(x_2, y_2)$ be two distinct points
given by their coordinates with reference to a rectangular system of axes.
The determinant

$$\begin{vmatrix} 1 & x & y \\ 1 & x_1 & y_1 \\ 1 & x_2 & y_2 \end{vmatrix} = 0, \tag{1}$$

when expanded, is an equation of the first degree in x and y and the
coefficients of x and y:

$$A = -\begin{vmatrix} 1 & y_1 \\ 1 & y_2 \end{vmatrix}, \qquad B = \begin{vmatrix} 1 & x_1 \\ 1 & x_2 \end{vmatrix}$$

are not both equal to 0. Hence, (1) represents the equation of a line
expressed in determinantal form. It is satisfied by $x = x_1$, $y = y_1$, and
$x = x_2$, $y = y_2$ so that the line determined by (1) goes through the points
A_1 and A_2. Three points $A_1(x_1, y_1)$, $A_2(x_2, y_2)$, $A_3(x_3, y_3)$ are therefore
collinear if and only if

$$\begin{vmatrix} 1 & x_1 & y_1 \\ 1 & x_2 & y_2 \\ 1 & x_3 & y_3 \end{vmatrix} = 0$$

Suppose now that $A_1(x_1, y_1, z_1)$, $A_2(x_2, y_2, z_2)$, $A_3(x_3, y_3, z_3)$ are three
noncollinear points in the space determined by their coordinates referred to a rectangular system of axes. The equation

$$\begin{vmatrix} 1 & x & y & z \\ 1 & x_1 & y_1 & z_1 \\ 1 & x_2 & y_2 & z_2 \\ 1 & x_3 & y_3 & z_3 \end{vmatrix} = 0 \tag{2}$$

is of the first degree in x, y, z and the coefficients of x, y, z:

$$A = -\begin{vmatrix} 1 & y_1 & z_1 \\ 1 & y_2 & z_2 \\ 1 & y_3 & z_3 \end{vmatrix}, \qquad B = \begin{vmatrix} 1 & x_1 & z_1 \\ 1 & x_2 & z_2 \\ 1 & x_3 & z_3 \end{vmatrix}, \qquad C = -\begin{vmatrix} 1 & x_1 & y_1 \\ 1 & x_2 & y_2 \\ 1 & x_3 & y_3 \end{vmatrix}$$

are not all three equal to 0. For the vanishing of all of them means that the projections of A_1, A_2, A_3 are collinear in each of the planes *YOZ*, *ZOX*, *XOY*, which implies that A_1, A_2, A_3 are collinear. Consequently, (2) represents a plane, and this plane goes through A_1, A_2, A_3 since the determinant vanishes for

$$\begin{array}{lll} x = x_1, & y = y_1, & z = z_1 \\ x = x_2, & y = y_2, & z = z_2 \\ x = x_3, & y = y_3, & z = z_3. \end{array}$$

It is just as easy to write in determinant form the equation of a circle through three noncollinear points $A_1(x_1,y_1)$, $A_2(x_2,y_2)$, $A_3(x_3,y_3)$. Equation

$$\begin{vmatrix} x^2 + y^2 & x & y & 1 \\ x_1^2 + y_1^2 & x_1 & y_1 & 1 \\ x_2^2 + y_2^2 & x_2 & y_2 & 1 \\ x_3^2 + y_3^2 & x_3 & y_3 & 1 \end{vmatrix} = 0 \tag{3}$$

has the form

$$A(x^2 + y^2) + Bx + Cy + D = 0,$$

with

$$A = \begin{vmatrix} x_1 & y_1 & 1 \\ x_2 & y_2 & 1 \\ x_3 & y_3 & 1 \end{vmatrix}$$

different from 0. Hence, (3) represents a circle and this circle goes through A_1, A_2, A_3. Similarly, the equation

$$\begin{vmatrix} x^2 + y^2 + z^2 & x & y & z & 1 \\ x_1^2 + y_1^2 + z_1^2 & x_1 & y_1 & z_1 & 1 \\ x_2^2 + y_2^2 + z_2^2 & x_2 & y_2 & z_2 & 1 \\ x_3^2 + y_3^2 + z_3^2 & x_3 & y_3 & z_3 & 1 \\ x_4^2 + y_4^2 + z_4^2 & x_4 & y_4 & z_4 & 1 \end{vmatrix} = 0 \tag{4}$$

represents a sphere passing through the four points $A_1(x_1, y_1, z_1)$, $A_2(x_2, y_2, z_2)$, $A_3(x_3, y_3, z_3)$, $A_4(x_4, y_4, z_4)$, which are not in the same plane.★

★**7. The Area of a Triangle, and the Volume of a Tetrahedron**   If

$$Ax + By + C = 0 \tag{1}$$

is the equation of a line l, and $P(X, Y)$ is an arbitrary point, the expression

$$d = \frac{AX + BY + C}{\sqrt{A^2 + B^2}}$$

represents the distance from P to l taken with a certain sign. Consider now a triangle with vertices $A_1(x_1, y_1)$, $A_2(x_2, y_2)$, $A_3(x_3, y_3)$. The equation of the side $A_1 A_2$ can be written in determinant form

$$\begin{vmatrix} 1 & x & y \\ 1 & x_1 & y_1 \\ 1 & x_2 & y_2 \end{vmatrix} = 0,$$

and the corresponding coefficients A and B in equation (1) are

$$A = y_1 - y_2, \qquad B = x_2 - x_1$$

so that

$$\sqrt{A^2 + B^2} = \sqrt{(x_2 - x_1)^2 + (y_2 - y_1)^2} = l_{12}$$

is the length of the side $A_1 A_2$. Hence,

$$\begin{vmatrix} 1 & x_3 & y_3 \\ 1 & x_1 & y_1 \\ 1 & x_2 & y_2 \end{vmatrix} : l_{12}$$

represents the altitude from the vertex A_3 taken positively or negatively, and therefore

$$\begin{vmatrix} 1 & x_3 & y_3 \\ 1 & x_1 & y_1 \\ 1 & x_2 & y_2 \end{vmatrix} = \begin{vmatrix} 1 & x_1 & y_1 \\ 1 & x_2 & y_2 \\ 1 & x_3 & y_3 \end{vmatrix}$$

is double the area of the triangle $A_1 A_2 A_3$ taken with sign $+$ or $-$. Denoting this area by A, we thus have

$$\pm 2A = \begin{vmatrix} 1 & x_1 & y_1 \\ 1 & x_2 & y_2 \\ 1 & x_3 & y_3 \end{vmatrix}$$

and it may be added, leaving an easy proof to the reader, that the sign will be $+$ or $-$ according as the sense of the triangle $A_1 A_2 A_3$ (with this order of vertices) is positive or negative.

The distance of a point $P(X, Y, Z)$ from the plane

$$Ax + By + Cz + D = 0,$$

taken with a certain sign, is given by the formula

$$d = \frac{AX + BY + CZ + D}{\sqrt{A^2 + B^2 + C^2}}.$$

Consider now the four vertices $A_1(x_1, y_1, z_1)$, $A_2(x_2, y_2, z_2)$, $A_3(x_3, y_3, z_3)$, $A_4(x_4, y_4, z_4)$ of a tetrahedron. The equation of the plane $A_1A_2A_3$, written in determinant form, is

$$\begin{vmatrix} 1 & x & y & z \\ 1 & x_1 & y_1 & z_1 \\ 1 & x_2 & y_2 & z_2 \\ 1 & x_3 & y_3 & z_3 \end{vmatrix} = 0,$$

and the corresponding coefficients

$$A = -\begin{vmatrix} 1 & y_1 & z_1 \\ 1 & y_2 & z_2 \\ 1 & y_3 & z_3 \end{vmatrix}, \qquad B = \begin{vmatrix} 1 & x_1 & z_1 \\ 1 & x_2 & z_2 \\ 1 & x_3 & z_3 \end{vmatrix}, \qquad C = -\begin{vmatrix} 1 & x_1 & y_1 \\ 1 & x_2 & y_2 \\ 1 & x_3 & y_3 \end{vmatrix}$$

represent numerically double the areas of the projections of the triangle $A_1A_2A_3$ on the coordinate planes YOZ, ZOX, XOY. As is known from analytic geometry, the sum of the squares of these areas is the square of the area of the triangle $A_1A_2A_3$. Calling this area Δ, we have therefore

$$A^2 + B^2 + C^2 = 4\Delta^2,$$

and the distance of the vertex A_4 of the tetrahedron from the opposite face, taken with the sign $+$ or $-$, is given by

$$d = \frac{1}{2\Delta} \begin{vmatrix} 1 & x_4 & y_4 & z_4 \\ 1 & x_1 & y_1 & z_1 \\ 1 & x_2 & y_2 & z_2 \\ 1 & x_3 & y_3 & z_3 \end{vmatrix} = -\frac{1}{2\Delta} \begin{vmatrix} 1 & x_1 & y_1 & z_1 \\ 1 & x_2 & y_2 & z_2 \\ 1 & x_3 & y_3 & z_3 \\ 1 & x_4 & y_4 & z_4 \end{vmatrix}.$$

Since the volume V of the tetrahedron is

$$V = \pm \tfrac{1}{3}\Delta d,$$

we have finally

$$\pm 6V = \begin{vmatrix} x_1 & y_1 & z_1 & 1 \\ x_2 & y_2 & z_2 & 1 \\ x_3 & y_3 & z_3 & 1 \\ x_4 & y_4 & z_4 & 1 \end{vmatrix}.$$

The sign $\pm$ depends on the sense attributed to the tetrahedron $A_1A_2A_3A_4$, the vertices being taken in this order, but it is not necessary for us to examine here this question of sign.★

★8. The Power of a Point with Respect to a Circle. Consider a circle Γ with center O and radius R, and a point P at a distance $PO = d$ from the center. The expression

$$p = d^2 - R^2$$

is called the *power of P with respect* to Γ. Let the equation of Γ be

$$x^2 + y^2 + Ax + By + C = 0,$$

which also can be written in the form

$$(x - a)^2 + (y - b)^2 - R^2 = 0,$$

a, b being coordinates of the center of Γ. Substituting here for x, y the coordinates X, Y of P, and taking into account that

$$(X - a)^2 + (Y - b)^2 = d^2,$$

we see that

$$(X - a)^2 + (Y - b)^2 - R^2 = p;$$

in other words, the power of the point $P(X, Y)$ with respect to the circle

$$x^2 + y^2 + Ax + By + C = 0$$

is

$$p = X^2 + Y^2 + AX + BY + C.$$

Let

$$\begin{vmatrix} x^2 + y^2 & x & y & 1 \\ x_1^2 + y_1^2 & x_1 & y_1 & 1 \\ x_2^2 + y_2^2 & x_2 & y_2 & 1 \\ x_3^2 + y_3^2 & x_3 & y_3 & 1 \end{vmatrix} = 0 \tag{1}$$

be the equation of the circle through the three points $A_1(x_1, y_1)$, $A_2(x_2, y_2)$, $A_3(x_3, y_3)$. Since the coefficient of $x^2 + y^2$ in the expanded determinant (1) is

$$\begin{vmatrix} x_1 & y_1 & 1 \\ x_2 & y_2 & 1 \\ x_3 & y_3 & 1 \end{vmatrix}$$

and represents the double area $2A$ of the triangle $A_1A_2A_3$, it is easy to see that the determinant

$$\begin{vmatrix} x_4^2 + y_4^2 & x_4 & y_4 & 1 \\ x_1^2 + y_1^2 & x_1 & y_1 & 1 \\ x_2^2 + y_2^2 & x_2 & y_2 & 1 \\ x_3^2 + y_3^2 & x_3 & y_3 & 1 \end{vmatrix}$$

is the product of $2A$ by the power p of an arbitrary point $A_4(x_4, y_4)$ with respect to the circle circumscribed about the triangle $A_1A_2A_3$. By a rearrangement of the rows in this determinant we have

$$- 2Ap = \begin{vmatrix} x_1^2 + y_1^2 & x_1 & y_1 & 1 \\ x_2^2 + y_2^2 & x_2 & y_2 & 1 \\ x_3^2 + y_3^2 & x_3 & y_3 & 1 \\ x_4^2 + y_4^2 & x_4 & y_4 & 1 \end{vmatrix}.$$

Combining this important formula in a proper manner with the rule for multiplication of determinants, we shall be able to obtain remarkable identities leading to some interesting geometrical theorems.★

★9. **Geometric Corollaries.** Multiplying the determinant

$$\begin{vmatrix} x_1^2 + y_1^2 & - 2x_1 & - 2y_1 & 1 \\ x_2^2 + y_2^2 & - 2x_2 & - 2y_2 & 1 \\ x_3^2 + y_3^2 & - 2x_3 & - 2y_3 & 1 \\ x_4^2 + y_4^2 & - 2x_4 & - 2y_4 & 1 \end{vmatrix} = - 8Ap$$

by

$$\begin{vmatrix} 1 & x_1 & y_1 & x_1^2 + y_1^2 \\ 1 & x_2 & y_2 & x_2^2 + y_2^2 \\ 1 & x_3 & y_3 & x_3^2 + y_3^2 \\ 1 & x_4 & y_4 & x_4^2 + y_4^2 \end{vmatrix} = 2Ap,$$

row by row, and setting for brevity

$$l_{ij}^2 = (x_i - x_j)^2 + (y_i - y_j)^2,$$

we get

$$- 16A^2p^2 = \begin{vmatrix} 0 & l_{12}^2 & l_{13}^2 & l_{14}^2 \\ l_{12}^2 & 0 & l_{23}^2 & l_{24}^2 \\ l_{13}^2 & l_{23}^2 & 0 & l_{34}^2 \\ l_{14}^2 & l_{24}^2 & l_{34}^2 & 0 \end{vmatrix}. \tag{1}$$

This is a relation between the area A of a triangle $A_1A_2A_3$, the power of the point A_4 with respect to the circle circumscribed about that triangle, and the six distances of the points A_1, A_2, A_3, A_4 taken two by two. Multiplying columns 2, 3, 4 in the determinant (1) by $(l_{13}l_{14})^2$, $(l_{12}l_{14})^2$, $(l_{12}l_{13})^2$, respectively, we get

$$- 16A^2p^2(l_{12}l_{13}l_{14})^4 = \begin{vmatrix} 0 & (l_{12}l_{13}l_{14})^2 & (l_{12}l_{13}l_{14})^2 & (l_{12}l_{13}l_{14})^2 \\ l_{12}^2 & 0 & (l_{12}l_{23}l_{14})^2 & (l_{12}l_{13}l_{24})^2 \\ l_{13}^2 & (l_{13}l_{14}l_{23})^2 & 0 & (l_{12}l_{13}l_{34})^2 \\ l_{14}^2 & (l_{13}l_{14}l_{24})^2 & (l_{12}l_{14}l_{34})^2 & 0 \end{vmatrix}$$

or

$$- 16A^2p^2(l_{12}l_{13}l_{14})^4 = (l_{12}l_{13}l_{14})^4 \begin{vmatrix} 0 & 1 & 1 & 1 \\ 1 & 0 & (l_{23}l_{14})^2 & (l_{13}l_{24})^2 \\ 1 & (l_{14}l_{23})^2 & 0 & (l_{12}l_{34})^2 \\ 1 & (l_{13}l_{24})^2 & (l_{12}l_{34})^2 & 0 \end{vmatrix}$$

and, finally,

$$- 16A^2p^2 = \begin{vmatrix} 0 & 1 & 1 & 1 \\ 1 & 0 & (l_{14}l_{23})^2 & (l_{13}l_{24})^2 \\ 1 & (l_{14}l_{23})^2 & 0 & (l_{12}l_{34})^2 \\ 1 & (l_{13}l_{24})^2 & (l_{12}l_{34})^2 & 0 \end{vmatrix}. \tag{2}$$

Suppose that A_4 is placed at the center of the circle circumscribed by the triangle $A_1A_2A_3$; then, $p = - R^2$ and $l_{14} = l_{24} = l_{34} = R$. Canceling the factor R^4 on both sides of (1), we come to the following relation between the sides and the area of a triangle in determinant form:

$$16A^2 = - \begin{vmatrix} 0 & l_{12}^2 & l_{13}^2 & 1 \\ l_{12}^2 & 0 & l_{22}^2 & 1 \\ l_{13}^2 & l_{23}^2 & 0 & 1 \\ 1 & 1 & 1 & 0 \end{vmatrix} = \begin{vmatrix} 1 & 0 & l_{12}^2 & l_{13}^2 \\ 1 & l_{12}^2 & 0 & l_{23}^2 \\ 1 & l_{13}^2 & l_{23}^2 & 0 \\ 0 & 1 & 1 & 1 \end{vmatrix}.$$

To use more familiar notations let a, b, c be the sides and let $l_{12} = a$, $l_{13} = b$, $l_{23} = c$. Then,

$$16A^2 = \begin{vmatrix} 1 & 0 & a^2 & b^2 \\ 1 & a^2 & 0 & c^2 \\ 1 & b^2 & c^2 & 0 \\ 0 & 1 & 1 & 1 \end{vmatrix},$$

or, on expanding the determinant,

$$16A^2 = (a + b + c)(a + b - c)(a + c - b)(b + c - a),$$

which is a well-known formula. On comparing the last determinant with that in the right-hand side of (2) we derive an interesting formula:

$$16A^2p^2 = (l_{23}l_{14} + l_{13}l_{24} + l_{12}l_{34})(l_{23}l_{14} + l_{13}l_{24} - l_{12}l_{34})(l_{23}l_{14} + l_{12}l_{34} - l_{13}l_{24})$$
$$(l_{13}l_{24} + l_{12}l_{34} - l_{23}l_{14}).$$

Suppose that the four points A_1, A_2, A_3, A_4 are on the same circle; then, $p = 0$ and, dropping the positive factor

$$l_{23}l_{14} + l_{13}l_{24} + l_{12}l_{34},$$

we have a relation between the distances of four concyclic points:

$$(l_{23}l_{14} + l_{13}l_{24} - l_{12}l_{34})(l_{23}l_{14} + l_{12}l_{34} - l_{13}l_{24})(l_{13}l_{24} + l_{12}l_{34} - l_{23}l_{14}) = 0,$$

or, writing for simplicity

$$l_{12} = a, \qquad l_{23} = b, \qquad l_{34} = c,$$
$$l_{14} = d, \qquad l_{13} = e, \qquad l_{24} = f,$$

then,

$$(bd + ef - ac)(bd + ac - ef)(ef + ac - bd) = 0.$$

If $A_1A_2A_3A_4$ is a convex quadrilateral with sides $A_1A_2 = a$, $A_2A_3 = b$, $A_3A_4 = c$, $A_4A_1 = d$ and diagonals $A_1A_3 = e$, $A_2A_4 = f$, then necessarily

$$ac + bd - ef = 0.$$

To prove this, notice that in $A_1A_2A_3A_4$, which is inscribed in a circle, the sums of opposite angles are 180°, and therefore there is one side to which two adjoining angles are obtuse. If nota-

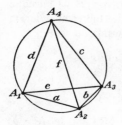

tions are so chosen that this side is b, then clearly

$$e > a, \qquad e > b, \qquad f > b, \qquad f > c$$

so that $ef > ac$ and hence

$$bd + ef - ac > 0.$$

On the other hand,

$$ef - bd = e(f - b) - b(d - e)$$

is positive in case $d < e$ since $f > b$. If $d > e$, it follows from the almost evident inequality

$$e + f > b + d$$

that

$$f - b > d - e,$$

and, again, that

$$ef - bd > (d - e)(e - b) > 0.$$

Thus, in all cases, $ef > bd$ and

$$ef + ac - bd > 0.$$

Therefore, if $A_1A_2A_3A_4$ is a convex polygon inscribed in a circle, there is a relation

$$ac + bd = ef$$

between the product of the diagonals and the sum of the products of the opposite sides known in elementary geometry as Ptolemy's theorem.★

★**10. Extension to Three Dimensions.** There is no difficulty in extending the considerations of Secs. 8 to 9 to three dimensions. In the first place, the equation of a sphere circumscribed to the tetrahedron with vertices

$$A_1(x_1, y_1, z_1), \quad A_2(x_2, y_2, z_2), \quad A_3(x_3, y_3, z_3), \quad A_4(x_4, y_4, z_4)$$

can be presented in determinant form thus:

$$\begin{vmatrix} x^2 + y^2 + z^2 & x & y & z & 1 \\ x_1^2 + y_1^2 + z_1^2 & x_1 & y_1 & z_1 & 1 \\ x_2^2 + y_2^2 + z_2^2 & x_2 & y_2 & z_2 & 1 \\ x_3^2 + y_3^2 + z_3^2 & x_3 & y_3 & z_3 & 1 \\ x_4^2 + y_4^2 + z_4^2 & x_4 & y_4 & z_4 & 1 \end{vmatrix} = 0.$$

Further, if R is the radius of this sphere and d the distance of some point $A_5(x_5, y_5, z_5)$ to its center, then the quantity

$$p = d^2 - R^2$$

is the power of A_5 with respect to that sphere, and as in Sec. 8 we easily establish the relation

$$6Vp = \begin{vmatrix} x_1^2 + y_1^2 + z_1^2 & x_1 & y_1 & z_1 & 1 \\ x_2^2 + y_2^2 + z_2^2 & x_2 & y_2 & z_2 & 1 \\ x_3^2 + y_3^2 + z_3^2 & x_3 & y_3 & z_3 & 1 \\ x_4^2 + y_4^2 + z_4^2 & x_4 & y_4 & z_4 & 1 \\ x_5^2 + y_5^2 + z_5^2 & x_5 & y_5 & z_5 & 1 \end{vmatrix},$$

where V is the volume of the tetrahedron $A_1 A_2 A_3 A_4$.

Multiplying now the determinant

$$\begin{vmatrix} x_1^2 + y_1^2 + z_1^2 & -2x_1 & -2y_1 & -2z_1 & 1 \\ x_2^2 + y_2^2 + z_2^2 & -2x_2 & -2y_2 & -2z_2 & 1 \\ \cdots & \cdots & \cdots & \cdots & \cdots \\ x_5^2 + y_5^2 + z_5^2 & -2x_5 & -2y_5 & -2z_5 & 1 \end{vmatrix} = -48Vp$$

by

$$\begin{vmatrix} 1 & x_1 & y_1 & z_1 & x_1^2 + y_1^2 + z_1^2 \\ 1 & x_2 & y_2 & z_2 & x_2^2 + y_2^2 + z_2^2 \\ \cdots & \cdots & \cdots & \cdots & \cdots \\ 1 & x_5 & y_5 & z_5 & x_5^2 + y_5^2 + z_5^2 \end{vmatrix} = -6Vp,$$

row by row, and setting for brevity

$$l_{ij}^2 = (x_i - x_j)^2 + (y_i - y_j)^2 + (z_i - z_j)^2,$$

we get

$$288V^2p^2 = \begin{vmatrix} 0 & l_{12}^2 & l_{13}^2 & l_{14}^2 & l_{15}^2 \\ l_{12}^2 & 0 & l_{23}^2 & l_{24}^2 & l_{25}^2 \\ l_{13}^2 & l_{23}^2 & 0 & l_{34}^2 & l_{35}^2 \\ l_{14}^2 & l_{24}^2 & l_{34}^2 & 0 & l_{45}^2 \\ l_{15}^2 & l_{25}^2 & l_{35}^2 & l_{45}^2 & 0 \end{vmatrix}.$$

The numbers l_{ij} are the ten distances of the five points A_1, A_2, A_3, A_4, A_5 taken two by two. If A_5 is placed at the center of the sphere circumscribed to the tetrahedron $A_1A_2A_3A_4$, then $p = -R^2$, and

$$l_{15} = l_{25} = l_{35} = l_{45} = R.$$

Canceling the factor R^4 on both sides, we come to the expression, in determinant form, of the volume of a tetrahedron in terms of its edges:

$$288V^2 = \begin{vmatrix} 0 & l_{12}^2 & l_{13}^2 & l_{14}^2 & 1 \\ l_{12}^2 & 0 & l_{23}^2 & l_{24}^2 & 1 \\ l_{13}^2 & l_{23}^2 & 0 & l_{34}^2 & 1 \\ l_{14}^2 & l_{24}^2 & l_{34}^2 & 0 & 1 \\ 1 & 1 & 1 & 1 & 0 \end{vmatrix}.$$

This determinant equated to 0 represents the relation between the mutual distances of any four coplanar points.★

CHAPTER XI

SYMMETRIC FUNCTIONS

1. Definition of Symmetric Functions. Sigma Functions. A polynomial in the variables $x_1, x_2, \ldots, x_n$ is called a *symmetric* polynomial or a symmetric function of these variables if it is not changed by changing the order of the variables in any manner whatsoever. For this it suffices that the interchange of any two variables leaves the polynomial unaltered. For example, the polynomials

$$a^2 + b^2 + c^2 - ab - ac - bc, \qquad a^3 + b^3 + c^3 - 3abc,$$
$$(a + b + c)abc$$

are symmetric polynomials, or symmetric functions, of the variables a, b, c. Similarly,

$$x_1^4 + x_2^4 + x_3^4 + x_4^4 + 3(x_1x_2 + x_2x_3)(x_1x_3 + x_2x_4)(x_1x_4 + x_2x_3)$$

and

$$(x_1 + x_2 - x_3 - x_4)(x_1 + x_3 - x_2 - x_4)(x_1 + x_4 - x_2 - x_3)$$

are symmetric functions of the four variables x_1, x_2, x_3, x_4. Taking the sum of all distinct terms obtained from a *typical* term

$$x_1^{\alpha_1} x_2^{\alpha_2} \cdots x_m^{\alpha_m},$$

where $\alpha_1, \alpha_2, \ldots, \alpha_m$ are positive integers, by replacing the indices $1, 2, \ldots, m$ by all possible arrangements of m numbers taken out of $1, 2, \ldots, n$, a symmetric function of the variables $x_1, x_2, \ldots, x_n$ is obtained that is called a *sigma function* of the type $(\alpha_1, \alpha_2, \ldots, \alpha_m)$ and is denoted by the sign [1]

$$\Sigma x_1^{\alpha_1} x_2^{\alpha_2} \cdots x_m^{\alpha_m}.$$

For example, for $n = 4$

$$\Sigma x_1^2 x_2^2 x_3 = x_1^2 x_2^2 x_3 + x_1^2 x_3^2 x_2 + x_2^2 x_3^2 x_1 + x_1^2 x_2^2 x_4 + x_1^2 x_4^2 x_2 + x_2^2 x_4^2 x_1$$
$$+ x_1^2 x_3^2 x_4 + x_1^2 x_4^2 x_3 + x_3^2 x_4^2 x_1 + x_2^2 x_3^2 x_4 + x_2^2 x_4^2 x_3 + x_3^2 x_4^2 x_2$$

is a sigma function of the type $(2, 2, 1)$ in the variables x_1, x_2, x_3, x_4.

Among the sigma functions especially important are the so-called *elementary symmetric functions:*

[1] More generally $\Sigma g(x_1, x_2, \ldots, x_m)$ means the sum of all distinct terms obtained by replacing indices $1, 2, \ldots, m$ by all possible arrangements of m numbers taken out of $1, 2, \ldots, n$.

257

$$\Sigma x_1 = f_1, \quad \Sigma x_1 x_2 = f_2, \quad \Sigma x_1 x_2 x_3 = f_3, \quad \ldots ,$$
$$\Sigma x_1 x_2 \cdots x_n = x_1 x_2 \cdots x_n = f_n.$$

Of importance are also the "sums of powers"

$$s_\alpha = \Sigma x_1^\alpha = x_1^\alpha + x_2^\alpha + \cdots + x_n^\alpha.$$

If a symmetric function $\phi(x_1, x_2, \ldots, x_n)$ contains a term

$$a x_1^{\alpha_1} x_2^{\alpha_2} \cdots x_m^{\alpha_m},$$

it will contain the aggregate of terms

$$a \Sigma x_1^{\alpha_1} x_2^{\alpha_2} \cdots x_m^{\alpha_m}$$

and if these do not exhaust all terms of $\phi(x_1, x_2, \ldots, x_n)$, then the difference

$$\phi(x_1, x_2, \ldots, x_n) - a \Sigma x_1^{\alpha_1} x_2^{\alpha_2} \cdots x_m^{\alpha_m}$$

will contain terms of the form

$$b x_1^{\beta_1} x_2^{\beta_2} \cdots x_r^{\beta_r}$$

and therefore also

$$b \Sigma x_1^{\beta_1} x_2^{\beta_2} \cdots x_r^{\beta_r}.$$

Continuing in the same way, we come to the conclusion that every symmetric function is a linear combination of sigma functions:

$$\phi(x_1, x_2, \ldots, x_n) = a \Sigma x_1^{\alpha_1} x_2^{\alpha_2} \cdots x_m^{\alpha_m} + b \Sigma x_1^{\beta_1} x_2^{\beta_2} \cdots x_r^{\beta_r}$$
$$+ c \Sigma x_1^{\gamma_1} x_2^{\gamma_2} \cdots x_s^{\gamma_s} + \cdots .$$

The process of splitting a symmetric function into sigma functions will now be shown by an example.

Example. Let

$$\phi = (x_1 + x_2 - x_3 - x_4)(x_1 + x_3 - x_2 - x_4)(x_1 + x_4 - x_2 - x_3).$$

This is a homogeneous symmetric polynomial of dimension 3. Typical terms of the component sigma functions will be

$$x_1^{\alpha_1} x_2^{\alpha_2} x_3^{\alpha_3} x_4^{\alpha_4},$$

where

$$\alpha_1 + \alpha_2 + \alpha_3 + \alpha_4 = 3,$$

and we may suppose that $\alpha_1 \geq \alpha_2 \geq \alpha_3 \geq \alpha_4 \geq 0$. The only possible sets of exponents are

α_1	α_2	α_3	α_4
3	0	0	0
2	1	0	0
1	1	1	0

and it remains only to determine the coefficients of the terms

$$a x_1^3, \quad b x_1^2 x_2, \quad c x_1 x_2 x_3.$$

To determine a and b we can set $x_3 = x_4 = 0$ in ϕ, which does not affect terms involving only x_1 and x_2. Since then ϕ becomes

$$(x_1 + x_2)(x_1 - x_2)(x_1 - x_2),$$

it is easy to see that $a = 1$, $b = -1$. To find c we may set $x_4 = 0$ in ϕ; then, ϕ becomes

$$(x_1 + x_2 - x_3)(x_1 - x_2 + x_3)(x_1 - x_2 - x_3).$$

To find the term involving $x_1x_2x_3$ we notice that it can be obtained in the following ways:

1. Picking x_1, x_2, x_3 from factors 1, 2, 3 we have their product $\quad + x_1x_2x_3$
2. Picking x_1, x_3, x_2 from factors 1, 2, 3 we have their product $\quad - x_1x_2x_3$
3. Picking x_2, x_1, x_3 from factors 1, 2, 3 we have their product $\quad - x_1x_2x_3$
4. Picking x_2, x_3, x_1 from factors 1, 2, 3 we have their product $\quad + x_1x_2x_3$
5. Picking x_3, x_1, x_2 from factors 1, 2, 3 we have their product $\quad + x_1x_2x_3$
6. Picking x_3, x_2, x_1 from factors 1, 2, 3 we have their product $\quad + x_1x_2x_3$

$$\text{Total} \quad \overline{2x_1x_2x_3}$$

Hence, $c = 2$ and

$$\phi = \Sigma x_1^3 - \Sigma x_1^2 x_2 + 2\Sigma x_1 x_2 x_3.$$

If in a symmetric function $\phi(x_1, x_2, \ldots, x_n)$ we substitute instead of the variables the numbers $\alpha_1, \alpha_2, \ldots, \alpha_n$, which are roots of an equation

$$x^n + p_1 x^{n-1} + p_2 x^{n-2} + \cdots + p_n = 0,$$

the resulting number $\phi(\alpha_1, \alpha_2, \ldots, \alpha_n)$ is called a *symmetric function of the roots of this equation*. This terminology is not correct since a definite number cannot be called a function, but it is consecrated by usage and we shall adhere to it. Thus, the elementary symmetric functions of the roots $\alpha_1, \alpha_2, \ldots, \alpha_n$ are

$$\Sigma \alpha_1 = -p_1, \quad \Sigma \alpha_1 \alpha_2 = p_2, \quad \Sigma \alpha_1 \alpha_2 \alpha_3 = -p_3, \quad \cdots,$$
$$\Sigma \alpha_1 \alpha_2 \cdots \alpha_n = (-1)^n p_n;$$

or

$$\Sigma \alpha_1 = -\frac{a_1}{a_0}, \quad \Sigma \alpha_1 \alpha_2 = \frac{a_2}{a_0}, \quad \Sigma \alpha_1 \alpha_2 \alpha_3 = -\frac{a_3}{a_0}, \quad \cdots,$$
$$\Sigma \alpha_1 \alpha_2 \cdots \alpha_n = (-1)^n \frac{a_n}{a_0}$$

if the equation is written in the form

$$a_0 x^n + a_1 x^{n-1} + \cdots + a_n = 0.$$

A rational function of the variables $x_1, x_2, \ldots, x_n$:

$$\frac{\phi(x_1, x_2, \ldots, x_n)}{\psi(x_1, x_2, \ldots, x_n)},$$

where the numerator and the denominator are polynomials, is called *symmetric* if it is not changed by all permutations of the variables. Thus,

$$\frac{a^2}{b+c} + \frac{b^2}{a+c} + \frac{c^2}{a+b}$$

is a rational symmetric function of the variables a, b, c. If the variables in a rational symmetric function are replaced by the roots of an equation, the resulting number is called again the symmetric function of roots of that equation—an incorrect terminology to which, however, it is customary to adhere.

Problems

Verify that the following functions are symmetric and break them up into sigma functions:

1. $(x_1 + x_2)(x_1 + x_3)(x_2 + x_3)$.

2. $(x_1 + x_2)^2 x_3 + (x_1 + x_3)^2 x_2 + (x_2 + x_3)^2 x_1$.

3. $(x_1 - x_2)^2 x_3 + (x_1 - x_3)^2 x_2 + (x_2 - x_3)^2 x_1$.

4. $(x_1 + x_2)^2 x_3^2 + (x_1 + x_3)^2 x_2^2 + (x_2 + x_3)^2 x_1^2$.

5. $(x_1 - x_2)^2(x_1 - x_3)^2 + (x_1 - x_2)^2(x_2 - x_3)^2 + (x_1 - x_3)^2(x_2 - x_3)^2$.

6. $(x_1 - x_2)^2(x_1 - x_3)^2(x_2 - x_3)^2$.

7. $(x_1 x_2 + x_3 x_4)^2 + (x_1 x_3 + x_2 x_4)^2 + (x_1 x_4 + x_2 x_3)^2$.

8. $(x_1 x_2 + x_3 x_4)(x_1 x_3 + x_2 x_4)(x_1 x_4 + x_2 x_3)$.

9. $(x_1 + x_2 - x_3 - x_4)^2 + (x_1 + x_3 - x_2 - x_4)^2 + (x_1 + x_4 - x_2 - x_3)^2$.

10. $(x_1 + x_2 - x_3 - x_4)^2(x_1 + x_3 - x_2 - x_4)^2$
$$+ (x_1 + x_2 - x_3 - x_4)^2(x_1 + x_4 - x_2 - x_3)^2$$
$$+ (x_1 + x_3 - x_2 - x_4)^2(x_1 + x_4 - x_2 - x_3)^2.$$

2. Newton's Formulas.

Sums of powers

$$s_\alpha = x_1^\alpha + x_2^\alpha + \cdots + x_n^\alpha$$

can be expressed as polynomials in the elementary symmetric functions or, which is the same, in terms of the coefficients of

$$f(x) = (x - x_1)(x - x_2) \cdots (x - x_n) = x^n + p_1 x^{n-1} + \cdots + p_n.$$

This can be shown in the following manner: On many occasions we made use of the identity

$$\frac{f'(x)}{f(x)} = \frac{1}{x - x_1} + \frac{1}{x - x_2} + \cdots + \frac{1}{x - x_n}.$$

Write it in the form

$$f'(x) = \frac{f(x)}{x - x_1} + \frac{f(x)}{x - x_2} + \cdots + \frac{f(x)}{x - x_n} \tag{1}$$

and notice that

$$\frac{f(x)}{x - x_i} = x^{n-1} + f_1(x_i)x^{n-2} + f_2(x_i)x^{n-3} + \cdots + f_{n-1}(x_i),$$

where

$$f_1(x) = x + p_1, \qquad f_2(x) = x^2 + p_1 x + p_2, \qquad f_3(x) = x^3 + p_1 x^2 + p_2 x + p_3,$$

and in general

$$f_k(x) = x^k + p_1 x^{k-1} + \cdots + p_k,$$

for $k = 1, 2, \ldots, n - 1$. On introducing the sums s_α we have

$$f_k(x_1) + f_k(x_2) + \cdots + f_k(x_n) = s_k + p_1 s_{k-1} + \cdots + n p_k,$$

so that the coefficient of x^{n-k-1} in the right-hand side of the identity (1) is

$$s_k + p_1 s_{k-1} + \cdots + n p_k.$$

On the left-hand side of the same identity the coefficient of x^{n-k-1} is $(n - k)p_k$, and on equating both expressions we get

$$s_k + p_1 s_{k-1} + \cdots + n p_k = (n - k)p_k,$$

or

$$s_k + p_1 s_{k-1} + \cdots + k p_k = 0$$

for $k = 1, 2, \ldots, n - 1$. This relation remains true for $k = n$. In fact,

$$f(x_i) = x_i^n + p_1 x_i^{n-1} + \cdots + p_n = 0$$

and, taking the sum of these identities for $i = 1, 2, \ldots, n$, we get

$$s_n + p_1 s_{n-1} + \cdots + n p_n = 0.$$

Thus,

$$s_1 + p_1 = 0,$$
$$s_2 + p_1 s_1 + 2 p_2 = 0,$$
$$s_3 + p_1 s_2 + p_2 s_1 + 3 p_3 = 0,$$
$$\cdots\cdots\cdots\cdots\cdots\cdots\cdots\cdots\cdots$$
$$s_n + p_1 s_{n-1} + p_2 s_{n-2} + \cdots + n p_n = 0.$$

These are the so-called *Newton's formulas*. From these identities we may calculate successively the sums of powers $s_1, s_2, \ldots, s_n$ in terms of the coefficients $p_1, p_2, \ldots, p_n$ of the polynomial

$$f(x) = (x - x_1)(x - x_2) \cdots (x - x_n) = x^n + p_1 x^{n-1} + \cdots + p_n,$$

and conversely these coefficients through $s_1, s_2, \ldots, s_n$. The expressions of $s_1, s_2, s_3, \ldots$ in terms of $p_1, p_2, p_3, \ldots$, being identities, will remain valid on replacing $p_1, p_2, \ldots$ by numbers, and $x_1, x_2, \ldots$ by the roots of the equation

$$x^n + p_1 x^{n-1} + \cdots + p_n = 0.$$

Thus, Newton's formulas allow us to compute by recurrence the sums of like powers of the roots of an equation, and conversely we may compute the coefficients of this equation if the sums of like powers of roots are known. Taking the sum of the obvious identities

$$x_i^\nu f(x_i) = x_i^{n+\nu} + p_1 x_i^{n+\nu-1} + \cdots + p_n x_i^\nu = 0$$

for $i = 1, 2, \ldots, n$, we obtain, taking $\nu = 1, 2, 3, \ldots$, the further recurrence relations

$$s_{n+1} + p_1 s_n + \cdots + p_n s_1 = 0,$$
$$s_{n+2} + p_1 s_{n+1} + \cdots + p_n s_2 = 0,$$
$$\cdots\cdots\cdots\cdots\cdots\cdots\cdots$$

which enable us to express s_{n+1}, s_{n+2}, ... in terms of p_1, p_2, ... The recurrence relations obtained by taking $\nu = -1, -2, \ldots$

$$s_{n-1} + p_1 s_{n-2} + \cdots + p_n s_{-1} = 0,$$
$$s_{n-2} + p_1 s_{n-3} + \cdots + p_n s_{-2} = 0,$$
$$\cdots\cdots\cdots\cdots\cdots\cdots\cdots$$

similarly will serve to compute the sums of the negative powers of the roots provided $p_n \neq 0$. The recurrence relations for s_{n+1}, s_{n+2}, ... look different from Newton's formulas but will coincide with them if we agree to set $p_r = 0$ whenever $r > n$. Here we give expressions for $s_1, s_2, \ldots, s_6$ in terms of $p_1, p_2, \ldots, p_6$:

$$s_1 = -p_1,$$
$$s_2 = p_1^2 - 2p_2,$$
$$s_3 = -p_1^3 + 3p_1 p_2 - 3p_3,$$
$$s_4 = p_1^4 - 4p_1^2 p_2 + 4p_1 p_3 - 4p_4 + 2p_2^2,$$
$$s_5 = -p_1^5 + 5p_1^3 p_2 + 5p_1 p_4 - 5p_3 p_1^2 - 5p_1 p_2^2 + 5p_2 p_3 - 5p_5,$$
$$s_6 = p_1^6 - 6p_1^4 p_2 + 6p_1^3 p_3 - 6p_1^2 p_4 - 12p_1 p_2 p_3 + 6p_2 p_4$$
$$+ 6p_1 p_5 - 6p_6 + 9p_1^2 p_2^2 - 2p_2^3 + 3p_3^2.$$

Observe that $p_r = 0$ if r is greater than the degree of the equation.

Problems

Compute by recurrence:

1. $s_1, s_2, s_3, s_4, s_5, s_6$ for $x^3 - 3x + 1 = 0$.

2. s_1, s_2, s_3, s_4 for $x^3 - 3x^2 + 2x - 1 = 0$.

3. s_1, s_2, s_3, s_4 for $x^4 - 4x - 1 = 0$.

4. s_{-2}, s_{-3} for $2x^4 - 6x^2 + x + 1 = 0$.

5. $s_{-1}, s_{-2}, s_{-3}, s_{-4}$ for $x^5 - x^3 + x^2 + 1 = 0$.

6. s_1, s_2, s_3 for $2x^4 + x^2 - x + 1 = 0$.

Find equations of lowest degree satisfying the conditions:

7. $s_1 = 0, s_2 = 5, s_3 = -6$.   **8.** $s_1 = 0, s_2 = 0, s_3 = s_4 = 1$.

9. $s_1 = 18, s_2 = 162, s_3 = 1701$.

10. $s_1 = 1, s_2 = -1, s_3 = 1, s_4 = -1, s_5 = 1$.

11. $s_{-1} = -1$, $s_{-2} = 2$, $s_{-3} = -3$.

12. $s_{-1} = 0$, $s_{-2} = 1$, $s_{-3} = 0$, $s_4 = 3$.

13. If a root r of an equation $f(x) = 0$ is larger in absolute value than the other roots, prove that the ratio

$$\frac{s_{n+1}}{s_n}$$

approaches r as a limit when n increases indefinitely. Hence, for large n

$$r = \frac{s_{n+1}}{s_n}$$

approximately; and, if all the roots are real and positive,

$$\frac{s_{n+1}}{s_n} < r < \sqrt[n]{s_n}.$$

This is the idea underlying Daniel Bernoulli's method for computing by approximation the largest (numerically) root of an equation. Can a similar method be devised for computing the numerically smallest root?

14. Apply this method to the equations

$$(a)\ x^2 - x - 1 = 0, \qquad (b)\ x^2 - 7x + 4 = 0,$$

going as far as $n = 10$, and examine the approximation obtained.

15. Compute the smallest root of $x^3 + 3x^2 + 2x - 1 = 0$ by Daniel Bernoulli's method going as far as $n = 10$, and examine the approximation.

*16. Let $y_1, y_2, \ldots, y_m$, where $m = n(n-1)/2$, be the sums of the roots of $f(x) = 0$ taken two at a time, and let

$$S_k = y_1^k + y_2^k + \cdots + y_m^k.$$

Show that

$$2S_k + 2^k s_k = s_0 s_k + \binom{k}{1} s_1 s_{k-1} + \binom{k}{2} s_2 s_{k-2} + \cdots + s_k s_0.$$

HINT: The right-hand side is

$$\sum_{j=1}^{n} \sum_{i=1}^{n} (x_j + x_i)^k.$$

*17. If the roots of the equation $x^3 - 3x + 1 = 0$ are α, β, γ, find an equation whose roots are $\alpha + \beta$, $\alpha + \gamma$, $\beta + \gamma$.

*18. Let $y_1, y_2, \ldots, y_m$, where $m = n(n-1)/2$, represent the squares of the differences between two roots of an equation of degree n and let

$$S_k = y_1^k + y_2^k + \cdots + y_m^k.$$

Show that

$$2S_k = s_0 s_{2k} - \binom{2k}{1} s_1 s_{2k-1} + \binom{2k}{2} s_2 s_{2k-2} - \cdots + s_{2k} s_0.$$

HINT: The right-hand side is

$$\sum_{j=1}^{n} \sum_{i=1}^{n} (x_j - x_i)^{2k}.$$

*19. If the roots of the equation $x^3 - 3x + 1 = 0$ are α, β, γ, find an equation with roots

$$(\alpha - \beta)^2, \quad (\alpha - \gamma)^2, \quad (\beta - \gamma)^2.$$

*20. Solve the same problem for the equation $x^3 - 7x + 7 = 0$.

*21. Find the lower bound of the distances between any two real roots of the equations (a) $x^3 - 3x + 1 = 0$, (b) $x^3 - 7x + 7 = 0$. How does a knowledge of this lower bound help to separate the roots?

★3. Fundamental Theorem on Symmetric Functions. The fact that the sums of powers are expressible as polynomials in the elementary symmetric functions is but a particular case of the following general theorem:

Fundamental Theorem on Symmetric Functions. Every symmetric function of the variables $x_1, x_2, \ldots, x_n$ can be expressed as a polynomial in the elementary symmetric functions. Moreover, coefficients of this polynomial are built up by additions and subtractions of the coefficients of the symmetric function. In particular, if the latter are integers, the former will be integers also.

PROOF. Among the various known proofs of this theorem we will select an elegant proof by Cauchy. First, the theorem will be proved in the case of two variables x_1, x_2 whose elementary symmetric functions are

$$f_1 = x_1 + x_2, \qquad f_2 = x_1 x_2.$$

Let $F(x_1, x_2)$ be a symmetric function. Arranged in powers of x_1, it can be written thus:

$$F(x_1, x_2) = A_0 x_1^m + A_1 x_1^{m-1} + \cdots + A_m,$$

where $A_0, A_1, \ldots, A_m$ are polynomials in x_2. Replace x_2 by $f_1 - x_1$ and arrange the resulting polynomial in powers of x_1, and f_1 again in powers of x_1 so that

$$F(x_1, x_2) = B_0 x_1^l + B_1 x_1^{l-1} + \cdots + B_l,$$

where now $B_0, B_1, \ldots, B_l$ are polynomials in f_1. Introducing a new variable t in place of x_1, divide

$$\phi(t) = B_0 t^l + B_1 t^{l-1} + \cdots + B_l$$

by

$$f(t) = t^2 - f_1 t + f_2.$$

Then, identically in t, we see that

$$\phi(t) = f(t)Q(t) + Ct + D,$$

where C and D are polynomials in f_1, f_2 with coefficients built up as stated in the theorem. In this identity set $t = x_1$; then, since $f(x_1) = 0$ and $\phi(x_1) = F(x_1, x_2)$,

$$F(x_1, x_2) = Cx_1 + D.$$

Exchange now x_1 and x_2. The left-hand side does not change and neither do C and D; hence,

$$Cx_1 + D = Cx_2 + D,$$

or

$$C(x_1 - x_2) = 0$$

identically in x_1, x_2, and therefore $C = 0$. Thus,

$$F(x_1, x_2) = D$$

and D is a polynomial in f_1 and f_2, which proves the theorem for two variables. We shall proceed further by induction. Assuming the theorem to be true in the case of symmetric functions of $n - 1$ variables, we shall prove its validity for symmetric functions in n variables. Denoting by ϕ_1, ϕ_2, . . . , ϕ_{n-1} the elementary symmetric functions of the $n - 1$ variables x_2, x_3, . . . , x_n, we obviously have

$$f_1 = x_1 + \phi_1, \qquad f_2 = x_1\phi_1 + \phi_2, \qquad \ldots, \qquad f_{n-1} = x_1\phi_{n-2} + \phi_{n-1},$$

whence conversely

$$\phi_1 = -x_1 + f_1, \qquad \phi_2 = x_1^2 - x_1 f_1 + f_2, \qquad \ldots,$$
$$\phi_{n-1} = (-1)^{n-1}[x_1^{n-1} - x_1^{n-2}f_1 + \cdots + (-1)^{n-1}f_{n-1}].$$

Now let $F(x_1, x_2, \ldots, x_n)$, or simply F, be a symmetric function in n variables. Arranging it in powers of x_1, we can write

$$F = A_0 x_1^m + A_1 x_1^{m-1} + \cdots + A_m$$

and here the coefficients are symmetric functions of x_2, x_3, . . . , x_n. Assuming that the theorem holds in the case of $n - 1$ variables, we can express A_0, A_1, . . . , A_m as polynomials in ϕ_1, ϕ_2, . . . , ϕ_{n-1}. If we substitute for ϕ_1, ϕ_2, . . . , ϕ_{n-1} their expressions through x_1, f_1, f_2, . . . , f_{n-1}, the coefficients A_0, A_1, . . . , A_m will be expressed as polynomials in x_1, f_1, f_2, . . . , f_{n-1}. On substituting these expressions into F and rearranging the resulting expression again in powers of x_1, we can write

$$F = B_0 x_1^l + B_1 x_1^{l-1} + \cdots + B_l$$

where B_0, B_1, . . . , B_l are polynomials in f_1, f_2, . . . , f_{n-1}. Introducing a new variable t, write

$$\phi(t) = B_0 t^l + B_1 t^{l-1} + \cdots + B_l,$$

and divide $\phi(t)$ by

$$f(t) = t^n - f_1 t^{n-1} + f_2 t^{n-2} - \cdots + (-1)^n f_n.$$

The remainder will be of degree not higher than $n - 1$, and we shall have the identity in t

$$\phi(t) = f(t)Q(t) + C_0 t^{n-1} + C_1 t^{n-2} + \cdots + C_{n-1}$$

where $C_0, C_1, \ldots, C_{n-1}$ are polynomials in $f_1, f_2, \ldots, f_n$ with coefficients built up as stated in the theorem. Now set in this identity $t = x_1$. Noticing that $f(x_1) = 0$, we obtain

$$F(x_1, x_2, \ldots, x_n) = C_0 x_1^{n-1} + C_1 x_1^{n-2} + \cdots + C_{n-1}.$$

Here the right-hand side does not change by interchanging x_1 with $x_2, x_3, \ldots, x_n$; neither do the coefficients $C_0, C_1, \ldots, C_{n-1}$ change so that the following identities in $x_1, x_2, \ldots, x_n$ hold:

$$C_0 x_1^{n-1} + C_1 x_1^{n-2} + \cdots + C_{n-1} - F = 0,$$
$$C_0 x_2^{n-1} + C_1 x_2^{n-2} + \cdots + C_{n-1} - F = 0,$$
$$\cdots \cdots \cdots \cdots \cdots \cdots \cdots \cdots$$
$$C_0 x_n^{n-1} + C_1 x_n^{n-2} + \cdots + C_{n-1} - F = 0.$$

This means that

$$C_0 t^{n-1} + C_1 t^{n-1} + \cdots + C_{n-1} - F$$

vanishes for $t = x_1, x_2, \ldots, x_n$, and this is possible only if

$$C_0 = C_1 = \cdots = C_{n-2} = 0, \qquad C_{n-1} - F = 0$$

so that

$$F = C_{n-1}.$$

But C_{n-1} is a polynomial in the elementary symmetric functions $f_1, f_2, \ldots, f_n$, which proves the theorem. The proof will be better understood by considering a particular example.

Example. Express
$$F = (x_1 + x_2)(x_1 + x_3)(x_2 + x_3)$$
in terms of
$$f_1 = x_1 + x_2 + x_3, \qquad f_2 = x_1 x_2 + x_1 x_3 + x_2 x_3, \qquad f_3 = x_1 x_2 x_3.$$
Arranging in powers of x_1, we have
$$F = (x_2 + x_3)x_1^2 + (x_2 + x_3)^2 x_1 + x_2 x_3(x_2 + x_3).$$
Substitute here
$$x_2 + x_3 = f_1 - x_1, \qquad x_2 x_3 = x_1^2 - f_1 x_1 + f_2$$
and arrange again in powers of x_1:
$$F = -x_1^3 + f_1 x_1^2 - f_2 x_1 + f_1 f_2.$$
Set
$$\phi(t) = -t^3 + f_1 t^2 - f_2 t + f_1 f_2$$
and divide by
$$f(t) = t^3 - f_1 t^2 + f_2 t - f_3.$$
The remainder will be
$$f_1 f_2 - f_3$$
and therefore
$$F = f_1 f_2 - f_3. \star$$

Problems

Express in terms of the elementary symmetric functions:

1. $x_1^2(x_2 + x_3) + x_2^2(x_1 + x_3) + x_3^2(x_1 + x_2)$.

2. $(x_1 - x_2)^2(x_1 - x_3)^2 + (x_1 - x_2)^2(x_2 - x_3)^2 + (x_1 - x_3)^2(x_2 - x_3)^2$.

3. $\Sigma x_1^4 x_2 x_3$ for $n = 3$.   **4.** $\Sigma x_1^2 x_2^2$ for $n = 4$.

4. Practical Methods. The proof of the fundamental theorem supplies also a method for the effective representation of a symmetric function as a polynomial in the elementary symmetric functions. In practice, however, it is more convenient to use for this purpose other more expedient methods two of which we shall briefly mention. Let us denote by the symbol

$$Sx_1^{\alpha_1} x_2^{\alpha_2} \cdots x_m^{\alpha_m}$$

the sum of all terms that are obtained from a typical term

$$x_1^{\alpha_1} x_2^{\alpha_2} \cdots x_m^{\alpha_m}$$

by substituting for the indices $1, 2, \ldots, m$ all possible arrangements of m numbers taken out of $1, 2, \ldots, n$. If among exponents α_1, $\alpha_2, \ldots, \alpha_m$ no two are equal, the new symbol, which may be called an *S-function*, coincides with the previously introduced sigma function with the same typical term. If, however, some of the exponents α_1, $\alpha_2, \ldots, \alpha_m$ are equal, then the new symbol contains a repetition of terms that in the sigma function occur only once. If among the exponents $\alpha_1, \alpha_2, \ldots, \alpha_m$ there are groups of $\lambda, \mu, \ldots, \sigma$ equal numbers, it is easy to see that each term of the sigma function occurs in the S-function $\lambda! \mu! \cdots \sigma!$ times so that

$$Sx_1^{\alpha_1} x_2^{\alpha_2} \cdots x_m^{\alpha_m} = \lambda! \mu! \cdots \sigma! \Sigma x_1^{\alpha_1} x_2^{\alpha_2} \cdots x_m^{\alpha_m}.$$

For instance,

$$\Sigma x_1^{\alpha} x_2^{\alpha} = \tfrac{1}{2} S x_1^{\alpha} x_2^{\alpha}, \qquad \Sigma x_1^{\alpha} x_2^{\alpha} x_3^{\beta} = \tfrac{1}{2} S x_1^{\alpha} x_2^{\alpha} x_3^{\beta},$$

provided $\beta \neq \alpha$. Also,

$$\Sigma x_1^{\alpha} x_2^{\alpha} x_3^{\alpha} = \tfrac{1}{6} S x_1^{\alpha} x_2^{\alpha} x_3^{\alpha}, \qquad \Sigma x_1^{\alpha} x_2^{\alpha} x_3^{\beta} x_4^{\beta} = \tfrac{1}{4} S x_1^{\alpha} x_2^{\alpha} x_3^{\beta} x_4^{\beta},$$

if $\beta \neq \alpha$. We have in all cases

$$Sx_1^{\alpha} \cdot Sx_1^{\beta} = Sx_1^{\alpha+\beta} + Sx_1^{\alpha} x_2^{\beta},$$

whence

$$Sx_1^{\alpha} x_2^{\beta} = s_{\alpha} s_{\beta} - s_{\alpha+\beta}$$

since in general

$$Sx_1^{m} = \Sigma x_1^{m} = s_m.$$

Thus, it is possible to express through sums of powers all double sigmas

$$\Sigma x_1^\alpha x_2^\beta.$$

To express similarly triple sigmas

$$\Sigma x_1^\alpha x_2^\beta x_3^\gamma$$

notice that under all circumstances

$$S x_1^\alpha x_2^\beta \cdot S x_1^\gamma = S x_1^{\alpha+\gamma} x_2^\beta + S x_1^\alpha x_2^{\beta+\gamma} + S x_1^\alpha x_2^\beta x_3^\gamma,$$

whence

$$S x_1^\alpha x_2^\beta x_3^\gamma = s_\alpha s_{\beta+\gamma} + s_\beta s_{\alpha+\gamma} + s_\gamma s_{\alpha+\beta} - 2 s_{\alpha+\beta+\gamma} - s_\alpha s_\beta s_\gamma.$$

The same method can be used to express through sums of powers quadruple sigmas, quintuple sigmas, etc. Since sums of powers can be expressed through the elementary symmetric functions, all sigmas ultimately can be expressed through these functions.

Example 1. Consider the symmetric function

$$F = (x_1 + x_2 - x_3 - x_4)(x_1 + x_3 - x_2 - x_4)(x_1 + x_4 - x_2 - x_3)$$
$$= \Sigma x_1^3 - \Sigma x_1^2 x_2 + 2\Sigma x_1 x_2 x_3.$$

Here

$$\Sigma x_1^2 x_2 = S x_1^2 x_2 = s_1 s_2 - s_3, \qquad \Sigma x_1^3 = s_3.$$

Denoting by

$$-p = \Sigma x_1, \qquad q = \Sigma x_1 x_2, \qquad -r = \Sigma x_1 x_2 x_3, \qquad s = x_1 x_2 x_3 x_4,$$

the elementary symmetric functions, we have

$$s_1 = -p, \qquad s_2 = p^2 - 2q, \qquad s_3 = -p^3 + 3pq - 3r$$

and so

$$F = -p^3 + 4pq - 8r.$$

Example 2. Let ω be an imaginary cube root of unity and

$$F = (x_1 + \omega x_2 + \omega^2 x_3)(x_1 + \omega^2 x_2 + \omega x_3)$$
$$= \Sigma x_1^2 - \Sigma x_1 x_2.$$

Keeping former notations,

$$\Sigma x_1^2 = p^2 - 2q, \qquad \Sigma x_1 x_2 = q$$

and so

$$F = p^2 - 3q.$$

Example 3. Let

$$F = (x_1 + \omega x_2 + \omega^2 x_3)^3 + (x_1 + \omega^2 x_2 + \omega x_3)^3.$$

By direct computation

$$F = 2\Sigma x_1^3 - 3\Sigma x_1^2 x_2 + 12 x_1 x_2 x_3.$$

On substituting here

$$\Sigma x_1^3 = -p^3 + 3pq - 3r, \qquad \Sigma x_1^2 x_2 = s_1 s_2 - s_3 = -pq + 3r, \qquad x_1 x_2 x_3 = -r,$$

we find

$$F = -2p^3 + 9pq - 27r.$$

The second method for the convenient representation of symmetric functions through the elementary symmetric functions will be explained by examples.

Example 4. Let

$$F = (x_1 + x_2 - x_3 - x_4)^2 + (x_1 + x_3 - x_2 - x_4)^2 + (x_1 + x_4 - x_2 - x_3)^2.$$

This polynomial is homogeneous of dimension 2, so that replacing x_1, x_2, x_3, x_4 by kx_1, kx_2, kx_3, kx_4, k being arbitrary, F becomes k^2F. If now F be expressed through the elementary symmetric functions, for which we keep the notations $-p$, q, $-r$, s, we notice that it will consist only of monomials of the form

$$p^\alpha q^\beta r^\gamma s^\delta$$

multiplied by constant factors. On replacing x_1, x_2, x_3, x_4 by kx_1, kx_2, kx_3, kx_4 it is clear that p, q, r, s are replaced by kp, k^2q, k^3r, k^4s, and the above monomial becomes

$$k^{\alpha+2\beta+3\gamma+4\delta}p^\alpha q^\beta r^\gamma s^\delta,$$

and since F is simply multiplied by k^2, the sum $\alpha + 2\beta + 3\gamma + 4\delta$ in all terms must have the same value 2. Thus,

$$\alpha + 2\beta + 3\gamma + 4\delta = 2.$$

Moreover, since 2 is the highest power of x_1 occurring in F, it follows from Sec. 7 below that

$$\alpha + \beta + \gamma + \delta \leq 2.$$

Both conditions on α, β, γ, δ are satisfied only for

$$\alpha = 0, \quad \beta = 1, \quad \gamma = 0, \quad \delta = 0,$$
$$\alpha = 2, \quad \beta = 0, \quad \gamma = 0, \quad \delta = 0.$$

The expression of F through p, q, r, s is therefore of the form

$$F = Ap^2 + Bq.$$

The unknown coefficients are determined by taking particular values for x_1, x_2, x_3, x_4 and equating the values of F found directly, with those resulting from the substitution of particular values of p and q. The two independent relations between A and B obtained in this manner will suffice to determine them. Let x_1, x_2, x_3, x_4 be the roots of the equation

$$x^4 - x^3 = 0$$

so that

$$x_1 = 1, \quad x_2 = x_3 = x_4 = 0.$$

Then,

$$F = 3, \quad p = 1, \quad q = 0,$$

and so $A = 3$. Now let x_1, x_2, x_3, x_4 be the roots of the equation

$$x^4 - 2x^3 + x^2 = 0$$

so that

$$x_1 = x_2 = 1, \quad x_3 = x_4 = 0.$$

Then, $F = 4$, $p = -2$, $q = 1$ and

$$4A - B = 4,$$

whence
$$B = 8,$$
and so
$$F = 3p^2 - 8q.$$

Example 5. Let
$$F = (x_1 + x_2 - x_3 - x_4)^2(x_1 + x_3 - x_2 - x_4)^2$$
$$+ (x_1 + x_2 - x_3 - x_4)^2(x_1 + x_4 - x_2 - x_3)^2$$
$$+ (x_1 + x_3 - x_2 - x_4)^2(x_1 + x_4 - x_2 - x_3)^2.$$

This polynomial is homogeneous of dimension 4; moreover, the exponent of the highest power of x_1 occurring in F is 4. For the same reasons as in the preceding example, for all monomials
$$p^\alpha q^\beta r^\gamma s^\delta$$
that enter into the expression of F we must have
$$\alpha + 2\beta + 3\gamma + 4\delta = 4,$$
$$\alpha + \beta + \gamma + \delta \leqq 4,$$
and the only numbers satisfying these requirements are

α	β	γ	δ
0	0	0	1
1	0	1	0
2	1	0	0
0	2	0	0
4	0	0	0

and so the literal form of F is
$$F = As + Bpr + Cp^2q + Dq^2 + Ep^4.$$

A comparison of the coefficients of x_1^4 in both members gives $E = 3$. Further, consider the particular equations

$x^4 - 1 = 0$. Roots $x_1 = 1, x_2 = -1, x_3 = i, x_4 = -i$;
$$p = q = r = 0, s = -1.$$

$(x^2 - 1)^2 = 0$. Roots $x_1 = 1, x_2 = -1, x_3 = 1, x_4 = -1$;
$$p = 0, q = -2, r = 0, s = 1.$$

$x(x - 1)(x^2 - 1) = 0$. Roots $x_1 = 1, x_2 = 1, x_3 = -1, x_4 = 0$;
$$p = -1, q = -1, r = 1, s = 0.$$

$(x + 1)^4 = 0$. Roots $x_1 = -1, \ x_2 = -1, \ x_3 = -1, \ x_4 = -1$;
$$p = 4, q = 6, r = 4, s = 1.$$

For these equations direct substitution gives, respectively,
$$F = 64, \quad F = 0, \quad F = 19, \quad F = 0.$$

On the other hand, taking into account the corresponding values of p, q, r, s, we have
$$-A = 64, \quad A + 4D = 0, \quad -B - C + D + E = 19,$$
$$A + 16B + 96C + 36D + 256E = 0,$$

which gives

$$A = -64, \quad D = 16, \quad B = 16, \quad C = -16, \quad E = 3.$$

The final result is therefore

$$F = 3p^4 - 16p^2q + 16pr + 16q^2 - 64s.$$

Problems

Express through the coefficients the following symmetric functions of the roots:

1. $\Sigma x_1^2 x_2$.

2. $\Sigma x_1^2 x_2^2$.

3. $\Sigma x_1^3 x_2$.

4. $\Sigma x_1^3 x_2^2$.

5. $\Sigma x_1^2 x_2 x_3$.

6. $\Sigma x_1^2 x_2^2 x_3$.

7. $\Sigma x_1^3 x_2^2 x_3$.

8. $\Sigma x_1^2 x_2^2 x_3 x_4$.

9. $(x_1 - x_2)^2 x_3 + (x_1 - x_3)^2 x_2 + (x_2 - x_3)^2 x_1$.

10. $(x_1 + x_2)^2 x_3^2 + (x_1 + x_3)^2 x_2^2 + (x_2 + x_3)^2 x_1^2$.

11. $(x_1 x_2 + x_3 x_4)(x_1 x_3 + x_2 x_4)(x_1 x_4 + x_2 x_3)$.

12. If roots of the equation $x^3 - 3x + 1 = 0$ be denoted by α, β, γ, find the equation whose roots are $\alpha + \alpha^{-1}$, $\beta + \beta^{-1}$, $\gamma + \gamma^{-1}$.

13. With the same notations and for the same equation, find the equation whose roots are $\alpha^2 + \beta^2$, $\alpha^2 + \gamma^2$, $\beta^2 + \gamma^2$.

14. Form the equation with roots $\alpha\beta$, $\alpha\gamma$, $\beta\gamma$ for the general cubic.

15. If α, β, γ are roots of $x^3 - x + 1 = 0$, form the equation with roots

$$\alpha(\beta - \gamma)^2, \qquad \beta(\alpha - \gamma)^2, \qquad \gamma(\alpha - \beta)^2.$$

*16. If

$$D_n = \Sigma \alpha^{n-1}\beta^{n-1}(\alpha - \beta)^2(\alpha + \beta)$$
$$E_n = \Sigma \alpha^{n-1}\beta^{n-1}(\alpha - \beta)^2,$$

show that

$$D_n = S_{n-1}S_{n+2} - S_n S_{n+1}, \qquad E_n = S_{n+1}S_{n-1} - S_n^2.$$

Hence, if α and β have greater moduli than the other roots of an equation,

$$\alpha + \beta = \lim \frac{D_n}{E_n}, \qquad \text{as } n \to \infty;$$

and

$$\alpha\beta = \lim \frac{E_{n+1}}{E_n}, \qquad \text{as } n \to \infty;$$

and for large n we have approximately

$$\alpha + \beta = \frac{D_n}{E_n}, \qquad \alpha\beta = \frac{E_{n+1}}{E_n}.$$

By this method evaluate approximately the two largest roots of the equation

$$(a) \quad x^3 - 5x^2 + 6x - 1 = 0.$$

Also, compute approximately the complex roots of the equation

$$(b) \quad x^3 + 6x^2 + 10x - 1 = 0.$$

17. Prove that rational symmetric functions of the type

$$\Sigma \frac{1}{x_1 - a} = \frac{1}{x_1 - a} + \frac{1}{x_2 - a} + \cdots + \frac{1}{x_n - a}$$

are expressible through the coefficients of the polynomial $f(x) = (x - x_1)(x - x_2)$ $\cdots (x - x_n)$ as follows:

$$\Sigma \frac{1}{x_1 - a} = -\frac{f'(a)}{f(a)}.$$

18. If a rational function $R(x)$ is split into simple elements:

$$R(x) = E(x) + \frac{A}{x - l} + \frac{B}{x - m} + \frac{C}{x - n} + \cdots$$

the rational symmetric function of the type

$$\Sigma R(x_1) = R(x_1) + R(x_2) + \cdots + R(x_n)$$

is expressible through the coefficients of $f(x) = (x - x_1)(x - x_2) \cdots (x - x_n)$ as follows:

$$\Sigma R(x_1) = \Sigma E(x_1) - A\frac{f'(l)}{f(l)} - B\frac{f'(m)}{f(m)} - C\frac{f'(n)}{f(n)} - \cdots,$$

and $\Sigma E(x_1)$, since $E(x)$ is a polynomial, can be expressed through $s_1, s_2, s_3, \ldots$

The letters α, β, γ in the following problems denote roots of a cubic equation specified in each problem. Calculate the symmetric functions:

19. $\Sigma \dfrac{\alpha}{\beta + \gamma} = \dfrac{\alpha}{\beta + \gamma} + \dfrac{\beta}{\alpha + \gamma} + \dfrac{\gamma}{\alpha + \beta}$ for $x^3 - 3x + 1 = 0$.

20. $\Sigma \dfrac{\alpha^2}{\beta + \gamma} = \dfrac{\alpha^2}{\beta + \gamma} + \dfrac{\beta^2}{\alpha + \gamma} + \dfrac{\gamma^2}{\alpha + \beta}$ for $x^3 + x^2 - 2x - 1 = 0$.

21. $\Sigma \dfrac{(\beta + \gamma)^2}{\alpha}$ for $x^3 - 3x^2 + 6x - 1 = 0$.

22. $\Sigma \dfrac{\alpha^2}{\beta^2 + \gamma^2}$ for $x^3 + x^2 - 4x + 2 = 0$.

23. $\Sigma \dfrac{(\alpha - \beta)^2}{(\alpha + \beta)^2}$ for $2x^3 - 2x - 1 = 0$.

24. $\Sigma \dfrac{\alpha^2 + \beta^2}{\alpha + \beta}$ for $2x^3 - 6x^2 + 1 = 0$.

25. $\Sigma \dfrac{\alpha^2 + \beta\gamma}{\beta + \gamma}$ for $x^3 + x^2 - 1 = 0$.

26. $\Sigma \dfrac{\alpha}{(\beta - \gamma)^2}$ for $x^3 - 1 = 0$.

5. Lagrange's Solution of Cubic Equations.

The algebraic solution of cubic and biquadratic equations as presented in Chap. V, although based on most elementary considerations, leaves the impression of a success achieved by the employment of ingenious artifices the reason of which is not clear. As in other cases, it is necessary to look at a problem from a higher standpoint in order to understand why an algebraic solution is possible for the cubic and the biquadratic equations. And not only that: the same principles properly generalized and subjected to a more profound examination show the reason why an algebraic solution is generally impossible for equations of degree higher than 4.

Consider at first a rational integral function in three variables x_1, x_2, x_3. If they are permuted in six possible manners, the function, in general, will acquire six distinct values. In particular cases it may happen, however, that the number of distinct values will be less than six, and then it will be either one (for symmetric functions) or two or three. Lagrange had shown that it is possible to find a linear function whose cube has only two different values.

Let ω be an imaginary cube root of unity and consider the linear function

$$x_1 + \omega x_2 + \omega^2 x_3.$$

To every even permutation of the indices 123, 231, 312 there correspond three values of this function:

$$y_1 = x_1 + \omega x_2 + \omega^2 x_3, \qquad y_2 = x_2 + \omega x_3 + \omega^2 x_1, \qquad y_3 = x_3 + \omega x_1 + \omega^2 x_2,$$

and to every odd permutation 132, 213, 321 there correspond three more:

$$y_4 = x_1 + \omega x_3 + \omega^2 x_2, \qquad y_5 = x_2 + \omega x_1 + \omega^2 x_3, \qquad y_6 = x_3 + \omega x_2 + \omega^2 x_1.$$

Notice that

$$y_2 = \omega^2 y_1, \qquad y_3 = \omega y_1, \qquad y_5 = \omega y_4, \qquad y_6 = \omega^2 y_4,$$

so that

$$y_1^3 = y_2^3 = y_3^3, \qquad y_4^3 = y_5^3 = y_6^3.$$

Hence,

$$(x_1 + \omega x_2 + \omega^2 x_3)^3$$

has only two distinct values

$$t_1 = (x_1 + \omega x_2 + \omega^2 x_3)^3, \qquad t_2 = (x_1 + \omega^2 x_2 + \omega x_3)^3,$$

and the combinations $t_1 + t_2$, $t_1 t_2$ are symmetric functions of x_1, x_2, x_3.

Supposing that x_1, x_2, x_3 are the roots of a cubic equation

$$x^3 + px^2 + qx + r = 0,$$

it was found (see Examples 2 and 3 in Sec. 4) that

$$t_1 + t_2 = -2p^3 + 9pq - 27r,$$
$$t_1 t_2 = (p^2 - 3q)^3.$$

Consequently, t_1 and t_2 are the roots of the quadratic equation

$$t^2 + (2p^3 - 9pq + 27r)t + (p^3 - 3q)^3 = 0$$

and can be found algebraically. Having found t_1 and t_2, on extracting cube roots, we get

$$x_1 + \omega x_2 + \omega^2 x_3 = \sqrt[3]{t_1}, \qquad x_1 + \omega^2 x_2 + \omega x_3 = \sqrt[3]{t_2},$$

and, in addition, we have the third equation

$$x_1 + x_2 + x_3 = -p.$$

It suffices to solve these equations in order to obtain the roots

$$x_1 = \tfrac{1}{3}(-p + \sqrt[3]{t_1} + \sqrt[3]{t_2}), \qquad x_2 = \tfrac{1}{3}(-p + \omega^2\sqrt[3]{t_1} + \omega\sqrt[3]{t_2}),$$
$$x_3 = \tfrac{1}{3}(-p + \omega\sqrt[3]{t_1} + \omega^2\sqrt[3]{t_2})$$

of the cubic equation. Notice that between the cube roots there exists the relation (Example 2, Sec. 4)

$$\sqrt[3]{t_1} \cdot \sqrt[3]{t_2} = p^2 - 3q.$$

6. Lagrange's Solution of Biquadratic Equation. The success in solving biquadratic equations algebraically depends on the fact that there are functions of the four variables that acquire only three distinct values on permuting these variables in the 24 possible ways. A function of this type is

$$(x_1 + x_2 - x_3 - x_4)^2,$$

which has only three distinct values:

$$\theta_1 = (x_1 + x_2 - x_3 - x_4)^2, \qquad \theta_2 = (x_1 + x_3 - x_2 - x_4)^2,$$
$$\theta_3 = (x_1 + x_4 - x_2 - x_3)^2$$

the symmetric combinations of which

$$\theta_1 + \theta_2 + \theta_3, \qquad \theta_1\theta_2 + \theta_1\theta_3 + \theta_2\theta_3, \qquad \theta_1\theta_2\theta_3$$

have been examined in Examples 1, 4, and 5 in Sec. 4. Through the coefficients of the equation

$$x^4 + px^3 + qx^2 + rx + s = 0$$

with roots x_1, x_2, x_3, x_4 these combinations are expressed as follows:

$$\theta_1 + \theta_2 + \theta_3 = 3p^2 - 8q,$$
$$\theta_1\theta_2 + \theta_1\theta_3 + \theta_2\theta_3 = 3p^4 - 16p^2q + 16pr + 16q^2 - 64s,$$
$$\theta_1\theta_2\theta_3 = (p^3 - 4pq + 8r)^2.$$

Thus, θ_1, θ_2, θ_3 are the roots of the cubic resolvent

$$\theta^3 - (3p^2 - 8q)\theta^2 + (3p^4 - 16p^2q + 16pr + 16q^2 - 64s)\theta$$
$$- (p^3 - 4pq + 8r)^2 = 0,$$

and θ_1, θ_2, θ_3 can be found by solving this equation. Once θ_1, θ_2, θ_3 are found, we may determine the roots x_1, x_2, x_3, x_4 by means of the linear equations

$$x_1 + x_2 + x_3 + x_4 = -p,$$
$$x_1 + x_2 - x_3 - x_4 = \sqrt{\theta_1},$$
$$x_1 - x_2 + x_3 - x_4 = \sqrt{\theta_2},$$
$$x_1 - x_2 - x_3 + x_4 = \sqrt{\theta_3},$$

whence

$$x_1 = \frac{-p + \sqrt{\theta_1} + \sqrt{\theta_2} + \sqrt{\theta_3}}{4}, \qquad x_2 = \frac{-p + \sqrt{\theta_1} - \sqrt{\theta_2} - \sqrt{\theta_3}}{4},$$

$$x_3 = \frac{-p - \sqrt{\theta_1} + \sqrt{\theta_2} - \sqrt{\theta_3}}{4}, \qquad x_4 = \frac{-p - \sqrt{\theta_1} - \sqrt{\theta_2} + \sqrt{\theta_3}}{4}.$$

Notice that the square roots $\sqrt{\theta_1}$, $\sqrt{\theta_2}$, $\sqrt{\theta_3}$ are not independent; between them there exists the relation (see Example 1, Sec. 4)

$$\sqrt{\theta_1}\sqrt{\theta_2}\sqrt{\theta_3} = -p + 4pq - 8r.$$

★7. The Gaussian Principle. For certain theoretical investigations the following proposition is of importance:

A nonidentically vanishing polynomial in the variables $f_1, f_2, \ldots, f_n$ cannot vanish identically in the variables $x_1, x_2, \ldots, x_n$ after replacing $f_1, f_2, \ldots, f_n$ by the elementary symmetric functions of $x_1, x_2, \ldots, x_n$.

To prove this proposition notice that the polynomial in question consists of terms of the form

$$A f_1^{\alpha_1} f_2^{\alpha_2} \cdots f_n^{\alpha_n}$$

where $A \neq 0$ and the exponents $\alpha_1, \alpha_2, \ldots, \alpha_n$ are nonnegative integers. Among these terms select that term in which the sums of the exponents

$$\alpha_1 + \alpha_2 + \cdots + \alpha_n, \qquad \alpha_2 + \alpha_3 + \cdots + \alpha_n,$$
$$\alpha_3 + \cdots + \alpha_n, \qquad \ldots, \qquad \alpha_{n-1} + \alpha_n, \qquad \alpha_n$$

have the greatest possible values. Such a term is unique. For if there were some other terms

$$B f_1^{\beta_1} f_2^{\beta_2} \cdots f_n^{\beta_n}$$

satisfying the same conditions, as to the exponents $\beta_1, \beta_2, \ldots, \beta_n$, we should have

$$\beta_n = \alpha_n, \qquad \beta_{n-1} + \beta_n = \alpha_{n-1} + \alpha_n, \qquad \ldots,$$
$$\beta_1 + \beta_2 + \cdots + \beta_n = \alpha_1 + \alpha_2 + \cdots + \alpha_n,$$

that is,

$$\beta_n = \alpha_n, \qquad \beta_{n-1} = \alpha_{n-1}, \qquad \ldots, \qquad \beta_1 = \alpha_1,$$

which is impossible since a polynomial is always written in such a way that similar monomials are combined.

To illustrate what has been said by an example, consider the polynomial

$$2f_1^6 f_2^3 - f_1^5 f_2^2 f_3^2 + 5f_1^4 f_2^2 f_3^2 f_4 - 7f_1^7 f_2 + f_1^3 f_2^3 f_3 f_4.$$

The sums of the exponents for the terms as they are written are

$$9, 3, 0, 0; \quad 9, 4, 2, 0; \quad 9, 5, 3, 1; \quad 8, 1, 0, 0; \quad 8, 5, 2, 1.$$

The unique term for which all these sums are the greatest is the term

$$5f_1^4 f_2^2 f_3^2 f_4.$$

In the unique term

$$A f_1^{\alpha_1} f_2^{\alpha_2} \cdots f_n^{\alpha_n}$$

thus selected replace $f_1, f_2, \ldots, f_n$ by the elementary symmetric functions of $x_1, x_2, \ldots, x_n$:

$$f_1 = x_1 + x_2 + \cdots, \quad f_2 = x_1 x_2 + x_2 x_3 + \cdots, \quad \ldots,$$
$$f_n = x_1 x_2 \cdots x_n$$

and expand. The expansion will contain the term

$$A x_1^{\alpha_1 + \alpha_2 + \cdots + \alpha_n} x_2^{\alpha_2 + \cdots + \alpha_n} \cdots x_n^{\alpha_n},$$

which cannot be canceled either by the other terms arising from the expansion of

$$A f_1^{\alpha_1} f_2^{\alpha_2} \cdots f_n^{\alpha_n},$$

or from the expansion of any other term

$$B f_1^{\beta_1} f_2^{\beta_2} \cdots f_n^{\beta_n}$$

that may be present in the polynomial. The fact that after replacing $f_1, f_2, \ldots, f_n$ by the elementary symmetric functions and expansion there is a term that cannot be canceled by the other terms proves the proposition. From this proposition two conclusions can be derived. First, *that a symmetric function of $x_1, x_2, \ldots, x_n$ can be represented as a polynominal in the elementary symmetric functions in one way only.* For, if there are two such representations, there would be a nonidentically vanishing polynomial in $f_1, f_2, \ldots, f_n$, which after the replacement of $f_1, f_2, \ldots, f_n$ by elementary symmetric functions and expanding, would become an identically vanishing polynomial in $x_1, x_2, \ldots, x_n$, an impossibility. Secondly, *the polynomial in the elementary symmetric functions that represents a symmetric polynomial is of degree equal to the highest power of x_1 occurring in that symmetric polynominal.* Hence, if r is the highest exponent of x_1 occurring in a symmetric function ϕ, and if the elementary symmetric functions are represented by

$$f_1 = -\frac{a_1}{a_0}, \quad f_2 = \frac{a_2}{a_0}, \quad \ldots, \quad f_n = (-1)^n \frac{a_n}{a_0}$$

then,

$$a_0^r \phi$$

can be expressed as a homogeneous polynomial in $a_0, a_1, a_2, \ldots, a_n$ of dimension r.★

CHAPTER XII

ELIMINATION

Resultant and Discriminants

1. Example of Elimination. It is proposed to solve simultaneously

$$x^2 + y^2 - 1 = 0, \qquad x^3 + y^3 - 1 = 0,$$

that is, to find such pairs of numbers x, y that satisfy both equations. To solve this problem we may seek to eliminate one of the unknowns, say x. To this end, solve the first equation for x^2:

$$x^2 = -(y^2 - 1);$$

then multiply both sides by x, which gives

$$x^3 = -x(y^2 - 1).$$

On the other hand, the value of x^3 found from the second equation is

$$x^3 = -(y^3 - 1)$$

and on equating both expressions for x^3 we obtain

$$x(y^2 - 1) = y^3 - 1.$$

Squaring this and replacing x^2 again by $-(y^2 - 1)$, we have

$$(y^2 - 1)^3 + (y^3 - 1)^2 = 0 \tag{1}$$

as a necessary condition that y must satisfy in order that the proposed equations may have solutions. But this condition is also sufficient; that is, if we take for y any root of (1), which is of the sixth degree, it is possible to satisfy the proposed equations by the same value of x.

The original equations are entirely equivalent to the equations

$$x^2 = -(y^2 - 1), \qquad x(y^2 - 1) = y^3 - 1.$$

Now, if $y^2 - 1 = 0$, it follows from (1) that $y^3 - 1 = 0$, and then the second equation is satisfied for no matter what value of x, and the first gives $x = 0$. If $y^2 - 1 \neq 0$, then,

$$x = \frac{y^3 - 1}{y^2 - 1},$$

and

$$x^2 = \frac{(y^3 - 1)^2}{(y^2 - 1)^2} = -(y^2 - 1)$$

by virtue of condition (1).

One may attempt to eliminate x in a different manner. Take the equation already obtained,

$$x(y^2 - 1) = y^3 - 1,$$

cube both members, and replace x^3 by $-(y^3 - 1)$; then, one has

$$(y^3 - 1)[(y^3 - 1)^2 + (y^2 - 1)^3] = 0$$

as a necessary consequence of the proposed equations. But this equation for y is satisfied by such values of y, on account of the factor $y^3 - 1$, for which the equations

$$x^2 + y^2 - 1 = 0, \qquad x^3 + y^3 - 1 = 0$$

have no common root x. In fact, if we take $y = \omega$, ω being an imaginary cube root of unity, then $y^3 - 1 = 0$ and from the second equation it follows that $x = 0$, whereas for $x = 0$

$$x^2 + y^2 - 1 = \omega^2 - 1 \neq 0.$$

This example shows that in carrying out the process of elimination one must proceed with great caution to avoid introducing extraneous factors, like $y^3 - 1$ in our case. To effect the elimination correctly without danger of introducing extraneous factors, the most natural way that presents itself is the following: For a given y the equation

$$x^2 + y^2 - 1 = 0$$

has roots x_1 and x_2. For this given value of y, suppose that one root satisfies also the equation

$$x^3 + y^3 - 1 = 0;$$

then,

$$(x_1^3 + y^3 - 1)(x_2^3 + y^3 - 1) = 0. \tag{2}$$

Conversely, if this condition is satisfied, then one of the factors is zero, say

$$x_1^3 + y^3 - 1 = 0,$$

and so the equations

$$x^2 + y^2 - 1 = 0, \qquad x^3 + y^3 - 1 = 0$$

have a common root $x = x_1$. Expanding the product in the left side of (2), we have

$$x_1^3 x_2^3 + (x_1^3 + x_2^3)(y^3 - 1) + (y^3 - 1)^2 = 0.$$

But

$$x_1 x_2 = y^2 - 1, \qquad x_1 + x_2 = 0, \qquad x_1^3 + x_2^3 = 0,$$

and so

$$(y^2 - 1)^3 + (y^3 - 1)^2 = 0$$

represents the condition necessary and sufficient that must be satisfied by y in order that the two equations

$$x^2 + y^2 - 1 = 0, \qquad x^3 + y^3 - 1 = 0$$

may have a common root x.

2. Resultant. The procedure just employed is quite general. Let

$$f(x) = a_0 x^n + a_1 x^{n-1} + \cdots + a_n = 0,$$
$$g(x) = b_0 x^m + b_1 x^{m-1} + \cdots + b_m = 0$$

be two equations whose coefficients may involve another unknown y. In order to find the necessary and sufficient condition for the existence of a root common to both, denote by $\alpha_1, \alpha_2, \ldots, \alpha_n$ the roots of $f(x) = 0$ supposing $a_0 \neq 0$. If one of these is also a root of the equation $g(x) = 0$, then,

$$g(\alpha_1)g(\alpha_2) \cdots g(\alpha_n) = 0.$$

Conversely, if this condition is satisfied, one of the factors is zero, say $g(\alpha_1) = 0$. Then, α_1 is a root common to both the equations $f(x) = 0$, $g(x) = 0$. The product

$$g(\alpha_1)g(\alpha_2) \cdots g(\alpha_n) = 0$$

is a symmetric function of roots $\alpha_1, \alpha_2, \ldots, \alpha_n$, and the highest power of α_1 that occurs in it is clearly α_1^m. According to the final remark in Chap. XI, Sec. 7,

$$a_0^m g(\alpha_1)g(\alpha_2) \cdots g(\alpha_n)$$

is a homogeneous polynomial of dimension m in the coefficients a_0, $a_1, \ldots, a_n$ and also, which is evident, a homogeneous polynomial of dimension n in the coefficients $b_0, b_1, \ldots, b_n$ of $g(x)$. This polynomial is called the *resultant* of $f(x)$ and $g(x)$ (in this order) and is denoted by $R(f,g)$ so that

$$R(f,g) = a_0^m g(\alpha_1)g(\alpha_2) \cdots g(\alpha_n).$$

The vanishing of the resultant $R(f,g)$ is therefore the necessary and sufficient condition for the two equations $f = 0$, $g = 0$ to have a common root, provided $a_0 \neq 0$. If $b_0 \neq 0$ and $\beta_1, \beta_2, \ldots, \beta_m$ are the roots of the equation $g(x) = 0$, the resultant $R(g,f)$ is

$$R(g,f) = b_0^n f(\beta_1)f(\beta_2) \cdots f(\beta_m),$$

and its vanishing is again the necessary and sufficient condition for the two equations $f = 0$ and $g = 0$ to have a common root.

There is a simple relation between two resultants $R(f,g)$ and $R(g,f)$. Taking into consideration the factorization

$$g(x) = b_0(x - \beta_1)(x - \beta_2) \cdots (x - \beta_m),$$

we can write

$$R(f,g) = a_0^m b_0^n (\alpha_1 - \beta_1)(\alpha_1 - \beta_2) \cdots (\alpha_1 - \beta_m)$$
$$(\alpha_2 - \beta_1)(\alpha_2 - \beta_2) \cdots (\alpha_2 - \beta_m)$$
$$\cdots\cdots\cdots\cdots\cdots\cdots$$
$$(\alpha_n - \beta_1)(\alpha_n - \beta_2) \cdots (\alpha_n - \beta_m).$$

Interchanging the letters α and β and rearranging the factors, we have also

$$R(f,g) = (-1)^{mn} a_0^m b_0^n (\beta_1 - \alpha_1)(\beta_1 - \alpha_2) \cdots (\beta_1 - \alpha_n)$$
$$(\beta_2 - \alpha_1)(\beta_2 - \alpha_2) \cdots (\beta_2 - \alpha_n)$$
$$\cdots\cdots\cdots\cdots\cdots\cdots$$
$$(\beta_m - \alpha_1)(\beta_m - \alpha_2) \cdots (\beta_m - \alpha_n).$$

The right-hand side here, except for the factor $(-1)^{mn}$, is $R(g,f)$ and so

$$R(g,f) = (-1)^{mn} R(f,g).$$

From this relation we may conclude that the vanishing of $R(f,g)$ is the necessary and sufficient condition for the equations $f = 0$ and $g = 0$ to have a root in common, if only not at the same time $a_0 = 0$ and $b_0 = 0$. For with $a_0 = 0$, by hypothesis $b_0 \neq 0$, and since $R(f,g) = 0$ implies $R(g,f) = 0$, it follows that $f = 0$ and $g = 0$ have a common root.

Example 1. To find the resultant of

$$f(x) = a_0 x^2 + a_1 x + a_2, \qquad g(x) = b_0 x^2 + b_1 x + b_2.$$

By definition

$$R(f,g) = a_0^2 (b_0 \alpha_1^2 + b_1 \alpha_1 + b_2)(b_0 \alpha_2^2 + b_1 \alpha_2 + b_2)$$
$$= a_0^2 [b_0^2 \alpha_1^2 \alpha_2^2 + b_0 b_1 (\alpha_1 \alpha_2^2 + \alpha_1^2 \alpha_2) + b_0 b_2 (\alpha_1^2 + \alpha_2^2) + b_1^2 \alpha_1 \alpha_2$$
$$+ b_1 b_2 (\alpha_1 + \alpha_2) + b_2^2].$$

On the other hand,

$$a_0^2 \alpha_1^2 \alpha_2^2 = a_2^2, \qquad a_0^2 \alpha_1 \alpha_2 (\alpha_1 + \alpha_2) = -a_1 a_2, \qquad a_0^2 (\alpha_1^2 + \alpha_2^2) = a_1^2 - 2a_0 a_2,$$
$$a_0^2 \alpha_1 \alpha_2 = a_0 a_2, \qquad a_0^2 (\alpha_1 + \alpha_2) = -a_0 a_1,$$

and therefore

$$R(f,g) = a_2^2 b_0^2 + a_0^2 b_2^2 - 2a_0 a_2 b_0 b_2 - a_0 a_1 b_1 b_2 - a_1 a_2 b_0 b_1 + a_1^2 b_0 b_2 + a_0 a_2 b_1^2.$$

This may be presented in a more elegant form, thus:

$$R(f,g) = (a_0 b_2 - a_2 b_0)^2 - (a_0 b_1 - a_1 b_0)(a_1 b_2 - a_2 b_1).$$

Example 2. To solve the equations

$$x^3 + 2x^2 y + 2y(y - 2)x + y^2 - 4 = 0, \qquad x^2 + 2xy + 2y^2 - 5y + 2 = 0.$$

It is convenient to arrange these equations in powers of y, since y occurs only in the second degree in both. From the equations thus arranged

$$(2x + 1)y^2 + (2x^2 - 4x)y + x^3 - 4 = 0,$$
$$2y^2 + (2x - 5)y + x^2 + 2 = 0,$$

eliminate y by forming their resultant according to the formula in Example 1:

$$R = x^4 + 13x^3 + 56x^2 + 80x = x(x + 4)^2(x + 5).$$

The values $x = 0, -4, -5$ are the only ones for which the equations have a common root y; this root can be found by the method for finding the highest common divisor. Thus, corresponding to $x = 0$ it is found that $y = 2$. Corresponding to $x = -4$, -5, it is found, respectively, that $y = 2$ and $y = 3$. All pairs x, y satisfying the proposed equations are

$$x = 0, \qquad y = 2$$
$$x = -4, \qquad y = 2$$
$$x = -5, \qquad y = 3.$$

Problems

Solve the following systems:

1. $x^2 - xy + y^2 = 1$,
$x^2 + xy - 3y^2 - 2x + 2y = -1$.

2. $3x^2 + 3y^2 - 8xy + 2x + 2y - 2 = 0$,
$xy + x - y + 1 = 0$.

3. $3x^2 + 3y^2 - 8xy + 2x + 2y - 2 = 0$,
$xy + x + y - 1 = 0$.

4. $(y - 1)x^2 + xy + y^2 - 2y = 0$,
$(y - 1)x + y = 0$.

5. $(y - 1)x^2 + 2x - 5y + 3 = 0$,
$x^2y + 9x - 10y = 0$.

6. $(y - 2)x^2 - 2x + 5y - 2 = 0$,
$yx^2 - 5x + 4y = 0$.

7. $(y - 1)x^3 + y(y + 1)x^2 + (3y^2 + y - 2)x + 2y = 0$,
$(y - 1)x^2 + y(y + 1)x + 3y^2 - 1 = 0$.

8. $x^2 + y^2 = 1$,
$x^3 + y^3 = 1$.

9. Eliminate x between $f(x) = 0$ and $y = x^2$. The resultant is

$$F(y) = f(\sqrt{y})f(-\sqrt{y}).$$

Hence, formulate the rule for forming the equation whose roots are the squares of the roots of $f(x) = 0$. Apply this rule to the particular example

$$f(x) = x^5 - 2x^4 + x^3 - x + 1 = 0.$$

10. To find the equation whose roots are the cubes of the roots of $f(x) = 0$ it suffices to eliminate x between $f(x) = 0$ and $y = x^3$. Show that the equation sought is

$$f(\sqrt[3]{y})f(\omega\sqrt[3]{y})f(\omega^2\sqrt[3]{y}) = 0,$$

ω being the imaginary cube root of unity. Apply this method to the example

$$f(x) = x^3 - 3x + 1 = 0.$$

11. From the definition of the resultant show that

$$a_0^{m-\mu}R(f, g + \lambda f) = R(f, l)$$

for an arbitrary polynomial λ, μ being the degree of $g + \lambda f$. Combining this property with the identity

$$R(f,g) = (-1)^{mn}R(g,f)$$

established in the text, show how the resultant can be computed by the process used to find the highest common divisor of two polynomials. Apply this method to the examples:

(a) $f(x) = x^4 - x^3 + 2x^2 - 3x + 1$, $g(x) = x^3 - 2x^2 + 3x - 1$.

(b) $f(x) = x^3 + 3yx^2 + (3y^2 - y + 1)x + y^3 - y^2 + 2y$,

$$g(x) = x^2 + 2yx + y^2 - y.$$

12. Show that

$$R(f, gg_1) = R(f, g)R(f, g_1).$$

3. Sylvester's Determinant. The resultant of two polynomials can be presented in a determinant form following an elegant method of elimination proposed by Sylvester. If two equations $f = 0$ and $g = 0$ have a common root, the polynomials f and g have a common divisor not a constant. Therefore, it is natural to ask the question: What is the necessary and sufficient condition that two polynomials

$$f(x) = a_0 x^n + a_1 x^{n-1} + \cdots + a_n,$$
$$g(x) = b_0 x^m + b_1 x^{m-1} + \cdots + b_m,$$

one of degree n (so that $a_0 \neq 0$) and another, not identically vanishing, of degree not exceeding m, should have a common divisor that is not a constant? The preliminary answer is given in the following:

LEMMA. *Polynomials f and g have a common divisor involving x if and only if two polynomials f_1 and g_1 of degrees not exceeding $n - 1$ and $m - 1$, respectively, and not identically vanishing, exist so that identically*

$$fg_1 + f_1 g = 0,$$

PROOF. (a) Let f and g have a common divisor d of degree > 0. Then,

$$f_1 = \frac{f}{d}, \qquad g_1 = -\frac{g}{d}$$

are two polynomials of degrees not exceeding $n - 1$ and $m - 1$, not identically vanishing, and such that

$$fg_1 + f_1 g = 0.$$

(b) Let the polynomials f_1 and g_1 as required in the lemma exist. Then, the highest common divisor of f and g cannot be a constant. For suppose it is a constant; then, the highest common divisor of

$$f_1 g \qquad \text{and} \qquad f_1 f$$

is f_1. But

$$f_1 g = -fg_1 \qquad \text{and} \qquad f_1 f$$

are both divisible by f; hence, f_1 is divisible by f, which is impossible since the degree of f_1 is lower than that of f. This contradiction shows that f and g have a common divisor that is not a constant. Assume now that

$$f_1 = \lambda_0 x^{n-1} + \lambda_1 x^{n-2} + \cdots + \lambda_{n-1},$$
$$g_1 = \mu_0 x^{m-1} + \mu_1 x^{m-2} + \cdots + \mu_{m-1},$$

and try to find $m + n$ constants $\lambda_0, \lambda_1, \ldots, \lambda_{n-1}; \mu_0, \mu_1, \ldots, \mu_{m-1}$, not all zero, so that identically

$$fg_1 + f_1g = 0.$$

This gives $m + n$ linear homogeneous equations to determine $\lambda_0, \lambda_1, \ldots, \lambda_{n-1}; \mu_0, \mu_1, \ldots, \mu_{m-1}$, these equations having a nontrivial solution if and only if their determinant vanishes. It cannot happen that all $\lambda_0, \lambda_1, \ldots, \lambda_{n-1}$ or all $\mu_0, \mu_1, \ldots, \mu_{m-1}$ vanish. Assume, for instance, that $\lambda_0 = \lambda_1 = \cdots = \lambda_{m-1} = 0$; then, not all $\mu_0, \mu_1, \ldots, \mu_{m-1}$ vanish. Since $f_1 = 0$ identically, it follows that

$$fg_1 = 0,$$

and this is impossible since neither f nor g_1 vanishes identically. From the lemma, it can be inferred that the vanishing of the determinant of the system serving to determine the coefficients of f_1 and g_1, f_1 and g_1 being nonidentically vanishing polynomials satisfying the identity

$$fg_1 + f_1g = 0,$$

is the necessary and sufficient condition for f and g to have their highest common divisor of degree > 0. We shall call this determinant Sylvester's determinant and denote it by $D(f,g)$.

In order to have a clear idea how this determinant is built we take the example

$$f = a_0 x^3 + a_1 x^2 + a_2 x + a_3,$$
$$g = b_0 x^2 + b_1 x + b_2.$$

Write the identities

$$xf(x) = a_0 x^4 + a_1 x^3 + a_2 x^2 + a_3 x,$$
$$f(x) = \qquad a_0 x^3 + a_1 x^2 + a_2 x + a_3,$$
$$x^2 g(x) = b_0 x^4 + b_1 x^3 + b_2 x^2,$$
$$xg(x) = \qquad b_0 x^3 + b_1 x^2 + b_2 x,$$
$$g(x) = \qquad\quad b_0 x^2 + b_1 x + b_2,$$

and multiply them by $\mu_0, \mu_1, \lambda_0, \lambda_1, \lambda_2$, respectively. Add and equate to 0 the coefficients of $x^4, x^3, x^2, x, 1$ in the right-hand member. This gives the system of equations

$$a_0\mu_0 + 0 \cdot \mu_1 + b_0\lambda_0 \qquad\qquad\qquad = 0,$$
$$a_1\mu_0 + a_0\mu_1 + b_1\lambda_0 + b_0\lambda_1 \qquad\qquad = 0,$$
$$a_2\mu_0 + a_1\mu_1 + b_2\lambda_0 + b_1\lambda_1 + b_0\lambda_2 = 0,$$
$$a_3\mu_0 + a_2\mu_1 + 0 \cdot \lambda_0 + b_2\lambda_1 + b_1\lambda_2 = 0,$$
$$0 \cdot \mu_0 + a_3\mu_1 + 0 \cdot \lambda_0 + 0 \cdot \lambda_1 + b_2\lambda_2 = 0,$$

whose determinant is

$$D(f,g) = \begin{vmatrix} a_0 & 0 & b_0 & 0 & 0 \\ a_1 & a_0 & b_1 & b_0 & 0 \\ a_2 & a_1 & b_2 & b_1 & b_0 \\ a_3 & a_2 & 0 & b_2 & b_1 \\ 0 & a_3 & 0 & 0 & b_2 \end{vmatrix},$$

or, exchanging rows and columns,

$$D(f,g) = \begin{vmatrix} a_0 & a_1 & a_2 & a_3 & 0 \\ 0 & a_0 & a_1 & a_2 & a_3 \\ b_0 & b_1 & b_2 & 0 & 0 \\ 0 & b_0 & b_1 & b_2 & 0 \\ 0 & 0 & b_0 & b_1 & b_2 \end{vmatrix}.$$

This is Sylvester's determinant in case $n = 3$, $m = 2$. It is clear that in the general case

$$D(f,g) = \left. \begin{vmatrix} a_0 & a_1 & a_2 & \cdots & a_n & & & \\ & a_0 & a_1 & \cdots & & a_n & & \\ & & \cdots & \cdots & \cdots & \cdots & & \\ & & & a_0 & a_1 & \cdots & \cdots & a_n \\ b_0 & b_1 & \cdots & b_m & & & & \\ & b_0 & b_1 & \cdots & b_m & & & \\ & & \cdots & \cdots & \cdots & \cdots & & \\ & & & b_0 & b_1 & \cdots & b_m & \end{vmatrix} \right\} \begin{matrix} n \text{ rows} \\ \\ \\ m \text{ rows} \end{matrix}$$

In this determinant the spaces left blank for typographical convenience should be filled with zeros. For example, if $n = 4$, $m = 3$,

$$D(f,g) = \begin{vmatrix} a_0 & a_1 & a_2 & a_3 & a_4 & 0 & 0 \\ 0 & a_0 & a_1 & a_2 & a_3 & a_4 & 0 \\ 0 & 0 & a_0 & a_1 & a_2 & a_3 & a_4 \\ b_0 & b_1 & b_2 & b_3 & 0 & 0 & 0 \\ 0 & b_0 & b_1 & b_2 & b_3 & 0 & 0 \\ 0 & 0 & b_0 & b_1 & b_2 & b_3 & 0 \\ 0 & 0 & 0 & b_0 & b_1 & b_2 & b_3 \end{vmatrix}.$$

Problems

Solve the following systems:

1. $x^3 + 3yx^2 + (3y^2 - y + 1)x + y^3 - y^2 + 2y = 0,$
$x^2 + 2yx + y^2 - y = 0.$

2. $x^3 + 2yx^2 + 2y(y - 2)x + y^2 - 4 = 0,$
$x^2 + 2xy + 2y^2 - 5y + 2 = 0.$

3. $x^3 + 3yx^2 - 3x^2 + 3y^2x - 6xy - x + y^3 - 3y^2 - y + 3 = 0,$
$x^3 - 3yx^2 + 3x^2 + 3y^2x - 6xy - x - y^3 + 3y^2 + y - 3 = 0.$

4. $x^3 + yx^2 - (y^2 + 1)x + y - y^3 = 0,$
$x^3 - yx^2 - (y^2 + 6y + 9)x + y^3 + 6y^2 + 9y = 0.$

5. $yx^3 - (y^3 - 3y - 1)x + y = 0,$
$x^2 - y^2 + 3 = 0.$

6. $5x^2 - 5y^2 - 3x + 9y = 0,$
$5x^3 + 5y^3 - 15x^2 - 13xy - y^2 = 0.$

7. $3x^3 + 9x^2y + 9xy^2 + 3y^3 + 2x^2 - 4xy + 2y^2 = 5,$
$4x^3 + 12x^2y + 12xy^2 + 4y^3 - x^2 + 2xy - y^2 = 3.$

8. $x^3 + yx^2 - 4 = 0,$
$x^3 + yx + 2 = 0.$

9. $x^3 + 4yx^2 - 3x + 2 = 0,$
$2x^3 + (y + 2)x^2 - 5x + 1 = 0.$

10. Find λ and μ so that equations
$x^3 - 6x^2 + \lambda x - 3 = 0,$
$x^3 - x^2 + \mu x + 2 = 0,$
have two common roots.

4. Identity of the Resultant and Sylvester's Determinant. The vanishing of Sylvester's determinant $D(f,g)$ is a necessary and sufficient condition that f and g have a highest common divisor of degree > 0, provided $a_0 \neq 0$. The vanishing of the resultant $R(f,g)$ is also a necessary and sufficient condition for that. This leads us to surmise the close relation between both. In fact, Sylvester's determinant and the resultant are identical polynomials in the coefficients of f and g. Not to interrupt the proof of this capital fact we shall prove first the following auxiliary proposition: *If* $\theta(x_1, x_2, \ldots, x_n)$ *is a polynomial in the variables* $x_1, x_2, \ldots, x_n$ *that is separately divisible by* $g(x_1), g(x_2), \ldots, g(x_n)$, *then it is divisible by the product* $g(x_1)g(x_2) \cdots g(x_n)$ *so that*

$$\theta(x_1, x_2, \ldots, x_n) = g(x_1)g(x_2) \cdots g(x_n)\Omega(x_1, x_2, \ldots, x_n)$$

where $\Omega(x_1, x_2, \ldots, x_n)$ *is a polynomial in* $x_1, x_2, \ldots, x_n$. To prove this statement assume that θ, being divisible by $g(x_1)g(x_2) \cdots g(x_{i-1})$, is also divisible by $g(x_i)$; then, we shall prove that θ is divisible by $g(x_1)g(x_2) \cdots g(x_i)$. By hypothesis

$$\theta(x_1, x_2, \ldots, x_n) = g(x_1)g(x_2) \cdots g(x_{i-1})\theta_1(x_1, x_2, \ldots, x_n)$$

θ_1 being a polynomial in $x_1, x_2, \ldots, x_n$. Arrange θ_1 in powers of x_i and divide by $g(x_i)$; then,

$$\theta_1 = g(x_i)\theta_2(x_1, x_2, \ldots, x_n) + p_0x_i^{m-1} + p_1x_i^{m-2} + \cdots + p_{m-1},$$

$p_0, p_1, \ldots \quad p_{m-1}$ being polynomials in $x_1, x_2, \ldots, x_{i-1}, x_{i+1}, \ldots, x_n$. Hence,

$$\theta = g(x_1)g(x_2) \cdots g(x_i)\theta_2 + P_0 x_i^{m-1} + P_1 x_i^{m-2} + \cdots + P_{m-1}$$

where the polynomials

$$P_k = g(x_1)g(x_2) \cdots g(x_{i-1})p_k, \qquad k = 0, 1, 2, \ldots, m-1,$$

do not contain x_i. Since θ is divisible by $g(x_i)$,

$$P_0 x_i^{m-1} + P_1 x_i^{m-2} + \cdots + P_{m-1}$$

will be divisible by $g(x_i)$ or

$$P_0 x_i^{m-1} + P_1 x_i^{m-2} + \cdots + P_{m-1} = g(x_i)T(x_1, x_2, \ldots, x_n).$$

But the right-hand side if not identically 0 has a degree in x_i of at least m; hence,

$$P_0 = P_1 = \cdots = P_{m-1} = 0$$

identically and the statement is proved. Since θ is divisible by $g(x_1)$ and $g(x_2)$, it will be divisible by $g(x_1)g(x_2)$ according to what has just been proved; θ, being divisible by $g(x_3)$, for the same reason will be divisible by $g(x_1)g(x_2)g(x_3)$, etc.

To save writing we shall show the identity of $D(f,g)$ and $R(f,g)$ in the particular case $n = 3$, $m = 2$, but the reasoning will be general. Let x_1, x_2, x_3 be variables and

$$f(x) = a_0(x - x_1)(x - x_2)(x - x_3).$$

Then, Sylvester's determinant

$$D(f,g) = \begin{vmatrix} a_0 & a_1 & a_2 & a_3 & 0 \\ 0 & a_0 & a_1 & a_2 & a_3 \\ b_0 & b_1 & b_2 & 0 & 0 \\ 0 & b_0 & b_1 & b_2 & 0 \\ 0 & 0 & b_0 & b_1 & b_2 \end{vmatrix}$$

is a polynomial in x_1, x_2, x_3. On multiplying the columns 1, 2, 3, 4 by x^4, x^3, x^2, x and adding to column 5 we can write $D(f,g)$ thus:

$$D(f,g) = \begin{vmatrix} a_0 & a_1 & a_2 & a_3 & xf(x) \\ 0 & a_0 & a_1 & a_2 & f(x) \\ b_0 & b_1 & b_2 & 0 & x^2 g(x) \\ 0 & b_0 & b_1 & b_2 & xg(x) \\ 0 & 0 & b_0 & b_1 & g(x) \end{vmatrix}.$$

Taking $x = x_1, x_2, x_3$ and noticing that $f(x_1) = f(x_2) = f(x_3) = 0$, we see that $D(f,g)$ is divisible by $g(x_1)$, $g(x_2)$, $g(x_3)$ and hence is divisible by $g(x_1)g(x_2)g(x_3)$ so that

$$D(f,g) = g(x_1)g(x_2)g(x_3)T(x_1, x_2, x_3)$$

where $T(x_1,x_2,x_3)$ is some polynomial in x_1, x_2, x_3. Writing $D(f,g)$ thus:

$$D(f,g) = a_3^2 \begin{vmatrix} \dfrac{a_0}{a_1} & \dfrac{a_1}{a_2} & \dfrac{a_2}{a_3} & 1 & 0 \\ 0 & \dfrac{a_0}{a_3} & \dfrac{a_1}{a_3} & \dfrac{a_2}{a_3} & 1 \\ b_0 & b_1 & b_2 & 0 & 0 \\ 0 & b_0 & b_1 & b_2 & 0 \\ 0 & 0 & b_0 & b_1 & b_2 \end{vmatrix},$$

and noticing that

$$-\frac{a_2}{a_3} = \frac{1}{x_1} + \frac{1}{x_2} + \frac{1}{x_3}, \qquad \frac{a_1}{a_3} = \frac{1}{x_1x_2} + \frac{1}{x_1x_3} + \frac{1}{x_2x_3}, \qquad -\frac{a_0}{a_3} = \frac{1}{x_1x_2x_3},$$

we see that the determinant when expanded contains, besides a constant term, only terms involving x_1, x_2, x_3 in the denominators. Hence, in the expansion of $D(f,g)$ the term of the highest degree in x_1, x_2, x_3 is of the form

$$Kx_1^2x_2^2x_3^2$$

with K a constant. A term of the same type is of highest degree in x_1, x_2, x_3 in the product

$$g(x_1)g(x_2)g(x_3),$$

whence one may conclude that $T(x_1,x_2,x_3)$ reduces to a constant. Call this constant C and to determine it set $x_1 = x_2 = x_3 = 0$ in both members of the identity

$$D(f,g) = Cg(x_1)g(x_2)g(x_3).$$

Since $a_1 = a_2 = a_3 = 0$ when $x_1 = x_2 = x_3 = 0$, we have

$$D(f,g) = \begin{vmatrix} a_0 & 0 & 0 & 0 & 0 \\ 0 & a_0 & 0 & 0 & 0 \\ b_0 & b_1 & b_2 & 0 & 0 \\ 0 & b_0 & b_1 & b_2 & 0 \\ 0 & 0 & b_0 & b_1 & b_2 \end{vmatrix} = a_0^2 b_2^3.$$

On the other hand,

$$g(0)g(0)g(0) = b_2^3,$$

and so

$$Cb_2^3 = a_0^2 b_2^3$$

whence, considering b_0, b_1, b_2 as variables,

$$C = a_0^2$$

and

$$D(f,g) = a_0^2 g(x_1)g(x_2)g(x_3) = R(f,g).$$

This is an identity in x_1, x_2, x_3; b_0, b_1, b_2 being considered as variables. But both members are symmetric functions of x_1, x_2, x_3. Being identical, they will remain identical when expressed as polynomials in a_0, a_1, a_2, a_3; b_0, b_1, b_2, according to the principle of Gauss proved in Chap. XI, Sec. 7. Thus, the identity of Sylvester's determinant and the resultant, considered as polynomials in coefficients of f and g, is proved.

5. Discriminant. The product of $\frac{1}{2}n(n-1)$ differences $x_\alpha - x_\beta$ corresponding to all possible selections of the two indices $\alpha < \beta$ out of the n numbers 1, 2, 3, . . . , n, namely,

$$(x_1 - x_2)(x_1 - x_3) \cdots (x_1 - x_n)(x_2 - x_3) \cdots (x_2 - x_n) \cdots (x_{n-1} - x_n)$$

merely changes its sign by a transposition of any two letters x_α and x_β and therefore by all permutations of them acquires only two values. Its square is therefore symmetric. If we set

$$f(x) = a_0(x - x_1)(x - x_2) \cdots (x - x_n) = a_0 x^n + a_1 x^{n-1} + \cdots + a_n,$$

the elementary symmetric functions are

$$-\frac{a_1}{a_0}, \frac{a_2}{a_0}, \ldots, (-1)^n \frac{a_n}{a_0}.$$

Hence,

$$(x_1 - x_2)^2(x_1 - x_3)^3 \cdots (x_1 - x_n)^2(x_2 - x_3)^2 \cdots (x_2 - x_n)^2 \cdots$$
$$(x_{n-1} - x_n)^2,$$

involving x_1, x_2, . . . , x_n symmetrically, can be expressed as a polynomial in

$$\frac{a_1}{a_0}, \frac{a_2}{a_0}, \ldots, \frac{a_n}{a_0}.$$

The highest power of x_1 occurring in the above product being $2n - 2$,

$$D = a_0^{2n-2}(x_1 - x_2)^2 \cdots (x_1 - x_n)^2(x_2 - x_3)^2 \cdots (x_2 - x_n)^2 \cdots$$
$$(x_{n-1} - x_n)^2$$

is a homogeneous polynomial of dimension $2n - 2$ in a_0, a_1, . . . , a_n. It is called the *discriminant* of $f(x)$. If x_1, x_2, . . . , x_n signify not variables but roots of the equation $f(x) = 0$, the same expression D is called the discriminant of that equation. Clearly the discriminant vanishes if and only if the equation has equal roots. The discriminant has a close relation to the resultant of $f(x)$ and its derivative. Since

$$f'(x_1) = a_0(x_1 - x_2) \cdots (x_1 - x_n),$$
$$f'(x_2) = a_0(x_2 - x_1) \cdots (x_2 - x_n),$$
$$\cdots \cdots \cdots \cdots \cdots \cdots \cdots \cdots$$
$$f'(x_n) = a_0(x_n - x_1) \cdots (x_n - x_{n-1}),$$

it is easy to see that

$$f'(x_1)f'(x_2) \cdots f'(x_n) = (-1)^{\frac{n(n-1)}{2}} a_0^n (x_1 - x_2)^2 \cdots (x_1 - x_n)^2 \cdots \\ (x_{n-1} - x_n)^2$$

$$= (-1)^{\frac{n(n-1)}{2}} a_0^{-n+2} D.$$

On the other hand,

$$R(f, f') = a_0^{n-1} f'(x_1) f'(x_2) \cdots f'(x_n),$$

and so

$$(-1)^{\frac{n(n-1)}{2}} a_0 D = R(f, f').$$

The explicit expression of the discriminant will be obtained by writing $R(f,f')$ as a Sylvester determinant.

Example 1. To find the discriminant of a quadratic polynomial

$$f(x) = a_0 x^2 + a_1 x + a_2.$$

The resultant of f and f' is

$$R(f, f') = \begin{vmatrix} a_0 & a_1 & a_2 \\ 2a_0 & a_1 & 0 \\ 0 & 2a_0 & a_1 \end{vmatrix}$$

whence

$$D = -\begin{vmatrix} 1 & a_1 & a_2 \\ 2 & a_1 & 0 \\ 0 & 2a_0 & a_1 \end{vmatrix} = a_1^2 - 4a_0 a_2.$$

Example 2. To find the discriminant of a cubic polynomial

$$f(x) = a_0 x^3 + a_1 x^2 + a_2 x + a_3.$$

Here

$$R(f, f') = \begin{vmatrix} a_0 & a_1 & a_2 & a_3 & 0 \\ 0 & a_0 & a_1 & a_2 & a_3 \\ 3a_0 & 2a_1 & a_2 & 0 & 0 \\ 0 & 3a_0 & 2a_1 & a_2 & 0 \\ 0 & 0 & 3a_0 & 2a_1 & a_2 \end{vmatrix}$$

whence

$$D = -\begin{vmatrix} 1 & a_1 & a_2 & a_3 & 0 \\ 0 & a_0 & a_1 & a_2 & a_3 \\ 3 & 2a_1 & a_2 & 0 & 0 \\ 0 & 3a_0 & 2a_1 & a_2 & 0 \\ 0 & 0 & 3a_0 & 2a_1 & a_2 \end{vmatrix} = \begin{vmatrix} 1 & a_1 & a_2 & a_3 & 0 \\ 0 & a_0 & a_1 & a_2 & a_3 \\ 0 & a_1 & 2a_2 & 3a_3 & 0 \\ 0 & 3a_0 & 2a_1 & a_2 & 0 \\ 0 & 0 & 3a_0 & 2a_1 & a_2 \end{vmatrix} = \begin{vmatrix} a_0 & a_1 & a_2 & a_3 \\ a_1 & 2a_2 & 3a_3 & 0 \\ 3a_0 & 2a_1 & a_2 & 0 \\ 0 & 3a_0 & 2a_1 & a_2 \end{vmatrix}.$$

After expanding this determinant the final expression of D is found to be

$$D = 18 a_0 a_1 a_2 a_3 - 4 a_1^3 a_3 + a_1^2 a_2^2 - 4 a_0 a_2^3 - 27 a_0^2 a_3^2.$$

In the case of an equation with numerical coefficients the computation of the discriminant can be reduced to the computation of a numerical determinant of the same order as the degree of the equation. If $\alpha_1, \alpha_2, \ldots, \alpha_r$ be the roots of this equation, then the square of Vandermonde's determinant

$$\begin{vmatrix} 1 & \alpha_1 & \alpha_1^2 & \cdots & \alpha_1^{n-1} \\ 1 & \alpha_2 & \alpha_2^2 & \cdots & \alpha_2^{n-1} \\ \cdots\cdots\cdots\cdots\cdots\cdots \\ 1 & \alpha_n & \alpha_n^2 & \cdots & \alpha_n^{n-1} \end{vmatrix} = (\alpha_n - \alpha_1) \cdots (\alpha_n - \alpha_{n-1}) \cdots (\alpha_2 - \alpha_1)$$

differs from D only by the factor a_0^{2n-2}. Now multiplying Vandermonde's determinant by itself, column by column, and denoting as usual by

$$s_i = \alpha_1^i + \alpha_2^i + \cdots + \alpha_n^i,$$

the sum of ith powers of roots, we have

$$\begin{vmatrix} 1 & \alpha_1 & \alpha_1^2 & \cdots & \alpha_1^{n-1} \\ 1 & \alpha_2 & \alpha_2^2 & \cdots & \alpha_2^{n-1} \\ \cdots\cdots\cdots\cdots\cdots\cdots \\ 1 & \alpha_n & \alpha_n^2 & \cdots & \alpha_n^{n-1} \end{vmatrix}^2 = \begin{vmatrix} s_0 & s_1 & \cdots & s_{n-1} \\ s_1 & s_2 & \cdots & s_n \\ \cdots\cdots\cdots\cdots\cdots \\ s_{n-1} & s_n & \cdots & s_{2n-2} \end{vmatrix},$$

and so

$$D = a_0^{2n-2} \begin{vmatrix} s_0 & s_1 & \cdots & s_{n-1} \\ s_1 & s_2 & \cdots & s_n \\ \cdots\cdots\cdots\cdots\cdots \\ s_{n-1} & s_n & \cdots & s_{2n-2} \end{vmatrix}.$$

The sums s_i are readily computed by means of Newton's formulas.

6. Imaginary Roots. The computation of imaginary roots can be reduced to that of real roots by means of elimination. Let the proposed equation

$$f(x) = a_0 x^n + a_1 x^{n-1} + \cdots + a_n = 0$$

have real coefficients and let $x = a + bi$ be one of its imaginary roots. Then, $f(a + bi) = 0$, and by Taylor's formula

$$f(a + bi) = f(a) + bi \frac{f'(a)}{1} - b^2 \frac{f''(a)}{1 \cdot 2} - ib^3 \frac{f'''(a)}{1 \cdot 2 \cdot 3} + \cdots .$$

The expression in the right-hand side vanishing by hypothesis, its real and imaginary parts vanish separately, that is,

$$f(a) - \frac{f''(a)}{1 \cdot 2} b^2 + \frac{f^{iv}(a)}{1 \cdot 2 \cdot 3 \cdot 4} b^4 - \cdots = 0,$$

$$f'(a) - \frac{f'''(a)}{1 \cdot 2 \cdot 3} b^2 + \frac{f^v(a)}{1 \cdot 2 \cdot 3 \cdot 4 \cdot 5} b^4 - \cdots = 0,$$

since for an imaginary root $b \neq 0$. On eliminating b^2 we get an equation of degree $n(n-1)/2$, which must be satisfied by the real parts of all the imaginary roots. The real roots of this equation leading to positive values of b^2 can be computed by one of the methods explained in Chap. VIII. The auxiliary equation serving to determine a is of degree 3, 6, 10, . . . corresponding to $n = 3, 4, 5, \ldots$. On account of the high degree of this equation and the labor involved in developing it, this method for the computation of the imaginary roots has only theoretical value except for the lowest degrees $n = 3$ and $n = 4$. By far the better method for this purpose is the so-called root squaring or Graeffe's method. The outline of this method can be found in Appendix V.

Example. Let the proposed equation be

$$f(x) = x^4 + 4x - 1 = 0.$$

The system of equations serving to determine a and b in this case is

$$b^4 - 6a^2b^2 + a^4 + 4a - 1 = 0,$$
$$ab^2 - a^3 - 1 = 0.$$

Clearly $a \neq 0$; hence,

$$b^2 = \frac{a^3 + 1}{a},$$

which after substituting into the first equation gives

$$4a^6 + a^2 - 1 = 0.$$

The only positive value of a^2 satisfying this equation is

$$a^2 = \tfrac{1}{2}$$

and correspondingly

$$a = \pm \frac{1}{\sqrt{2}}.$$

We must take

$$a = \frac{1}{\sqrt{2}}$$

in order to have a positive value of b^2. Then,

$$b^2 = \frac{1 + \sqrt{8}}{2}, \qquad b = \pm \frac{\sqrt{1 + \sqrt{8}}}{\sqrt{2}},$$

and the imaginary roots of our equation are

$$\frac{1 + i\sqrt{1 + \sqrt{8}}}{\sqrt{2}}, \qquad \frac{1 - i\sqrt{1 + \sqrt{8}}}{\sqrt{2}}.$$

Problems

Find the imaginary roots of the following equations:

1. $x^3 + x + 10 = 0.$
2. $x^3 - 2x - 2 = 0.$
3. $x^3 - 2x - 5 = 0.$
4. $x^4 + x^2 - 2x + 6 = 0.$
5. $x^4 - 2x^3 + 6x^2 - 2x + 5 = 0.$
6. $x^4 - 4x^2 + 8x - 4 = 0.$
7. $x^4 - x + 1 = 0.$
8. $x^4 - 2x^3 - 1 = 0.$

APPENDIX I

THE FUNDAMENTAL THEOREM OF ALGEBRA

1. It was assumed throughout this book that every algebraic equation with real or imaginary coefficients has at least one complex root. This truth is known as the *fundamental theorem of algebra*. The empirical evidence in favor of this proposition, gathered from innumerable particular examples, is so strong that for a long time it was regarded as something evident. In 1746 D'Alembert first stated the fact of the existence of roots as a theorem and made an attempt to prove it. Judging by the standards of rigor accepted today, D'Alembert's proof is defective in many respects, but it contains a sound germ and can be elaborated into a rigorous proof. For instance, one of the proofs proposed by Weierstrass is based on D'Alembert's idea. The first complete proofs of the fundamental theorem were given by Gauss in the beginning of the past century, and since that time many others have been added. Among the many existing proofs perhaps the first and the fourth (which is only another presentation of the first) proofs by Gauss show in the clearest intuitive way why any equation should have a root, and although partisans of extreme rigor may insist on the necessity of various additions, we shall present here the fourth Gaussian proof as the most suitable for the purposes of this book.

2. The coefficients of the polynomial

$$f(x) = x^n + ax^{n-1} + bx^{n-2} + \cdots + l$$

are complex numbers. Written in trigonometric form let them be

$$a = A(\cos \alpha + i \sin \alpha), \qquad b = B(\cos \beta + i \sin \beta), \quad \ldots,$$
$$l = L(\cos \lambda + i \sin \lambda).$$

To x we shall attribute also a complex value

$$x = r(\cos \phi + i \sin \phi).$$

On substituting it into $f(x)$ and separating real and imaginary parts with the help of de Moivre's theorem we can write

$$f(x) = T + iU,$$

where

$$T = r^n \cos n\phi + Ar^{n-1} \cos[(n - 1)\phi + \alpha] + \cdots + L \cos \lambda,$$
$$U = r^n \sin n\phi + Ar^{n-1} \sin[(n - 1)\phi + \alpha] + \cdots + L \sin \lambda.$$

The fundamental theorem will be proved if we can show the existence of a point with polar coordinates r and ϕ at which both T and U vanish, and this will be achieved by resorting to geometrical intuition.

3. Before entering into the heart of the proof some simple preliminary considerations must be developed. In the first place, it is possible to determine such a number R that for $r > R$

$$r^n - \sqrt{2}(Ar^{n-1} + Br^{n-2} + \cdots + L) > 0. \tag{1}$$

If C is a positive number larger than all of the numbers $A, B, \ldots, L$, the inequality (1) will certainly be fulfilled if

$$r^n - \sqrt{2}C(r^{n-1} + r^{n-2} + \cdots + 1) > 0$$

or

$$r^n\left[1 - \sqrt{2}C\left(\frac{1}{r} + \frac{1}{r^2} + \cdots + \frac{1}{r^n}\right)\right] > 0. \tag{2}$$

But, supposing $r > 1$,

$$\frac{1}{r} + \frac{1}{r^2} + \cdots + \frac{1}{r^n} < \frac{1}{r-1}$$

and consequently (2) and still more (1) will hold if $r > 1$ and

$$1 - \frac{\sqrt{2}C}{r-1} > 0,$$

that is, if

$$r > 1 + \sqrt{2}C.$$

It suffices to take

$$R = 1 + \sqrt{2}C$$

to have the inequality (1) fulfilled for $r > R$.

4. In the second place, it is possible to show that the circumference of a circle of radius $r > R$ consists of $2n$ arcs inside of which T takes alternately positive and negative values. To this end introduce the angle

$$\omega = \frac{\pi}{4n}$$

and consider on the circumference $4n$ points with amplitudes

$$\omega, 3\omega, 5\omega, \ldots, (8n - 3)\omega, (8n - 1)\omega.$$

These points will be denoted by $P_0, P_1, P_2, \ldots, P_{4n-1}$. Combining them into $2n$ pairs $P_0, P_1; P_2, P_3; \ldots; P_{4n-2}, P_{4n-1}$, it will be easy to show that at points of each pair T has values of opposite signs, and that these signs will be

$$(-1)^k, \quad (-1)^{k+1}$$

for the pair P_{2k}, P_{2k+1}. In fact, amplitudes corresponding to P_{2k} and P_{2k+1} are

$$\phi = (4k + 1)\frac{\pi}{4n} \quad \text{and} \quad \phi' = (4k + 3)\frac{\pi}{4n},$$

and therefore

$$\cos n\phi = (-1)^k \frac{1}{\sqrt{2}} \quad \text{and} \quad \cos n\phi' = (-1)^{k+1}\frac{1}{\sqrt{2}}.$$

Multiplying the corresponding values of T by $(-1)^k$ and $(-1)^{k+1}$, respectively, we have

$$(-1)^k T = \frac{r^n}{\sqrt{2}} + (-1)^k A r^{n-1} \cos [(n-1)\phi + \alpha] + \cdots$$
$$+ (-1)^k L \cos \lambda,$$

$$(-1)^{k+1} T = \frac{r^n}{\sqrt{2}} + (-1)^{k+1} A r^{n-1} \cos [(n-1)\phi' + \alpha] + \cdots$$
$$+ (-1)^{k+1} L \cos \lambda,$$

and hence, replacing

$$(-1)^k \cos [(n-1)\phi + \alpha], \ldots, (-1)^k \cos \lambda$$
$$(-1)^{k+1} \cos [(n-1)\phi' + \alpha], \ldots, (-1)^{k+1} \cos \lambda$$

by -1, the following inequalities are derived:

$$(-1)^k T \geqq \frac{r^n}{\sqrt{2}} - A r^{n-1} - \cdots - L,$$

$$(-1)^{k+1} T \geqq \frac{r^n}{\sqrt{2}} - A r^{n-1} - \cdots - L,$$

the right-hand sides of which, by the choice of r, are positive, and this proves the statement. Since T varies continuously with ϕ, it will vanish $2n$ times at points (0), (1), (2), $\ldots$, $(2n-1)$ whose amplitudes are, respectively, contained between

$$\omega, 3\omega; 5\omega, 7\omega; 9\omega, 11\omega; \ldots; (8n-3)\omega, (8n-1)\omega.$$

It is important to show that these points are the only ones where T vanishes. To show this express $\cos \phi$ and $\sin \phi$ through $\xi = \tan \phi/2$ thus:

$$\cos \phi = \frac{1 - \xi^2}{1 + \xi^2}, \quad \sin \phi = \frac{2\xi}{1 + \xi^2},$$

and introduce the corresponding expression

$$x = r\left(\frac{1 - \xi^2}{1 + \xi^2} + i\frac{2\xi}{1 + \xi^2}\right)$$

into $f(x)$. Then, the real part T appears in the form

$$T = \frac{P_{2n}(\xi)}{(1 + \xi^2)^n},$$

$P_{2n}(\xi)$ being real polynomial of degree not higher than $2n$. Since this polynomial vanishes for $2n$ distinct values of ξ, as has been proved, its degree is $2n$ and it cannot vanish for any other values of ξ. Besides, the $2n$ roots whose existence is shown must be simple roots and therefore on describing the circumference of the circle of radius r the positive and negative values of T alternate. Since at the point with amplitude ω the value of T is positive and changes sign on passing through the point (0), on the $2n$ arcs between

(0) and (1); (1) and (2); . . . ; $(2n - 2)$ and $(2n - 1)$;

$$(2n - 1) \text{ and } (0);$$

the signs of T will be alternately $- + - +$, etc.

5. Thirdly, we shall show that U has positive values at the points (0), (2), . . . , $(2n - 2)$, and negative values at the points (1), (3), . . . , $(2n - 1)$. The amplitude ϕ of the point (k) lies between

$$(4k + 1)\frac{\pi}{4n} \quad \text{and} \quad (4k + 3)\frac{\pi}{4n};$$

hence, $(-1)^k \sin n\phi$ is not only positive but greater than $1/\sqrt{2}$. Multiplying U by $(-1)^k$ and replacing

$$(-1)^k \sin [(n - 1)\phi + \alpha], \; . . . , \; (-1)^k \sin \lambda$$

by -1, we shall have

$$(-1)^k U \geqq r^n(-1)^k \sin n\phi - Ar^{n-1} - \cdots - L,$$

and also the inequality

$$(-1)^k U > \frac{r^n}{\sqrt{2}} - Ar^{n-1} - \cdots - L;$$

but the right-hand side here is positive by the choice of r. Therefore, the sign of U at the point (k) is $(-1)^k$.

6. A circle Γ of radius $> R$ being selected, its circumference is divided by the points (0), (1), (2), . . . , $(2n - 1)$ into $2n$ arcs on which the sign of T is alternately negative and positive. When the circle Γ expands, the arcs $(0)(1)$; $(1)(2)$; etc., sweep out $2n$ regions, extending to infinity, within which T has alternately negative and positive values, and these regions are separated from one another by lines on which $T = 0$. To describe the situation more intuitively we shall call those n regions outside of Γ, where $T < 0$, "seas," and the other n regions, where $T > 0$, "lands." Lines on which $T = 0$ will be then "seashores."

Now the n seas and the n lands existing in the exterior of Γ extend themselves into the interior of Γ across the arcs $(0)(1)$; $(1)(2)$; etc. Starting from the end point (1) of the arc $(0)(1)$ through which a sea penetrates into the interior of Γ, we imagine that we walk along the seashore so that the land is always on our right, heading inward. We must eventually come out of Γ, and when we cross it again, heading outward, the land must still be on our right. If the circumference is followed in a counterclockwise direction, lands and seas alternate, whence it follows that we cross Γ, heading outward, at a point (k) with k even, that is either at (2), which is the simplest case, or at (4), (6), etc. Thus, there is a continuous line L leading from (1) to some point (k) with even k. On the line L constantly $T = 0$, and at the point (1), according to Sec. 5, $U < 0$, whereas $U > 0$ at the point (k). Since U varies continuously, at some point of L it must take the value 0, so that there is a point within Γ at which both $T = 0$ and $U = 0$, which proves the existence of a root.

The annexed figure, borrowed from Gauss and made for a special equation of the fifth degree, illustrates beautifully what has just been said. Areas covered by seas are shaded, and those representing lands are left white. Now, if we start from (1) and walk along the shore we reach (2) on the circle Γ, and in between on this shore there is a point a representing a root. We can just as well start from points (3), (5), (7), (9) and follow the shore as prescribed. From (3) we reach (8) passing through another root b. From (5) we reach (6), passing through a root c, and from (7) we reach (4) passing again through c. Finally, from (9) we reach (0) passing through a root d. This accounts for four roots a, b, c, d of which c is a double root.

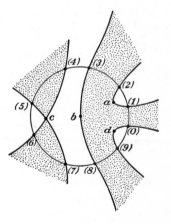

The fundamental theorem by its nature belongs rather to analysis than to algebra. It is natural, therefore, that this Gaussian proof contains transcendental elements of analytical and topological character. There are proofs, like the second proof by Gauss and others modeled after it, in which transcendental elements are reduced to a minimum. Only one fact belonging to analysis is required in these proofs, namely, that a real equation of an odd degree has a real root.

APPENDIX II

ON THE THEOREM OF VINCENT

1. If an equation without multiple roots is transformed successively by the substitutions

$$x = a + \frac{1}{y}, \qquad y = b + \frac{1}{z}, \qquad z = c + \frac{1}{t}; \qquad \cdots$$

where $a, b, c, \ldots$ are arbitrary positive integers, the transformed equation, after a sufficiently large number of such transformations, will present either no variations or just one. This remarkable theorem was published by Vincent in 1836 in the first volume of *Liouville's Journal*, but later so completely forgotten that no mention of it is found even in such a capital work as the *Enzyclopädie der mathematischen Wissenschaften*. Yet Vincent's theorem is the base of the very efficient method for separating real roots explained in Chap. VI. In this appendix we shall give a proof of a slightly sharper proposition, which is stated as follows:

Let N_k be the kth term of the series

$$1, 1, 2, 3, 5, 8, 13, 21, \ldots$$

in which each term is the sum of the preceding two and where $\Delta > 0$ is the smallest distance between any two roots of the equation $f(x) = 0$ of degree n and without multiple roots. Let the number m be so chosen that

$$N_{m-1}\Delta > \frac{1}{2}, \qquad \Delta N_m N_{m-1} > 1 + \frac{1}{\epsilon} \tag{1}$$

where

$$\epsilon = \left(1 + \frac{1}{n}\right)^{\frac{1}{n-1}} - 1. \tag{2}$$

Then, the substitution

$$x = a_1 + \cfrac{1}{a_2 + \cfrac{\cdots}{\cdots + \cfrac{1}{a_m + \cfrac{1}{\xi}}}} \tag{3}$$

presented in the form of a continued fraction with arbitrary positive integral elements $a_1, a_2, \ldots, a_m$ transforms the equation $f(x) = 0$ into the equation $F(\xi) = 0$, which has not more than one variation.

2. Let in general P_k/Q_k be the kth convergent to the continued fraction

$$a_1 + \cfrac{1}{a_2 + \cfrac{1}{a_3 +}}$$

$$\cdot$$
$$\cdot$$
$$\cdot$$

From the law of convergents:

$$P_{k+1} = a_{k+1}P_k + P_{k-1},$$
$$Q_{k+1} = a_{k+1}Q_k + Q_{k-1},$$

considering that $Q_1 = 1$, $Q_2 = a_2 \geqq 1$, it follows at once that $Q_k \geqq N_k$. Further, the relation (3) appears in the form

$$x = \frac{P_m\xi + P_{m-1}}{Q_m\xi + Q_{m-1}},$$

whence

$$\xi = -\frac{P_{m-1} - Q_{m-1}x}{P_m - Q_mx}. \tag{4}$$

If x is any root of the equation $f(x) = 0$, the quantity ξ, determined by (4), is the corresponding root of the transformed equation $F(\xi) = 0$.

Suppose that x is an imaginary root

$$x = a + bi, \qquad b \neq 0.$$

Then, the real part of the corresponding root ξ is

$$R(\xi) = -\frac{(P_{m-1} - Q_{m-1}a)(P_m - Q_ma) + Q_{m-1}Q_mb^2}{(P_m - Q_ma)^2 + Q_m^2b^2}.$$

It is certainly negative if

$$(P_{m-1} - Q_{m-1}a)(P_m - Q_ma) \geqq 0.$$

If, on the contrary,

$$(P_{m-1} - Q_{m-1}a)(P_m - Q_ma) < 0,$$

it means that a is contained between two consecutive convergents

$$\frac{P_{m-1}}{Q_{m-1}}, \quad \frac{P_m}{Q_m},$$

the difference between which is in absolute value equal to

$$\frac{1}{Q_{m-1}Q_m}.$$

Hence,

$$\left|\frac{P_{m-1}}{Q_{m-1}} - a\right| < \frac{1}{Q_{m-1}Q_m}, \quad \left|\frac{P_m}{Q_m} - a\right| < \frac{1}{Q_{m-1}Q_m}$$

whence

$$|(P_{m-1} - aQ_{m-1})(P_m - aQ_m)| < \frac{1}{Q_{m-1}Q_m} \leqq 1.$$

Consequently, the real part of ξ will be negative if

$$Q_{m-1}Q_m b^2 > 1.$$

But $Q_m \geqq Q_{m-1} \geqq N_{m-1}$ and $|b| > \Delta/2$; the inequality just written follows therefore from

$$N_{m-1}\Delta > \tfrac{1}{2}$$

and this by virtue of (1) is true. Thus, imaginary roots of the transformed equation have negative real parts.

3. Suppose that x denotes indefinitely real roots of the equation $f(x) = 0$ if there are any. Two cases must be considered. Suppose at first that corresponding to all real roots x

$$(P_{m-1} - Q_{m-1}x)(P_m - Q_m x) > 0.$$

Then, it follows from (4) that all real roots of the transformed equation $F(\xi) = 0$ will be negative. It has been proved already that all imaginary roots of this equation have negative real parts. Consequently, $F(\xi)$, apart from a constant factor, consists of real linear factors $\xi + a$ with positive a, and quadratic factors $\xi^2 + b\xi + c$ with positive b and c. Hence, the polynomial $F(\xi)$ presents no variation.

Let us suppose in the second place that for some real root x

$$(P_{m-1} - Q_{m-1}x)(P_m - Q_m x) \leqq 0.$$

Then, x belongs to the interval

$$\frac{P_{m-1}}{Q_{m-1}}, \quad \frac{P_m}{Q_m}$$

and therefore

$$\left|x - \frac{P_m}{Q_m}\right| \leqq \frac{1}{Q_{m-1}Q_m}.$$

Let x' be any other root, real or imaginary, of $f(x) = 0$, and ξ' the corresponding root of the transformed equation. Then, keeping in mind that

$$P_m Q_{m-1} - P_{m-1} Q_m = (-1)^m,$$

it follows from (4) that

$$\xi' + \frac{Q_{m-1}}{Q_m} = \frac{(-1)^m}{Q_m(P_m - Q_m x')}$$

or

$$\xi' = -\frac{Q_{m-1}}{Q_m}\left[1 - \frac{(-1)^m}{Q_m Q_{m-1}\left(\dfrac{P_m}{Q_m} - x'\right)}\right] = -\frac{Q_{m-1}}{Q_m}(1 + \alpha)$$

with

$$\alpha = \frac{(-1)^{m-1}}{Q_m Q_{m-1}\left(\dfrac{P_m}{Q_{-,}} - x'\right)}.$$

Now

$$\left|\frac{P_m}{Q_m} - x'\right| = \left|\frac{P_m}{Q_m} - x + x - x'\right| \geqq \Delta - \frac{1}{Q_m Q_{m-1}} > 0,$$

and

$$|\alpha| \leqq \frac{1}{\Delta Q_m Q_{m-1} - 1} \leqq \frac{1}{\Delta N_m N_{m-1} - 1}.$$

But by the second of the inequalities (1)

$$\Delta N_m N_{m-1} - 1 > \frac{1}{\epsilon}$$

and so

$$|\alpha| < \epsilon.$$

Thus, the roots of the transformed equation corresponding to the roots $x_1, x_2, \ldots, x_{n-1}$ of the equation $f(x) = 0$ different from x are of the form

$$\xi_k = -\frac{Q_{m-1}}{Q_m}(1 + \alpha_k)$$

and

$$|\alpha_k| < \epsilon.$$

Let us examine now the product

$$(t + 1 + \alpha_1)(t + 1 + \alpha_2) \cdots (t + 1 + \alpha_{n-1}) = t^{n-1} + R_1 t^{n-2} + \cdots + R_{n-1}.$$

We have

$$R_k = \Sigma(1 + \alpha_1)(1 + \alpha_2) \cdots (1 + \alpha_k),$$

the sum consisting of $\binom{n-1}{k}$ terms. Since

$$|(1 + \alpha_1)(1 + \alpha_2) \cdots (1 + \alpha_k) - 1| \leq (1 + |\alpha_1|)(1 + |\alpha_2|) \cdots$$
$$(1 + |\alpha_k|) - 1$$

and

$$|\alpha_h| < \epsilon, \qquad h = 1, 2, \ldots, n - 1,$$

we have

$$|(1 + \alpha_1)(1 + \alpha_2) \cdots (1 + \alpha_k) - 1| \leq (1 + \epsilon)^k - 1$$
$$\leq (1 + \epsilon)^{n-1} - 1 = \frac{1}{n}.$$

It is easy then to understand that R_k can be presented in the form

$$R_k = \binom{n-1}{k}(1 + \delta_k)$$

where

$$|\delta_k| < \frac{1}{n}.$$

Hence, $R_k > 0$.

4. It is of the utmost importance to show that the ratio (we set $R_0 = 1$)

$$\frac{R_k}{R_{k-1}}$$

diminishes with increasing k. Since

$$\frac{R_k}{R_{k-1}} = \frac{n-k}{k}\frac{1 + \delta_k}{1 + \delta_{k-1}}, \qquad \frac{R_{k+1}}{R_k} = \frac{n-k-1}{k+1}\frac{1 + \delta_{k+1}}{1 + \delta_k},$$

it suffices to show that

$$\frac{k(n - k - 1)}{(k + 1)(n - k)} < \frac{(1 + \delta_k)^2}{(1 + \delta_{k-1})(1 + \delta_{k+1})}. \tag{5}$$

Now

$$\frac{k(n - k - 1)}{(k + 1)(n - k)} = 1 - \frac{n}{(k + 1)(n - k)} \leq 1 - \frac{4n}{(n + 1)^2} = \left(\frac{n - 1}{n + 1}\right)^2.$$

On the other hand,

$$\frac{(1 + \delta_k)^2}{(1 + \delta_{k-1})(1 + \delta_{k+1})} > \frac{\left(1 - \dfrac{1}{n}\right)^2}{\left(1 + \dfrac{1}{n}\right)^2} = \frac{(n - 1)^2}{(n + 1)^2},$$

and so the inequality (5) is proved.

5. Suppose that for the root x the strict inequality

$$(P_{m-1} - Q_{m-1}x)(P_m - Q_m x) < 0$$

holds. Then, the root ξ of the transformed equation corresponding to x will be positive, and it can be shown that the polynomial

$$(u - \xi)(u - \xi_1) \cdots (u - \xi_{n-1}) = \phi(u),$$

differing from $F(u)$ only by a constant factor, presents just one variation. On substituting

$$u = \frac{Q_{m-1}}{Q_m}v, \qquad \xi = \frac{Q_{m-1}}{Q_m}\omega, \qquad \omega > 0$$

we have

$$\phi(u) = \left(\frac{Q_{m-1}}{Q_m}\right)^n (v - \omega)(v + 1 + \alpha_1) \cdots (v + 1 + \alpha_{n-1}),$$

so that it suffices to prove the statement for the polynomial

$$
\begin{aligned}
(v - \omega)(v + 1 + \alpha_1) &\cdots (v + 1 + \alpha_{n-1}) \\
&= (v - \omega)(v^{n-1} + R_1 v^{n-2} + \cdots + R_{n-1}) \\
&= v^n + (R_1 - \omega)v^{n-1} + (R_2 - R_1\omega)v^{n-2} + \cdots \\
&\qquad + (R_{n-1} - R_{n-2}\omega)v - R_{n-1}\omega.
\end{aligned}
$$

This polynomial certainly presents one variation and not more than one since $R_1, R_2, \ldots, R_{n-1}, \omega$ are positive and

$$R_1 - \omega, \qquad \frac{R_2}{R_1} - \omega, \qquad \ldots, \qquad \frac{R_{n-1}}{R_{n-2}} - \omega, \qquad - \omega$$

is a decreasing sequence of numbers.

It remains to consider the case

$$(P_{m-1} - Q_{m-1}x)(P_m - Q_m x) = 0.$$

All conclusions as to the roots of the transformed equation corresponding to the roots differing from x remain in force, and, in particular, the conclusion that $R_k > 0$. Only in case

$$P_{m-1} - Q_{m-1}x = 0$$

is the quantity denoted by ξ equal to 0. Clearly then the transformed equation has no variation. In case

$$P_m - Q_m x = 0$$

we have $\xi = \infty$, which shows that the transformed equation reduces to degree $n - 1$. The polynomial $F(u)$ differs only by a constant factor from the product

$$(u - \xi_1)(u - \xi_2) \cdots (u - \xi_{n-1}),$$

which in expanded form presents no variation. Thus, Vincent's theorem is proved completely.

APPENDIX III

ON EQUATIONS WHOSE ROOTS HAVE NEGATIVE
REAL PART

1. In the investigation of dynamical stability of mechanical systems it is necessary to have a simple criterion to decide whether all roots of an algebraic equation with real coefficients

$$f(x) = p_0 + p_1 x + p_2 x^2 + \cdots + p_n x^n = 0 \tag{1}$$

have negative real part. This question was raised by Maxwell, investigated further by Routh, and solved in a very elegant way by Hurwitz. Hurwitz found the following criterion: *Equation (1) has all roots with negative real part if, and only if, the n determinants*

$$D_1 = p_1, \qquad D_2 = \begin{vmatrix} p_1 & p_0 \\ p_3 & p_2 \end{vmatrix}, \qquad D_3 = \begin{vmatrix} p_1 & p_0 & 0 \\ p_3 & p_2 & p_1 \\ p_5 & p_4 & p_3 \end{vmatrix}, \qquad \ldots,$$

$$D_n = \begin{vmatrix} p_1 & p_0 & \cdots & 0 \\ p_3 & p_2 & \cdots & 0 \\ \cdots\cdots\cdots\cdots\cdots\cdots \\ p_{2n-1} & p_{2n-2} & \cdots & p_n \end{vmatrix}$$

are positive, provided, as can legitimately be assumed, $p_0 > 0$. The indices of the letters p in each row of every determinant decrease by 1, and the letters with negative indices as well as those with indices $> n$ are replaced by zeros.

In case $n = 2$ this criterion is verified directly. For the conditions then are

$$p_0 > 0, \qquad p_1 > 0, \qquad \begin{vmatrix} p_1 & p_0 \\ 0 & p_2 \end{vmatrix} = p_1 p_2 > 0,$$

or $p_0 > 0$, $p_1 > 0$, $p_2 > 0$, and then it is evident that

$$p_0 + p_1 x + p_2 x^2 = 0$$

has roots with negative real part. Assuming Hurwitz's criterion to be true for equations of degree $n - 1$, we are going to show that it remains true for equations of degree n, thus completing the induction. We shall follow a simple and elegant method of I. Schur.

The following proposition can be established without difficulty: *If all roots of $f(x)$ have negative real part and ξ is any complex number, then*

$$\begin{array}{ll} |f(\xi)| > |f(-\xi)| & \text{if} \quad R\xi > 0, \\ |f(\xi)| = |f(-\xi)| & \text{if} \quad R\xi = 0, \\ |f(\xi)| < |f(-\xi)| & \text{if} \quad R\xi < 0. \end{array} \tag{2}$$

Consider the factorization of $f(x)$:

$$f(x) = p_n(x - \alpha_1)(x - \alpha_2) \cdots (x - \alpha_n).$$

Supposing $\xi = a + bi$ and considering real linear factor $x - \alpha_k$, we have

$$|\xi - \alpha_k|^2 = (a - \alpha_k)^2 + b^2;$$

hence, since by hypothesis $\alpha_k < 0$,

$$|\xi - \alpha_k| \gtreqless |-\xi - \alpha_k| \tag{3}$$

according as

$$a = R\xi \gtreqless 0.$$

Imaginary linear factors occur in conjugate pairs. Let $x - \alpha_r$ and $x - \alpha_s$ be one such pair and

$$\alpha_r = \gamma_r + i\delta_r, \qquad \alpha_s = \gamma_r - i\delta_r, \qquad \gamma_r < 0, \qquad \delta_r > 0.$$

Then,

$$|\xi - \alpha_r|^2 = (a - \gamma_r)^2 + (b - \delta_r)^2, \qquad |\xi - \alpha_s|^2 = (a - \gamma_r)^2 + (b + \delta_r)^2,$$

and

$$|-\xi - \alpha_r|^2 = (a + \gamma_r)^2 + (b + \delta_r)^2, \qquad |-\xi - \alpha_s|^2 = (a + \gamma_r)^2 + (b - \delta_r)^2.$$

But

$$(a - \gamma_r)^2 \gtreqless (a + \gamma_r)^2$$

according as

$$a \gtreqless 0.$$

Consequently,

$$|(\xi - \alpha_r)(\xi - \alpha_s)| \gtreqless |(-\xi - \alpha_r)(-\xi - \alpha_s)| \tag{4}$$

according as

$$R\xi \gtreqless 0.$$

The inequalities (2) now follow easily from (3) and (4).

2. If we set

$$\tfrac{1}{2}\psi = p_0 p_1 + \omega(p_1 p_2 - p_0 p_3)x + p_0 p_3 x^2 + \omega(p_1 p_4 - p_0 p_5)x^3 + \cdots, \tag{5}$$

it is easy to verify that

$$x\psi = \left[p_0\left(1 - \frac{\omega}{x}\right) + p_1\omega\right]f(x) - \left[p_0\left(1 - \frac{\omega}{x}\right) - p_1\omega\right]f(-x)$$

or

$$x\psi = A(x)f(x) - B(x)f(-x) \tag{6}$$

where

$$A(x) = p_0\left(1 - \frac{\omega}{x}\right) + p_1\omega, \qquad B(x) = p_0\left(1 - \frac{\omega}{x}\right) - p_1\omega.$$

Suppose that $p_0 > 0$, $p_1 > 0$, $\omega > 0$ and let x be such a complex number $\neq 0$ that

$$R\frac{\omega}{x} < 1.$$

Then, it is easy to see that

$$|A(x)| > |B(x)|. \tag{7}$$

3. Roots of the equation $\psi = 0$ vary with ω, but their reciprocals will be bounded if ω is bounded, for example, if $\omega < 1$. For this equation, as is easily seen, is of degree $n - 1$, and the reciprocals of its roots satisfy the equation

$$x^{n-1} + \omega\frac{p_1p_2 - p_0p_3}{p_0p_1}x^{n-2} + \frac{p_3}{p_1}x^{n-3} + \cdots = 0$$

whose coefficients are bounded. Hence, ω can be taken positive and so small that for all roots of $\psi = 0$

$$R\frac{\omega}{x} < 1.$$

Then, the real parts of all the roots of $\psi = 0$ will be negative if the real parts of the roots of $f(x) = 0$ are negative. For suppose that for some root ξ of $\psi = 0$

$$R\xi \geqq 0.$$

Then, according to (6)

$$A(\xi)f(\xi) = B(\xi)f(-\xi),$$

and according to (7)

$$|A(\xi)| > |B(\xi)|$$

since

$$R\frac{\omega}{\xi} < 1.$$

Consequently,

$$|f(\xi)| = \frac{|B(\xi)|}{|A(\xi)|}\, |f(-\xi)| < |f(-\xi)|,$$

which is impossible if $R\xi \geq 0$.

4. The converse is also true: If $p_0 > 0$, $p_1 > 0$, and all roots of $\psi = 0$ have negative real part, then all roots of $f(x) = 0$ have negative real part. Changing, in the identity

$$x\psi(x) = A(x)f(x) - B(x)f(-x),$$

x into $-x$, we have

$$x\psi(-x) = B(-x)f(x) - A(-x)f(-x)$$

and, solving for $f(x)$,

$$f(x) = \frac{xA(-x)}{A(x)A(-x) - B(x)B(-x)}\psi(x) - \frac{xB(x)}{A(x)A(-x) - B(x)B(-x)}\psi(-x).$$

But

$$A(x)A(-x) - B(x)B(-x) = 4p_0p_1\omega$$

and so

$$4p_0p_1\omega f(x) = xA(-x)\psi(x) - xB(x)\psi(-x). \tag{8}$$

Now, suppose that $\xi \neq 0$ is a root of $f(x) = 0$ and that $R\xi \geq 0$. It follows then from (8) that

$$A(-\xi)\psi(\xi) = B(\xi)\psi(-\xi). \tag{9}$$

Setting $\xi = a + bi$, we have

$$A(-\xi) = p_0(1 + a\omega + ib\omega) + p_1\omega,$$
$$B(\xi) = p_0(1 - a\omega - ib\omega) - p_1\omega,$$

and

$$|A(-\xi)|^2 = [p_0(1 + a\omega) + p_1\omega]^2 + p_0^2b^2\omega^2,$$
$$|B(\xi)|^2 = [p_0(1 - a\omega) - p_1\omega]^2 + p_0^2b^2\omega^2,$$

whence

$$|A(-\xi)|^2 - |B(\xi)|^2 = 4\omega p_0^2 a + 4p_0p_1\omega > 0,$$

that is,

$$|A(-\xi)| > |B(\xi)|.$$

Then, from (9)

$$|\psi(\xi)| = \frac{|B(\xi)|}{|A(-\xi)|}\, |\psi(-\xi)| < |\psi(-\xi)|$$

and this is impossible for $R\xi \geq 0$ since all roots of $\psi = 0$, by hypothesis, have negative real part.

5. Corresponding to the determinants D_1, D_2, . . . for the equation $\psi = 0$ we have the determinants

$$\Delta_1 = \omega(p_1p_2 - p_0p_3), \qquad \Delta_2 = \begin{vmatrix} \omega(p_1p_2 - p_0p_3) & p_0p_1 \\ \omega(p_1p_4 - p_0p_5) & p_0p_3 \end{vmatrix},$$

$$\Delta_3 = \begin{vmatrix} \omega(p_1p_2 - p_0p_3) & p_0p_1 & 0 \\ \omega(p_1p_4 - p_0p_5) & p_0p_3 & \omega(p_1p_2 - p_0p_3) \\ \omega(p_1p_6 - p_0p_7) & p_0p_5 & \omega(p_1p_4 - p_0p_5) \end{vmatrix}, \qquad \ldots$$

Now

$$\Delta_1 = \omega D_2, \qquad \Delta_2 = \omega p_0 \begin{vmatrix} p_1p_2 - p_0p_3 & p_1 \\ p_1p_4 - p_0p_5 & p_3 \end{vmatrix} = \omega p_0 p_1^{-1} \begin{vmatrix} p_1 & 0 & 0 \\ p_3 & p_1p_2 - p_0p_3 & p_1 \\ p_5 & p_1p_4 - p_0p_5 & p_3 \end{vmatrix}$$

or

$$\Delta_2 = \omega p_0 \begin{vmatrix} p_1 & p_0 & 0 \\ p_3 & p_2 & p_1 \\ p_5 & p_4 & p_3 \end{vmatrix} = \omega p_0 D_3.$$

Similarly,

$$\Delta_3 = \omega^2 p_1^{-1} \begin{vmatrix} p_1 & 0 & 0 \\ p_3 & p_1p_2 - p_0p_3 & p_0p_1 \\ p_5 & p_1p_4 - p_0p_5 & p_0p_3 \\ p_7 & p_1p_6 - p_0p_7 & p_0p_5 \end{vmatrix} = \omega^2 p_0 p_1 \begin{vmatrix} p_1 & p_0 & 0 & 0 \\ p_3 & p_2 & p_1 & p_0 \\ p_5 & p_4 & p_3 & p_2 \\ p_7 & p_6 & p_5 & p_4 \end{vmatrix}$$

or

$$\Delta_3 = \omega^2 p_0 p_1 D_4.$$

In general, every determinant Δ_k differs from D_{k+1} only by a positive factor. Now we can prove that the conditions

$$D_1 > 0, \qquad D_2 > 0, \qquad \ldots, \qquad D_n > 0$$

are sufficient for an equation $f(x) = 0$ of degree n to have all roots with negative real part. For then we have $p_0 > 0$, $p_1 > 0$ and

$$\Delta_1 > 0, \qquad \Delta_2 > 0, \qquad \ldots, \qquad \Delta_{n-1} > 0.$$

Hence, assuming that the criterion holds for equations of degree $n - 1$, equation $\psi = 0$ will have all roots with negative real part and then (Sec. 4) the same will be true for $f(x) = 0$. Conditions

$$D_1 > 0, \qquad D_2 > 0, \qquad \ldots, \qquad D_n > 0$$

are also necessary. For if $f(x) = 0$ has all roots with negative real part, the polynomial $f(x)$ consists of linear factors $x + \alpha$ with positive α and quadratic factors $x^2 + 2\beta x + \gamma$ with positive β and γ. Hence, all coefficients of $f(x)$ are of the same sign, and supposing $p_0 > 0$, we shall have also $D_1 = p_1 > 0$. Equation $\psi = 0$ having all roots with negative real part (Sec. 3), it is necessary that

$$\Delta_1 > 0, \qquad \Delta_2 > 0, \qquad \ldots, \qquad \Delta_{n-1} > 0$$

by hypothesis; consequently, also

$$D_2 > 0, \qquad D_3 > 0, \qquad \ldots, \qquad D_n > 0.$$

Thus,

$$D_1 > 0, \qquad D_2 > 0, \qquad \ldots, \qquad D_n > 0$$

if all roots of $f(x) = 0$ have negative real part provided $p_0 > 0$.

APPENDIX IV

ITERATIVE SOLUTION OF THE FREQUENCY EQUATION

1. In the study of small vibrations of mechanical systems one has to deal with the system of linear differential equations of the second order with constant coefficients

$$\ddot{q}_1 + a_{11}q_1 + a_{12}q_2 + \cdots + a_{1n}q_n = 0,$$
$$\ddot{q}_2 + a_{21}q_1 + a_{22}q_2 + \cdots + a_{2n}q_n = 0,$$
$$\cdots \cdots \cdots \cdots \cdots \cdots \cdots \cdots \cdots$$
$$\ddot{q}_n + a_{n1}q_1 + a_{n2}q_2 + \cdots + a_{nn}q_n = 0,$$

$$(1)$$

which determines the variation with time of the n independent parameters $q_1, q_2, \ldots, q_n$ defining the configuration of the system. The real constants a_{ij} satisfy the condition

$$a_{ij} = a_{ji}$$

for all indices i, j so that the matrix of the linear forms on the left of (1) is symmetrical. To solve equations (1) solutions of the form

$$q_k = A_k \sin{(pt + \alpha)} \qquad (2)$$

are sought representing natural harmonic oscillations with the same frequency p and phase α, but different amplitudes for different parameters. Substituting expressions (2) into equations (1) and canceling the factor $\sin{(pt + \alpha)}$, we get the following equations:

$$(a_{11} - p^2)A_1 + a_{12}A_2 + \cdots + a_{1n}A_n = 0,$$
$$a_{21}A_1 + (a_{22} - p^2)A_2 + \cdots + a_{2n}A_n = 0,$$
$$\cdots \cdots \cdots \cdots \cdots \cdots \cdots \cdots \cdots$$
$$a_{n1}A_1 + a_{n2}A_2 + \cdots + (a_{nn} - p^2)A_n = 0.$$

$$(3)$$

Writing $p^2 = \lambda$ and considering that not all the amplitudes vanish, we conclude that the determinant of system (3) must vanish. This gives the following equation for λ in determinant form:

$$F(\lambda) = \begin{vmatrix} a_{11} - \lambda & a_{12} & \cdots & a_{1n} \\ a_{21} & a_{22} - \lambda & \cdots & a_{2n} \\ \cdots & \cdots & \cdots & \cdots \\ a_{n1} & a_{n2} & \cdots & a_{nn} - \lambda \end{vmatrix} = 0 \qquad (4)$$

whose degree is n. It is called the *frequency equation*. The same equation occurs in the theory of secular perturbations of planetary orbits and for this reason is often called the *secular equation*. Considered independently of physical and astronomical problems, the same equation is known as the *characteristic equation* of the matrix

$$A = \begin{pmatrix} a_{11} & a_{21} & \cdots & a_{1n} \\ a_{21} & a_{22} & \cdots & a_{2n} \\ \cdots\cdots\cdots\cdots\cdots \\ a_{n1} & a_{n2} & \cdots & a_{nn} \end{pmatrix}.$$

As was mentioned already this matrix will be supposed symmetrical.

2. Two sets of amplitudes $A_1, A_2, \ldots, A_n$ and $A'_1, A'_2, \ldots, A'_n$ corresponding to two different roots λ and λ' of the frequency equation satisfy the condition of *orthogonality*

$$A_1A'_1 + A_2A'_2 + \cdots + A_nA'_n = 0.$$

The origin of this expression is explained by the fact that two vectors

$$A = (A_1, A_2, \ldots, A_n)$$
$$A' = (A'_1, A'_2, \ldots, A'_n)$$

are called orthogonal if their scalar product

$$A \cdot A' = A_1A'_1 + A_2A'_2 + \cdots + A_nA'_n$$

vanishes. To prove the statement consider two systems of linear equations

$$
\begin{aligned}
a_{11}A_1 + a_{12}A_2 + \cdots + a_{1n}A_n &= \lambda A_1, \\
a_{21}A_1 + a_{22}A_2 + \cdots + a_{2n}A_n &= \lambda A_2, \\
\cdots\cdots\cdots\cdots\cdots\cdots\cdots\cdots \\
a_{n1}A_1 + a_{n2}A_2 + \cdots + a_{nn}A_n &= \lambda A_n,
\end{aligned}
\tag{5}
$$

and

$$
\begin{aligned}
a_{11}A'_1 + a_{12}A'_2 + \cdots + a_{1n}A'_n &= \lambda' A_1, \\
a_{21}A'_1 + a_{22}A'_2 + \cdots + a_{2n}A'_n &= \lambda' A_2, \\
\cdots\cdots\cdots\cdots\cdots\cdots\cdots\cdots \\
a_{n1}A'_1 + a_{n2}A'_2 + \cdots + a_{nn}A'_n &= \lambda' A_n.
\end{aligned}
\tag{6}
$$

Equations (5) multiplied, respectively, by $A'_1, A'_2, \ldots, A'_n$ and added give the result

$$\sum_{i,j} a_{ij}A'_iA_j = \lambda(A_1A'_1 + A_2A'_2 + \cdots + A_nA'_n) \tag{7}$$

where on the left the double summation refers to indices i, j independently running through $1, 2, \ldots, n$. Equations (6) multiplied by $A_1, A_2, \ldots, A_n$ and added give, on the other hand,

$$\sum_{i,j} a_{ji}A'_iA_j = \lambda'(A_1A'_1 + A_2A'_2 + \cdots + A_nA'_n). \tag{8}$$

But because $a_{ij} = a_{ji}$ both sums on the left of (7) and (8) are equal, and on subtracting we get

$$(\lambda - \lambda')(A_1A_1' + A_2A_2' + \cdots + A_nA_n') = 0,$$

and so

$$A_1A_1' + A_2A_2' + \cdots + A_nA_n' = 0$$

since $\lambda' \neq \lambda$.

The orthogonality relation may be used to prove, following Lagrange, that the roots of the frequency equation are real. For suppose that there is an imaginary root $\lambda = \alpha + i\beta$, $\beta \neq 0$; then there will be a conjugate imaginary root $\lambda' = \alpha - i\beta$ and $\lambda' \neq \lambda$. Equations (5) are satisfied by complex numbers $A_1, A_2, \ldots, A_n$ not all equal to zero. Equations (6) corresponding to the conjugate root will be satisfied by numbers $\bar{A}_1, \bar{A}_2, \ldots, \bar{A}_n$ conjugate, respectively, to $A_1, A_2, \ldots, A_n$, and the orthogonality relation gives

$$A_1\bar{A}_1 + A_2\bar{A}_2 + \cdots + A_n\bar{A}_n = 0$$

or

$$|A_1|^2 + |A_2|^2 + \cdots + |A_n|^2 = 0,$$

which requires that

$$A_1 = A_2 = \cdots = A_n = 0,$$

contrary to hypothesis.

In mechanical applications the roots λ are not only real but positive and unequal, barring very exceptional cases of no interest in applications. The formation of the frequency equation requires the expansion of the determinant (4), which might be of a high order in some cases (as in planetary theory), and that is a very tedious task. When roots λ differ considerably in magnitude, the engineers prefer approximate evaluation of them by a certain iteration process. The purpose of this appendix is to explain the outlines of this process.

3. The exposition will gain considerably in simplicity and elegance by making use of the algebra of matrices. To this end we must add some notions not given in the text in Chap. IX, Secs. 12 to 15.

A matrix A* obtained from the matrix A by interchanging rows and columns is called the transposed of, or conjugate to, A. A symmetric matrix is self-conjugate so that

$$A^* = A,$$

and, vice versa, any matrix satisfying this condition is symmetric. Determinants of conjugate matrices are equal. By the rule of multiplication of matrices one verifies the following equation:

$$(AB)^* = B^*A^*,$$

which can be interpreted by saying that the conjugate matrix of the product is equal to the product of the conjugates of the factors taken in the reverse order. A matrix S is called *orthogonal* if it satisfies the equation

$$SS^* = E, \tag{9}$$

from which it follows, in the first place, by the rule of multiplication of determinants that

$$(\det S)^2 = 1,$$

so that

$$\det S = \pm 1;$$

and, in the second place, that

$$S^* = S^{-1},$$

and hence also

$$S^*S = E. \tag{10}$$

If c_{ij} are elements of an orthogonal matrix, it follows from the equations (9) and (10) that they satisfy the relations

$$c_{i1}^2 + c_{i2}^2 + \cdots + c_{in}^2 = 1$$
$$c_{i1}c_{j1} + c_{i2}c_{j2} + \cdots + c_{in}c_{jn} = 0 \quad \text{for} \quad i \neq j \tag{11}$$

and

$$c_{1i}^2 + c_{2i}^2 + \cdots + c_{ni}^2 = 1$$
$$c_{1i}c_{1j} + c_{2i}c_{2j} + \cdots + c_{ni}c_{nj} = 0 \quad \text{for} \quad i \neq j. \tag{12}$$

4. Let $A_1, A_2, \ldots, A_n$ be the amplitudes corresponding to a root λ of the frequency equation. On multiplying them by a convenient factor it can be assumed that

$$A_1^2 + A_2^2 + \cdots + A_n^2 = 1.$$

Amplitudes satisfying this condition will be called *normalized*. Let

$$\alpha_1, \quad \alpha_2, \quad \ldots, \quad \alpha_n,$$
$$\beta_1, \quad \beta_2, \quad \ldots, \quad \beta_n,$$
$$\ldots \ldots \ldots \ldots \ldots$$
$$\nu_1, \quad \nu_2, \quad \ldots, \quad \nu_n$$

be normalized amplitudes corresponding, respectively, to the roots $\lambda_1, \lambda_2, \ldots, \lambda_n$ of the frequency equation. If we suppose for simplicity that these roots are simple, normalized amplitudes corresponding to them will satisfy the orthogonality relation whence, by virtue of (11), it follows that the matrix

$$S = \begin{pmatrix} \alpha_1 & \alpha_2 & \cdots & \alpha_n \\ \beta_1 & \beta_2 & \cdots & \beta_n \\ \cdots & \cdots & \cdots & \cdots \\ \nu_1 & \nu_2 & \cdots & \nu_n \end{pmatrix}$$

is orthogonal. The amplitudes corresponding to the root λ_1 satisfy the equations

$$a_{11}\alpha_1 + a_{12}\alpha_2 + \cdots + a_{1n}\alpha_n = \lambda_1\alpha_1,$$
$$a_{21}\alpha_1 + a_{22}\alpha_2 + \cdots + a_{2n}\alpha_n = \lambda_1\alpha_2,$$
$$\cdots\cdots\cdots\cdots\cdots\cdots\cdots\cdots$$
$$a_{n1}\alpha_1 + a_{n2}\alpha_2 + \cdots + a_{nn}\alpha_n = \lambda_1\alpha_n,$$

which, by introducing the column matrix,

$$(\alpha) = \begin{pmatrix} \alpha_1 \\ \alpha_2 \\ \cdot \\ \cdot \\ \cdot \\ \alpha_n \end{pmatrix}$$

can be condensed into one matrix equation

$$A(\alpha) = \lambda_1(\alpha).$$

From this equation it follows that

$$A^2(\alpha) = AA(\alpha) = \lambda_1 A(\alpha) = \lambda_1^2(\alpha),$$
$$A^3(\alpha) = AA^2(\alpha) = \lambda_1^2 A(\alpha) = \lambda_1^3(\alpha),$$

and in general

$$A^m(\alpha) = \lambda_1^m(\alpha).$$

Introducing similarly column matrices (β), . . . , (ν) corresponding to the roots λ_2, . . . , λ_n, we have thus n matrix equations

$$A^m(\alpha) = \lambda_1^m(\alpha), \qquad A^m(\beta) = \lambda_2^m(\beta), \qquad \ldots, \qquad A^m(\nu) = \lambda_n^m(\nu). \quad (13)$$

Let

$$C = \begin{pmatrix} c_1 \\ c_2 \\ \cdot \\ \cdot \\ \cdot \\ c_n \end{pmatrix}$$

be any column matrix and write

$$AC = C_1, \qquad AC_1 = C_2, \qquad AC_2 = C_3, \qquad \ldots ;$$

then in general

$$C_m = AC_{m-1} = A^m C. \tag{14}$$

On the other hand, constants γ_1, γ_2, . . . , γ_n can be so determined that

$$C = \gamma_1(\alpha) + \gamma_2(\beta) + \cdots + \gamma_n(\nu).$$

From (14) and (13) it follows then that

$$C_m = \lambda_1^m \gamma_1(\alpha) + \gamma_2^m \lambda_2(\beta) + \cdots + \lambda_n^m \gamma_n(\nu).$$

If $c_1^{(m)}, c_2^{(m)}, \ldots, c_n^{(m)}$ are elements of C_m, the last equation shows that

$$c_i^{(m)} = \gamma_1 \lambda_1^m \alpha_i + \gamma_2 \lambda_2^m \beta_i + \cdots + \gamma_n \lambda_n^m \nu_i \qquad (15)$$

for $i = 1, 2, \ldots, n$. Notice that owing to the orthogonality of S

$$\gamma_1 = c_1 \alpha_1 + c_2 \alpha_2 + \cdots + c_n \alpha_n. \qquad (16)$$

From (15) it follows that

$$\frac{c_i^{(m+1)}}{c_i^{(m)}} = \frac{\gamma_1 \lambda_1^{m+1} \alpha_i + \gamma_2 \lambda_2^{m+1} \beta_i + \cdots + \gamma_n \lambda_n^{m+1} \nu_i}{\gamma_1 \lambda_1^m \alpha_i + \gamma_2 \lambda_2^m \beta_i + \cdots + \gamma_n \lambda_n^m \nu_i},$$

or

$$\frac{c_i^{(m+1)}}{c_i^{(m)}} = \lambda_1 \frac{\gamma_1 \alpha_i + \gamma_2 \beta_i \left(\frac{\lambda_2}{\lambda_1}\right)^{m+1} + \cdots + \gamma_n \nu_i \left(\frac{\lambda_n}{\lambda_1}\right)^{m+1}}{\gamma_1 \alpha_i + \gamma_2 \beta_i \left(\frac{\lambda_2}{\lambda_1}\right)^{m} + \cdots + \gamma_n \nu_i \left(\frac{\lambda_n}{\lambda_1}\right)^{m}}.$$

Suppose now that λ_1 is the largest root; then, the ratios

$$\frac{\lambda_2}{\lambda_1}, \ldots, \frac{\lambda_n}{\lambda_1}$$

being less than 1, their powers will converge to 0 when the exponents increase indefinitely. Provided therefore that $\gamma_1 \alpha_i \neq 0$, the limit of the ratio

$$\frac{c_i^{(m+1)}}{c_i^{(m)}}$$

as $m \to \infty$ will be λ_1, and for large m the largest root (corresponding to the highest frequency) may be calculated approximately from the equation

$$\lambda_1 = \frac{c_i^{(m+1)}}{c_i^{(m)}},$$

where, in general, i may be any of the numbers $1, 2, \ldots, n$. Similar considerations show also that for large m the ratios

$$\frac{c_2^{(m)}}{c_1^{(m)}}, \ldots, \frac{c_n^{(m)}}{c_1^{(m)}}$$

will be nearly equal to the ratios of the amplitudes

$$\frac{\alpha_2}{\alpha_1}, \ldots, \frac{\alpha_n}{\alpha_1}.$$

The computation of $c_1^{(m)}$, $c_2^{(m)}$, . . . , $c_n^{(m)}$ can be made by a simple recurrent process using the formulas

$$c_1^{(m+1)} = a_{11}c_1^{(m)} + a_{12}c_2^{(m)} + \cdots + a_{1n}c_n^{(m)},$$
$$c_2^{(m+1)} = a_{21}c_1^{(m)} + a_{22}c_2^{(m)} + \cdots + a_{2n}c_n^{(m)},$$
$$\cdots \cdots \cdots \cdots \cdots \cdots \cdots \cdots \cdots \cdots$$
$$c_n^{(m+1)} = a_{n1}c_1^{(m)} + a_{n2}c_2^{(m)} + \cdots + a_{nn}c_n^{(m)}.$$

As to the initial values c_1, c_2, . . . , c_n they can be taken arbitrarily with the sole condition that

$$\gamma_1 = c_1\alpha_1 + c_2\alpha_2 + \cdots + c_n\alpha_n \neq 0.$$

If c_1, c_2, . . . , c_n are taken at random, this condition in general will be satisfied.

To find the least root of the frequency equation, or the lowest frequency, the same method can be applied to the reciprocal matrix A^{-1}. To prove this, notice that the characteristic equation of A is

$$F(\lambda) = \det (A - \lambda E) = 0.$$

Now

$$A(A^{-1} - \lambda^{-1}E) = - \lambda^{-1}(A - \lambda E)$$

whence by the rule of multiplication of determinants

$$\det A \cdot \det (A^{-1} - \lambda^{-1}E) = (- 1)^n\lambda^{-n} \det (A - \lambda E),$$

and from this it follows that the roots of the characteristic equation for the reciprocal matrix A^{-1} are

$$\lambda_1^{-1}, \lambda_2^{-1}, \ldots , \lambda_n^{-1},$$

and the largest of them is the reciprocal of the smallest root corresponding to A. Hence, the smallest root can be computed by the above process. Once the largest root is computed with sufficient accuracy, the other roots can be computed approximately by the same method supplemented with certain other considerations, but we shall not consider this question. The method explained in this appendix is based on the same idea as the old method proposed by Daniel Bernoulli for any algebraic equation and has the same defects of slow convergence when the roots are numerically not very different. The slow convergence that sometimes occurs is compensated by the fact that accidental errors of computation only retard the process but have no influence on the final result. The simplicity of calculation is another advantage especially appreciable when the work is entrusted to a staff of computers. Probably these are the reasons why many engineers use this method in preference to others.

The defect of slow convergence can be remedied to a certain extent by the following procedure, which amounts to using $2m$ iterations instead of m: Referring to the expression

$$c_i^{(m)} = \gamma_1 \alpha_i \lambda_1^m + \gamma_2 \beta_i \lambda_2^m + \cdots + \gamma_n \nu_i \lambda_n^m$$

and the orthogonality relations (12), it is easily found that

$$\Gamma_m = c_1^{(m)} c_1^{(m+1)} + c_2^{(m)} c_2^{(m+1)} + \cdots + c_n^{(m)} c_n^{(m+1)}$$
$$\Delta_m = c_1^{(m)} c_1^{(m)} + c_2^{(m)} c_2^{(m)} + \cdots + c_n^{(m)} c_n^{(m)}$$

will have the following simple expressions:

$$\Gamma_m = \gamma_1^2 \lambda_1^{2m+1} + \gamma_2^2 \lambda_2^{2m+1} + \cdots + \gamma_n^2 \lambda_n^{2m+1}$$
$$\Delta_m = \gamma_1^2 \lambda_1^{2m} + \gamma_2^2 \lambda_2^{2m} + \cdots + \gamma_n^2 \lambda_n^{2m}$$

whence the quotient

$$\frac{\Gamma_m}{\Delta_m}$$

will be approximately equal to the largest root λ_1 and it is easy to see that

$$\frac{\Gamma_m}{\Delta_m} < \lambda_1.$$

Example. Let the matrix be

$$A = \begin{pmatrix} 2 & -1 & 0 \\ -1 & 2 & -1 \\ 0 & -1 & 1 \end{pmatrix}.$$

Starting with

$$c_1^{(0)} = 1, \qquad c_2^{(0)} = 0, \qquad c_3^{(0)} = 0$$

we readily compute by recurrence

$$c_1^{(10)} = 14041; \qquad c_2^{(10)} = -17460; \qquad c_3^{(10)} = 7753;$$
$$c_1^{(11)} = 45542; \qquad c_2^{(11)} = -56714; \qquad c_3^{(11)} = 25213;$$

whence

$$\Gamma_{10} = 1825158051,$$
$$\Delta_{10} = 562110290,$$

and

$$\frac{\Gamma_{10}}{\Delta_{10}} = 3.246975,$$

whereas the largest characteristic root of the matrix A is

$$3.2469795$$

so that the approximation even after such a small number of steps is quite satisfactory.

APPENDIX V

GRAEFFE'S METHOD

1. When it is necessary to compute all the roots of an equation, including imaginary roots, the so-called root-squaring or Graeffe's method is preferable to others, and it is the only practical method for the computation of imaginary roots. The first idea of this method occurs in the writings of Waring in the eighteenth century. Later a method for computation of roots based on the same idea was proposed independently by Dandelin (1826) and Lobatchevsky (1834) but was developed in all details only by Graeffe (1837).

Let

$$f(x) = p_0 + p_1x + p_2x^2 + \cdots + p_nx^n = 0$$

be the proposed equation with roots $\alpha_1, \alpha_2, \ldots, \alpha_n$. The first part in Graeffe's method is an algorithm for the formation of the equation with roots $\alpha_1^2, \alpha_2^2, \ldots, \alpha_n^2$. Since

$$f(x) = p_n(x - \alpha_1)(x - \alpha_2) \cdots (x - \alpha_n),$$
$$(-1)^n f(-x) = p_n(x + \alpha_1)(x + \alpha_2) \cdots (x + \alpha_n),$$

and on multiplying these equations member by member we have

$$(-1)^n f(x)f(-x) = p_n^2(x^2 - \alpha_1^2)(x^2 - \alpha_2^2) \cdots (x^2 - \alpha_n^2),$$

so that, replacing x by $\sqrt{x}$, the requested equation with roots $\alpha_1^2, \alpha_2^2, \ldots, \alpha_n^2$ may be written thus:

$$F(x) = f(\sqrt{x})f(-\sqrt{x}) = 0.$$

Now

$$f(\sqrt{x}) = p_0 + p_2x + p_4x^2 + \cdots + \sqrt{x}(p_1 + p_3x + p_5x^2 + \cdots),$$
$$f(-\sqrt{x}) = p_0 + p_2x + p_4x^2 + \cdots - \sqrt{x}(p_1 + p_3x + p_5x^2 + \cdots),$$

and so

$$F(x) = (p_0 + p_2x + p_4x^2 + \cdots)^2 - x(p_1 + p_3x + p_5x^2 + \cdots)^2,$$

whence it is easy to see that the coefficient of x^i in $F(x)$ will be

$$(-1)^i[p_i^2 - 2p_{i-1}p_{i+1} + 2p_{i-2}p_{i+2} - \cdots],$$

the terms in the brackets being continued as long as the indices of the letters p do not become negative or greater than n. Some simplification results by changing x into $-x$ in $F(x)$, that is, considering the equation whose roots are $-\alpha_1^2,\ -\alpha_2^2,\ \ldots,\ -\alpha_n^2$. In this case, we have the following uniform rule: Writing the original equation in the form

$$a_0 x^n + a_1 x^{n-1} + \cdots + a_n = 0,$$

the coefficient of x^{n-i} in the transformed equation, whose roots are $-\alpha_1^2,\ -\alpha_2^2,\ \ldots,\ -\alpha_n^2$, is expressed by the sum

$$a_i^2 - 2a_{i-1}a_{i+1} + 2a_{i-2}a_{i+2} - \cdots,$$

which is continued as long as indices do not become negative or greater than n.

By a repetition of the same process, transformed equations are obtained whose roots are

$$-\alpha_1^4,\quad -\alpha_2^4,\quad \cdots,\quad -\alpha_n^4,$$
$$-\alpha_1^8,\quad -\alpha_2^8,\quad \ldots,\quad -\alpha_n^8,$$
$$-\alpha_1^{16},\quad -\alpha_2^{16},\quad \ldots,\quad -\alpha_n^{16},$$
$$\text{etc.,}$$

that is, the negatives of the roots $\alpha_1, \alpha_2, \ldots, \alpha_n$ raised to exponents that are rapidly increasing powers of 2. After a few such operations the coefficients of the transformed equations become exceedingly large and it is necessary to replace them by approximate values retaining a certain fixed number of significant digits, for example, five or six. Generally retaining, for instance, six digits, one may expect to obtain roots by Graeffe's method with also six significant digits. The computation is carried out conveniently with the help of a calculating machine. Another way to avoid dealing with very large numbers is to replace the coefficients of the transformed equations by their logarithms (or rather the logarithms of their absolute values) appending the letter n at the end of the logarithm to indicate a negative coefficient. The coefficients in the transformed equations will be all positive if the proposed equation has only real roots, so that negative coefficients indicate the presence of imaginary roots, though the converse is not necessarily true. When using logarithmic values of the coefficients, it is very helpful to have tables of Gaussian logarithms that permit taking directly from the table the logarithm of a sum or difference of two numbers whose logarithms are given. The excellent six-place tables of logarithms by Bremiker contain also tables of Gaussian logarithms.

Example 1. Let the proposed equation be

$$x^4 + 2x^3 - 6x^2 - 2x + 1 = 0.$$

The coefficients of the first three transformed equations are computed exactly by the above given rule and exhibited in the following table:

2^0	1	2	-6	-2	1
2^1	1	16	46	16	1
2^2	1	164	1606	164	1
2^3	1	23684	2525446	23684	1

Numbers in the column on the left show the exponents of the powers to which the roots are raised. Passing to the following transformed equations, we retain only six significant digits. The results are as follows:

2^4	1	555881×10^3	637676×10^7	555881×10^3	1
2^5	1	308991×10^{12}	406631×10^{20}	308991×10^{12}	1

For the next transformed equation the coefficients within the assumed accuracy will be nearly equal to the squares of the preceding coefficients—an important circumstance connected with Graeffe's method that shows that it is not necessary to continue the computation further.

Example 2. Let the proposed equation be

$$x^4 + 8x^3 - 7x^2 - 50x - 30 = 0.$$

The coefficients of the first transformed equation are computed directly:

$$2^1 : \quad 1 \qquad 78 \qquad 789 \qquad 2080 \qquad 900$$

are replaced by their logarithms:

$$2^1 : \quad 0 \qquad 1.892095 \qquad 2.897077 \qquad 3.318063 \qquad 2.954243.$$

Further computation made logarithmically with the help of Bremiker's tables yields

2^2	0	3.653792	5.476891	6.463324	5.908486
2^3	0	7.294564	10.804246	12.900923	11.816972
2^4	0	14.588985	21.573568	25.801272	23.633944
2^5	0	29.177970	43.145614	51.602544	47.267888

At the next step all logarithms would be doubled so that with the assumed accuracy it is needless to continue further.

2. In order to explain how the process described in Sec. 1 can be used for the computation of roots, we begin with the simplest case when the roots $\alpha_1, \alpha_2, \ldots, \alpha_n$ are all real and of different absolute value. Denoting in general $|\alpha_k|$ by ρ_k, suppose that

$$\rho_1 > \rho_2 > \cdots > \rho_n.$$

Denote by $A_0, A_1, \ldots, A_n$ the coefficients of the transformed equation corresponding to the degree $m = 2^h$ of the roots. Then, clearly

$$\frac{A_1}{A_0} = \Sigma \rho_1^m, \qquad \frac{A_2}{A_0} = \Sigma \rho_1^m \rho_2^m, \qquad \frac{A_3}{A_0} = \Sigma \rho_1^m \rho_2^m \rho_3^m, \qquad \ldots \ldots$$

In these sums the terms ρ_1^m, $\rho_1^m\rho_2^m$, $\rho_1^m\rho_2^m\rho_3^m$, . . . are the dominant terms, and the other terms will be incomparably smaller if m is a sufficiently large number. Hence, with small relative errors ϵ_1, ϵ_2, ϵ_3, . . . we can write

$$\frac{A_1}{A_0} = \rho_1^m(1 + \epsilon_1), \qquad \frac{A_2}{A_0} = \rho_1^m\rho_2^m(1 + \epsilon_2), \qquad \frac{A_3}{A_0} = \rho_1^m\rho_2^m\rho_3^m(1 + \epsilon_3), \qquad \cdots$$

whence with small errors

$$m \log \rho_1 = \log \frac{A_1}{A_0}, \qquad m \log \rho_1\rho_2 = \log \frac{A_2}{A_0}, \qquad m \log \rho_1\rho_2\rho_3 = \log \frac{A_3}{A_0}$$

and with still smaller errors (on account of the large divisor m)

$$\log \rho_1 = \frac{1}{m} \log \frac{A_1}{A_0}, \qquad \log \rho_1\rho_2 = \frac{1}{m} \log \frac{A_2}{A_0},$$

$$\log \rho_1\rho_2\rho_3 = \frac{1}{m} \log \frac{A_3}{A_0}, \qquad \cdots \cdot \quad (1)$$

Passing to the next transformed equation whose coefficients are A_0', A_1', A_2', . . . we have also

$$\frac{A_1'}{A_0'} = \rho_1^{2m}(1 + \epsilon_1'), \qquad \frac{A_2'}{A_0'} = (\rho_1\rho_2)^{2m}(1 + \epsilon_2'),$$

$$\frac{A_3'}{A_0'} = (\rho_1\rho_2\rho_3)^{2m}(1 + \epsilon_3'), \qquad \cdots$$

or with small relative errors

$$\frac{A_1'}{A_0'} = \left(\frac{A_1}{A_0}\right)^2, \qquad \frac{A_2'}{A_0'} = \left(\frac{A_2}{A_0}\right)^2, \qquad \frac{A_3'}{A_0'} = \left(\frac{A_3}{A_0}\right)^2, \qquad \cdots \,,$$

which explains in general the phenomenon observed in the two preceding examples. Conversely, when these equations are satisfied, within the prescribed degree of accuracy, it may be safely assumed that at least with the same degree of accuracy equations (1) will be fulfilled. The logarithms of the moduli ρ_1, ρ_2, ρ_3, . . . having been found from (1), these moduli are found immediately by logarithmic tables. As to the signs of the roots they will be found without difficulty if the location of the roots is roughly known.

Example 1. To solve the equation

$$x^4 + 2x^3 - 6x^2 - 2x + 1 = 0$$

examined in Example 1, Sec. 1. The coefficients of the fifth transformed equation are

| 1 | 308991×10^{12} | 406631×10^{20} | 308991×10^{12} | 1 |

and their logarithms

| 0 | 17.489945 | 25.609200 | 17.489945 | 0 |

whence

$$32 \log \rho_1 = 17.489945, \qquad 32 \log \rho_1\rho_2 = 25.609200,$$
$$32 \log \rho_1\rho_2\rho_3 = 17.489945, \qquad 32 \log \rho_1\rho_2\rho_3\rho_4 = 0$$

and consequently

$$\log \rho_1 = 0.546561, \qquad \log \rho_2 = 0.253727, \qquad \log \rho_3 = \overline{1}.746273,$$
$$\log \rho_4 = \overline{1}.453439$$

$$\rho_1 = 3.520150, \qquad \rho_2 = 1.793604, \qquad \rho_3 = 0.557536, \qquad \rho_4 = 0.284079.$$

With the rough knowledge of the location of roots it is found that these roots are

$$
\begin{array}{r}
- 3.520150 \\
+ 1.793604 \\
- 0.557536 \\
+ 0.284079 \\
\hline
\text{Sum} \quad - 2.000003
\end{array}
$$

instead of $- 2$. The equation being presented in the form

$$(x^2 + x - 1)^2 = 5x^2$$

is readily solved and its roots are

$$\frac{\sqrt{5} - 1 \pm \sqrt{10 - \sqrt{20}}}{2}, \qquad \frac{- \sqrt{5} - 1 \pm \sqrt{10 + \sqrt{20}}}{2}$$

or approximately

$$
\begin{array}{r}
- 3.5201470 \\
+ 1.7936045 \\
- 0.5575365 \\
+ 0.2840790
\end{array}
$$

The above-found values are as close to these as can be expected using six-place tables of logarithms.

Example 2. To solve the equation

$$x^4 + 8x^3 - 7x^2 - 50x - 30 = 0$$

examined in Example 2, Sec. 1. The logarithms of the coefficients of the fifth transformed equation are

$$\begin{array}{ccccc} 0 & 29.177970 & 43.145609 & 51.602544 & 47.267888 \end{array}$$

whence

$$
\begin{array}{r}
43.145609 \qquad\qquad 51.602544 \\
- 29.177970 \qquad\quad - 43.145609 \\
\end{array}
$$

$$32 \log \rho_1 = 29.177970 \quad 32 \log \rho_2 = \overline{13.967639} \quad 32 \log \rho_3 = \overline{8.456935}$$

$$
\begin{array}{r}
47.267888 \\
- 51.602544 \\
\hline
32 \log \rho_4 = \overline{5}.665344
\end{array}
$$

and hence

$$
\begin{array}{ll}
\log \rho_1 = 0.911812; & \rho_1 = 8.16228 \\
\log \rho_2 = 0.436489; & \rho_2 = 2.73204 \\
\log \rho_3 = 0.264279; & \rho_3 = 1.83772 \\
\log \rho_4 = \overline{1}.864542; & \rho_4 = 0.73205
\end{array}
$$

With little trouble it is found that the roots themselves are

$$
\begin{array}{r}
- \ 8.16228 \\
+ \ 2.73205 \\
- \ 1.83772 \\
- \ 0.73205 \\
\hline
\text{Sum} \quad - \ 8.00000
\end{array}
$$

3. In general let the moduli of the roots $\rho_1, \rho_2, \ldots, \rho_n$ be divided in groups such that

$$
\begin{aligned}
\rho_1 &\geqq \rho_2 \geqq \cdots \geqq \rho_\lambda, \\
\rho_{\lambda+1} &\geqq \rho_{\lambda+2} \geqq \cdots \geqq \rho_\mu, \\
\rho_{\mu+1} &\geqq \rho_{\mu+2} \geqq \cdots \geqq \rho_\nu, \\
&\cdots \cdots \cdots \cdots \cdots
\end{aligned}
$$

but

$$
\rho_\lambda > \rho_{\lambda+1}, \qquad \rho_\mu > \rho_{\mu+1}, \qquad \rho_\nu > \rho_{\nu+1}, \qquad \cdots
$$

In the expressions

$$
\frac{A_\lambda}{A_0} = \Sigma \alpha_1^m \alpha_2^m \cdots \alpha_\lambda^m, \qquad \frac{A_\mu}{A_0} = \Sigma \alpha_1^m \alpha_2^m \cdots \alpha_\mu^m,
$$

$$
\frac{A_\nu}{A_0} = \Sigma \alpha_1^m \alpha_2^m \cdots \alpha_\nu^m, \qquad \cdots
$$

the terms $\alpha_1^m \alpha_2^m \cdots \alpha_\lambda^m$, $\alpha_1^m \alpha_2^m \cdots \alpha_\mu^m$, $\alpha_1^m \alpha_2^m \cdots \alpha_\nu^m$, $\ldots$ are dominant and the other terms are incomparably smaller for large m. Hence, with arbitrarily small relative errors $\epsilon_1, \epsilon_2, \epsilon_3, \ldots$

$$
\frac{A_\lambda}{A_0} = (\alpha_1 \alpha_2 \cdots \alpha_\lambda)^m (1 + \epsilon_1), \qquad \frac{A_\mu}{A_0} = (\alpha_1 \alpha_2 \cdots \alpha_\mu)^m (1 + \epsilon_2),
$$

$$
\frac{A_\nu}{A_0} = (\alpha_1 \alpha_2 \cdots \alpha_\nu)^m (1 + \epsilon_3), \qquad \cdots
$$

or, which is the same,

$$
\frac{A_\lambda}{A_0} = (\rho_1 \rho_2 \cdots \rho_\lambda)^m (1 + \epsilon_1),
$$

$$
\frac{A_\mu}{A_0} = (\rho_1 \rho_2 \cdots \rho_\mu)^m (1 + \epsilon_2), \qquad \cdots,
$$

whence again with small errors

$$
\rho_1 \rho_2 \cdots \rho_\lambda = \frac{1}{m} \log \frac{A_\lambda}{A_0}, \qquad \rho_{\lambda+1} \cdots \rho_\mu = \frac{1}{m} \log \frac{A_\mu}{A_\lambda},
$$

$$
\rho_{\mu+1} \cdots \rho_\nu = \frac{1}{m} \log \frac{A_\nu}{A_\mu}, \qquad \cdots \quad (2)
$$

and these relations will be satisfied within the prescribed accuracy when the transformations are carried so far that the logarithms of the ratios

$$
\frac{A_\lambda}{A_0}, \quad \frac{A_\mu}{A_0}, \quad \frac{A_\nu}{A_0}, \quad \cdots
$$

begin to double on passing from one transformed equation to the next. The series of coefficients

$$A_0, \quad A_1, \quad A_2, \quad \ldots$$

splits into segments

$$A_0, \quad \ldots, \quad A_\lambda$$
$$A_\lambda, \quad \ldots, \quad A_\mu$$
$$A_\mu, \quad \ldots, \quad A_\nu$$
$$\ldots \ldots \ldots$$

the extreme terms of which show a regularity of behavior in that the ratios

$$\frac{A_\lambda}{A_0}, \quad \frac{A_\mu}{A_0}, \quad \frac{A_\nu}{A_0}, \quad \ldots$$

show the tendency to be squared or their logarithms to be doubled when passing from one transform to the next, while intermediate terms may not manifest such a regular behavior. In practice, therefore, it is always easy to find, while carrying out transformations, values of the indices λ, μ, ν, ... From equations (2) one can find approximate values of the products of the moduli $\rho_1 \ldots \rho_\lambda$; $\rho_{\lambda+1} \ldots \rho_\mu$; $\ldots$ of each group. To see how this helps to find approximately the moduli of the real and imaginary roots suppose, for instance, that among the roots

$$\alpha_1, \quad \alpha_2, \quad \alpha_3, \quad \alpha_4, \quad \alpha_5, \quad \alpha_6,$$

of an equation of the sixth degree, arranged in a decreasing order of moduli, α_1 is a real root, α_2 and α_3 conjugate imaginary roots, α_4 real, and α_5 and α_6 conjugate imaginary, no three roots having the same moduli. Then,

$$\rho_1 > \rho_2 = \rho_3 > \rho_4 > \rho_5 = \rho_6,$$

and for large m the sequence of coefficients

$$A_0, A_1, A_2, A_3, A_4, A_5, A_6$$

will be found to split into segments

$$A_0, \quad A_1$$
$$A_1, \quad A_2, \quad A_3$$
$$A_3, \quad A_4$$
$$A_4, \quad A_5, \quad A_6$$

Supposing for simplicity that $A_0 = 1$, the coefficients A_1, A_3, A_4, A_6 will show a regular behavior as explained above, but A_2 and A_5 in general will vary erratically, changing sign and showing the considerable influence of the neighboring coefficients A_1, A_3 and A_4, A_6. Once the segmentation becomes manifest, the moduli of the real roots ρ_1, ρ_4 are found approximately from

$$\log \rho_1 = \frac{1}{m} \log \frac{A_1}{A_0}, \qquad \log \rho_4 = \frac{1}{m} \log \frac{A_4}{A_3},$$

and the squares of the moduli of the imaginary roots from

$$\log \rho_2^2 = \frac{1}{m} \log \frac{A_3}{A_1}, \qquad \log \rho_5^2 = \frac{1}{m} \log \frac{A_6}{A_4}.$$

The procedure illustrated in this example will serve to find the moduli of the real and imaginary roots excluding the case of more than two roots of equal (or very nearly equal) moduli, which requires application of special artifices.

4. When the moduli of the roots are found by approximation, the signs of the real roots are easily determined and the arguments (or real parts) of the imaginary roots can be found in the following manner: Suppose at first that there is just one pair of imaginary roots $a \pm bi$ whose modulus is ρ. If the sum of the real roots is denoted by σ, the real and imaginary parts are determined from

$$2a = -\frac{a_1}{a_0} - \sigma, \qquad b = \sqrt{\rho^2 - a^2}.$$

When there are two pairs of imaginary roots $a_1 \pm b_1 i$, $a_2 \pm b_2 i$ of unequal moduli ρ_1, ρ_2, consider the sum of all the roots and the sum of their reciprocals; calling σ and σ' the sum of the real roots and of their reciprocals, we have the two equations

$$2a_1 + 2a_2 = -\frac{a_1}{a_0} - \sigma, \qquad \frac{2a_1}{\rho_1^2} + \frac{2a_2}{\rho_2^2} = -\frac{a_{n-1}}{a_n} - \sigma',$$

which serve to determine a_1 and a_2, after which b_1 and b_2 are found from

$$b_1 = \sqrt{\rho_1^2 - a_1^2}, \qquad b_2 = \sqrt{\rho_2^2 - a_2^2}.$$

In case there are three or more pairs of imaginary roots of unequal moduli the following general procedure can be used: We may suppose that the degree of the proposed equation $f(x) = 0$ is even, equal to 2ν; for in the contrary case it may be replaced by $xf(x) = 0$. Let ρ be the modulus of some imaginary root and ϕ its argument. Substituting $x = \rho(\cos \phi + i \sin \phi)$ into

$$x^{-\nu}f(x) = 0$$

and equating to 0 the real and imaginary parts, we have two relations of the form

$$C_0 \cos \nu\phi + C_1 \cos (\nu - 1)\phi + \cdots + C_{\nu-1} \cos \phi + C_\nu = 0,$$
$$D_0 \sin \nu\phi + D_1 \sin (\nu - 1)\phi + \cdots + D_{\nu-1} \sin \phi = 0,$$

where the coefficients are known quantities. Replacing in general

$$\cos k\phi, \quad \frac{\sin k\phi}{\sin \phi}$$

by their expressions through $t = \cos \phi$, we find that t is a unique common root of two equations

$$F(t) = 0, \qquad \Phi(t) = 0$$

one of degree ν and the other of degree $\nu - 1$. This common root can be found by the method used in finding the highest common divisor. The computation is carried out conveniently if a calculating machine is available.

There is another method for computing the real parts of imaginary roots. Take a real number k (for practical reasons it should not be taken too small or too large) and transform the proposed equation by the substitution

$$y = x - k.$$

Writing

$$|\alpha_\nu| = \rho_\nu, \qquad |\alpha_\nu - k| = r_\nu, \qquad \alpha_\nu = a_\nu + b_\nu i,$$

we have

$$a_\nu^2 + b_\nu^2 = \rho_\nu^2, \qquad (a_\nu - k)^2 + b_\nu^2 = r_\nu^2$$

whence

$$a_\nu = \frac{\rho_\nu^2 - r_\nu^2 + k^2}{2k}.$$

The squares of the moduli of the equation in y will be found by another application of Graeffe's method; let them be

$$r^2, \quad s^2, \quad t^2, \quad \ldots .$$

Since the signs of the real roots α_ν are known, there is no difficulty in finding the corresponding r_ν. If $2k$ in absolute value does not exceed the smallest difference between the moduli of the nonconjugate imaginary roots, it is easy to see that with $\rho_\nu > \rho_\mu$ we have also $r_\nu > r_\mu$, and so the values r_ν for imaginary roots can be found without ambiguity among $r^2, s^2, t^2, \ldots$. But values of k satisfying the above-mentioned requirement may be so small as to cause considerable loss of accuracy in determining a_ν. With larger k the identification of r_ν corresponding to ρ_ν may be achieved by a tentative process, easily carried out in practice, which is based on the inequality

$$|\rho_\nu - r_\nu| > |k|,$$

and the relation

$$\sum_{\nu=1}^{n} \frac{r_\nu^2 - k^2}{\rho_\nu^2} = 2k\frac{a_{n-1}}{a_n} + n,$$

which follows from the expression of the sum of the reciprocals of the roots taking into account the above expression of a_ν.

5. It is well to illustrate these general considerations by numerical examples.

Example 1. To solve the equation

$$x^5 + x^4 - 7x^3 - 22x^2 + x + 1 = 0.$$

The results of the computation, which the reader is urged to repeat, are given in the following table:

2^0	1	1	-7	-22	1	1
2^1	1	15	95	500	45	1
2^2	1	35	-5885	241480	1025	1
2^3	1	12995	177317×10^2	583247×10^5	567665	1
2^4	1	133407×10^3	-120145×10^{10}	340177×10^{16}	205594×10^6	1
2^5	1	202003×10^{11}	53584×10^{25}	115720×10^{38}	354654×10^{17}	1
2^6	1	406980×10^{27}	-180391×10^{54}	133911×10^{81}	123465×10^{40}	1
2^7	*	*	*	*	152409×10^{85}	1

The irregular behavior of the third coefficient indicates the presence of imaginary roots. After six transformations the segmentation becomes manifest and the group into which the coefficients are split are the following:

$$
\begin{array}{cc}
1 & 406980 \times 10^{27} \\
406980 \times 10^{27}, & -180391 \times 10^{54}, \quad 133911 \times 10^{81} \\
133911 \times 10^{81}, & 123468 \times 10^{40} \\
123468 \times 10^{40}, & 1
\end{array}
$$

This indicates the presence of three real and two imaginary roots. Resorting to logarithms, we find

$$\log 406980 \times 10^{27} = 32.609573, \qquad \log 133911 \times 10^{81} = 86.126816,$$
$$\log 123468 \times 10^{40} = 45.091554,$$

whence

$$64 \log \rho_1 = 32.609573, \qquad 64 \log \rho_2^2 = 53.517243,$$
$$64 \log \rho_4 = -41.035262, \qquad 64 \log \rho_5 = -45.091554,$$

and

$$\log \rho_1 = 0.509524, \qquad \log \rho_4 = \bar{1}.358824, \qquad \log \rho_5 = \bar{1}.295444,$$
$$\rho_1 = 3.23239, \qquad \rho_4 = 0.228467, \qquad \rho_5 = 0.197445.$$

which lead to the following real roots:

$$+ 3.23239, \qquad + 0.22847, \qquad - 0.19745.$$

The logarithm of the square of the moduli of the two imaginary roots is

$$\log \rho_2^2 = 0.836207,$$

and so, denoting it simply by ρ^2, we have

$$\rho^2 = 6.85815.$$

Considering the sum of all the roots, we have

$$3.26341 + 2a = -1,$$

whence

$$a = -2.131705$$

and

$$b = 1.52117$$

so that the imaginary roots are

$$-2.13171 \pm 1.52117i$$

Example 2. To solve the equation

$$x^5 - x^3 - 2x^2 - 2x - 1 = 0.$$

The results of the computation are given in the table:

2^0	1	0	-1	-2	-2	-1
2^1	1	2	-3	0	0	1
2^2	1	10	9	4	0	1
2^3	1	82	1	36	-8	1
2^4	1	6722	-5919	1476	-8	1
2^5	1	451971×10^2	151912×10^2	209732×10	-2888	1
2^6	1	204278×10^{10}	41187×10^9	448658×10^7	414590×10	1
2^7	*	*	*	201291×10^{20}	*	*

The irregular behavior of the numbers in the third and fifth columns indicates two pairs of imaginary roots. The segmentation is almost completed after six transformations, and the seventh is needed only to make a slight adjustment of the number in the fourth column. Passing to logarithms and dividing the logarithm of the number in the fourth column by 2, we have the following segments of the logarithmic coefficients:

$$\begin{array}{ccc} 0 & 15.310222 & \\ 15.310222 & * & 12.651912 \\ 12.651904 & * & 0 \end{array}$$

whence

$$64 \log \rho_1 = 15.310222, \quad 64 \log \rho_2^2 = 61.341690 - 64, \quad 64 \log \rho_4^2 = 51.348096 - 64,$$

$$\log \rho_1 = 0.239222, \qquad \log \rho_2^2 = \bar{1}.958464, \qquad \log \rho_4^2 = \bar{1}.802314,$$

$$\rho_1 = 1.73469, \qquad \rho_2^2 = 0.908790, \qquad \rho_4^2 = 0.634329$$

The real root is

$$1.73469,$$

and squares of the moduli of the imaginary roots, which we denote simply by ρ^2 and ρ'^2, are

$$\rho^2 = 0.908790, \qquad \rho'^2 = 0.634329.$$

Considering the sum of all the roots and their reciprocals, we have the two equations

$$a + a' = -\ 0.86735$$
$$\frac{a}{\rho^2} + \frac{a'}{\rho'^2} = -\ 1.28824$$

the solution of which yields

$$a = -\ 0.16616, \qquad a' = -\ 0.70119,$$

and further

$$b = 0.93871, \qquad b' = 0.37770,$$

so that the imaginary roots are

$$-\ 0.16616 \pm 0.93871i,$$
$$-\ 0.70119 \pm 0.37770i.$$

Example 3.   To solve the equation

$$x^6 - x^5 + 2x^4 - 3x^3 + 2x^2 + x + 1 = 0.$$

After the third transformation the computations are made logarithmically, and the results are exhibited in the table:

2^0	1	-1	2	-3	2	1	1
2^1	1	-3	2	-3	14	-3	1
2^2	1	5	14	-31	182	-19	1
2^3	1	-3	870	-4327	31974	-3	1
2^3	0	0.477121_n	2.939519	3.636187_n	4.504797	0.477121_n	0
2^4	0	3.238297_n	5.900305	7.567165_n	9.009583	4.805766_n	0
2^5	0	6.148167	11.704236	14.419561_n	18.019162	9.310391	0
2^6	0	11.985109	23.409727	29.995115_n	36.038324	*	0
2^7	*	*	46.819467	*	*	*	*

The segmentation is almost complete after six transformations, and the seventh is required only to make a slight adjustment in the third column. Dividing the last number in this column by 2, we have the following segments after six transformations:

0	*	23.409733
23.409733	*	36.038324
36.038324	*	0

This indicates the presence of six imaginary roots. The logarithms of the squares of their moduli are

$$\log \rho_1^2 = 0.365777, \qquad \log \rho_2^2 = 0.197321, \qquad \log \rho_3^2 = 1.436901$$
$$\rho_1^2 = 2.32154, \qquad \rho_2^2 = 1.57515, \qquad \rho_3^2 = 0.273464.$$

To find arguments we write the proposed equation in the form

$$x^3 + \frac{1}{x^3} - x^2 + \frac{1}{x^2} + 2x + \frac{2}{x} - 3 = 0,$$

and introduce

$$x = \rho(\cos \phi + i \sin \phi).$$

Equating to zero the real and imaginary parts, we get

$$\left(\rho^3 + \frac{1}{\rho^3}\right) \cos 3\phi + \left(\frac{1}{\rho^2} - \rho^2\right) \cos 2\phi + 2\left(\rho + \frac{1}{\rho}\right) \cos \phi - 3 = 0,$$

$$\left(\rho^3 - \frac{1}{\rho^3}\right) \sin 3\phi - \left(\frac{1}{\rho^2} + \rho^2\right) \sin 2\phi + 2\left(\rho - \frac{1}{\rho}\right) \sin \phi = 0,$$

and, introducing $t = \cos \phi$,

$$F(t) = 4\left(\rho^3 + \frac{1}{\rho^3}\right)t^3 - 2\left(\rho^2 - \frac{1}{\rho^2}\right)t^2 + \left[2\left(\rho + \frac{1}{\rho}\right) - 3\left(\rho^3 + \frac{1}{\rho^3}\right)\right]t + \rho^2 - \frac{1}{\rho^2} - 3 = 0,$$

$$\phi(t) = 4\left(\rho^3 - \frac{1}{\rho^3}\right)t^2 - 2\left(\rho^2 + \frac{1}{\rho^2}\right)t + 2\left(\rho - \frac{1}{\rho}\right) - \left(\rho^3 - \frac{1}{\rho^3}\right) = 0.$$

Substituting here $\rho = \rho_1$ and reducing the coefficients of the highest power of t to 1, we have

$$t^3 - 0.247490t^2 - 0.464659t - 0.072593 = 0$$
$$t^2 - 0.422840t - 0.116748 = 0$$

and the common root of these equations, found by the method of the highest common divisor, is

$$t = \cos \phi_1 = -0.190333$$

and

$$a_1 = \rho_1 \cos \phi_1 = -0.290083.$$

Similarly, it is found for the two other pairs of complex roots:

$$a_2 = 1.08018, \qquad a_3 = -0.290093.$$

Finally, the imaginary parts of these roots are found to be

$$1.49579, \quad 0.63903, \quad 0.43510$$

so that the requested six imaginary roots are

$$-0.29008 \pm 1.49579i$$
$$1.08018 \pm 0.63903i$$
$$-0.29009 \pm 0.43510i$$

and their sum turns out to be 1.00002 instead of 1, which suggests that only the last decimals may be somewhat in error.

To verify these numbers we may use the second method explained on p. 326. To this end take $k = 1$ and transform the equation

$$x^6 - x^5 + 2x^4 - 3x^3 + 2x^2 + x + 1 = 0$$

by the substitution $y = x - 1$; the transformed equation in y will be

$$y^6 + 5y^5 + 12y^4 + 15y^3 + 10y^2 + 5y + 3 = 0.$$

After five transformations the coefficients of the transformed equation are split into the following segments:

$$1 \qquad , \qquad -408929 \times 10^4, \qquad 832061 \times 10^{13}$$
$$832061 \times 10^{13}, \qquad -178969 \times 10^{18}, \qquad 314087 \times 10^{22}$$
$$314087 \times 10^{22}, \qquad -317600 \times 10^{16}, \qquad 185302 \times 10^{10}$$

whence it is found that the squares of the moduli of the imaginary roots are

$$3.90171, \quad 1.85365, \quad 0.414800.$$

With the help of the relation

$$\frac{r_1^2 - 1}{\rho_1^2} + \frac{r_2^2 - 1}{\rho_2^2} + \frac{r_3^2 - 1}{\rho_3^2} = 4$$

it is easily found that corresponding to

$$\rho_1^2 = 2.32154, \qquad \rho_2^2 = 1.57515, \qquad \rho_3^2 = 0.273464$$

we must take

$$r_1^2 = 3.90171, \qquad r_2^2 = 0.41480, \qquad r_3^2 = 1.85365.$$

Hence,

$$a_1 = \frac{3.32154 - 3.90171}{2} = -0.290085,$$

$$a_2 = \frac{2.57515 - 0.41480}{2} = 1.080175,$$

$$a_3 = \frac{1.27346 - 1.85365}{2} = -0.290095,$$

that is, practically the same numbers as obtained by another method. A sharper calculation yields the following values of the roots with seven decimals:

$$-0.2900809 \pm 1.4957939i$$
$$1.0801744 \pm 0.6390402i$$
$$-0.2900935 \pm 0.4350983i$$

and

$$\rho_1^2 = 2.3215463$$
$$\rho_2^2 = 1.5751490$$
$$\rho_3^2 = 0.2734647.$$

In this exposition we confined ourselves only to the main features of Graeffe's method, leaving aside such questions as computation of roots with the same or nearly the same moduli, improvements of values found, as well as the deeper theoretical investigation of the whole subject. For this the reader may consult the lengthy paper by A. Ostrowski: *Recherches sur la méthode de Graeffe* in *Acta Mathematica*, vol. 72, 1940.

ANSWERS

Chapter I

§6. **1.** $-3-i$.   **2.** $3+2i$.   **3.** $5i$.

4. $-i$.   **5.** $-1+i$.   **6.** $\frac{1}{2}-\frac{1}{2}i$.

7. $\frac{3}{2}-\frac{1}{2}i$.   **8.** $\frac{3}{2}-\frac{1}{2}i$.   **9.** -2.

10. -1.   **11.** -1.   **12.** $\frac{3}{8}+\frac{3}{8}i$.

13. $x=1.\ y=2$.   **14.** 1.   **15.** $x=\pm 1$.

§7. **1.** 1.   **2.** 1.   **3.** 5.

4. 1.   **5.** x^2+1.   **6.** $2x^2-2x+1$.

7. x^3-x^2+x-1 if $x \geqq 1$; and $-x^3+x^2-x+1$ if $x<1$.

8. 0.   **9.** $\pm\dfrac{1+i}{\sqrt{2}},\ \pm\dfrac{1-i}{\sqrt{2}}$.

10. (a) $0, 1, -\dfrac{1}{2}\pm i\dfrac{\sqrt{3}}{2};$ (b) $0, \pm 1, \pm i$.

§8. **2.** 1.   **3.** (a) 1; (b) <1.

§9. **2.** 15 for $z=2$.

§10. **1.** $\pm\dfrac{1+i}{\sqrt{2}}$.   **2.** $\pm\left(\dfrac{1}{2}+i\dfrac{\sqrt{3}}{2}\right)$.   **3.** $\pm\dfrac{\sqrt{3}+i}{2}$.

4. $\pm(1-2i)$.   **5.** $\pm(6-7i)$.   **6.** $\pm(\sqrt{2}+i\sqrt{3})$.

7. $\pm(x+i)$.   **8.** $\dfrac{-1\pm i\sqrt{3}}{2}$.   **9.** $\dfrac{3}{4}\pm i\dfrac{\sqrt{7}}{4}$.

10. $1+i, 1+2i$.   **11.** $-1+i, \frac{3}{2}+4i$.   **12.** $\pm 1, \pm i$.

13. $\pm\dfrac{1+i}{\sqrt{2}},\ \pm\dfrac{1-i}{\sqrt{2}}$.   **14.** $\pm(1+i), \pm(1-i)$.

15. $\pm\dfrac{5-i}{\sqrt{2}},\ \pm\dfrac{1+5i}{\sqrt{2}}$.   **16.** $1, -\dfrac{1}{2}\pm i\dfrac{\sqrt{3}}{2}$.

17. $-i, \dfrac{\pm\sqrt{3}+i}{2}$.   **18.** $\pm 1, -\dfrac{1}{2}\pm i\dfrac{\sqrt{3}}{2}, \dfrac{1}{2}\pm i\dfrac{\sqrt{3}}{2}$.

19. $\pm i, \pm\dfrac{\sqrt{3}+i}{2}, \pm\dfrac{\sqrt{3}-i}{2}$.

20. $\dfrac{-1+i}{\sqrt[3]{2}}, \dfrac{1\pm\sqrt{3}+(-1\pm\sqrt{3})i}{2\sqrt[3]{2}}$.

22. (a) x^2+x+1, (b) $x^2+1=0$.

§12. **1.** (a) $-60°$, (b) $+60°$, (c) $+120°$.

2. (a) $-90°$, (b) $+90°$, (c) $+45°$.

3. $+30°$ **4.** $-80°$.

§13. **1.** $4(\cos\pi+i\sin\pi)$.   **2.** $\cos\dfrac{\pi}{2}+i\sin\dfrac{\pi}{2}$.

3. $6(\cos\frac{3}{2}\pi+i\sin\frac{3}{2}\pi)$.   **4.** $\sqrt{2}\left(\cos\dfrac{3\pi}{4}+i\sin\dfrac{3\pi}{4}\right)$.

5. $\cos 300° + i \sin 300°$.   **6.** $\cos 120° + i \sin 120°$.

7. $2(\cos 330° + i \sin 330°)$.

8. $2\sqrt{2}\ (\cos 255° + i \sin 255°)$.

9. $r = 5, \phi = 216°52'12''$.

10. $r = \sqrt{5}, \phi = 153°26'6''$.

11. $2 \cos \dfrac{\alpha}{2}\left(\cos \dfrac{\alpha}{2} + i \sin \dfrac{\alpha}{2}\right)$ if $-\pi < \alpha < \pi$.

12. $2 \cos \dfrac{\alpha - \beta}{2}\left(\cos \dfrac{\alpha + \beta}{2} + i \sin \dfrac{\alpha + \beta}{2}\right)$ if $\cos \dfrac{\alpha - \beta}{2} > 0$ and

$-2 \cos \dfrac{\alpha - \beta}{2}\left[\cos\left(\pi + \dfrac{\alpha + \beta}{2}\right) + i \sin\left(\pi + \dfrac{\alpha + \beta}{2}\right)\right]$ if $\cos \dfrac{\alpha - \beta}{2} < 0$.

§14.   **1.** $2^n\left(\cos \dfrac{n\pi}{6} + i \sin \dfrac{n\pi}{6}\right)$.   **2.** $8^{\frac{n}{2}}\left(\cos \dfrac{n\pi}{12} - i \sin \dfrac{n\pi}{12}\right)$.

3. $\cos\left(\dfrac{n\pi}{2} - n\phi\right) + i \sin\left(\dfrac{n\pi}{2} - n\phi\right)$.

4. $2^n\left(\sin \dfrac{\theta - \phi}{2}\right)^n\left[\cos \dfrac{n(\theta + \phi)}{2} - i \sin \dfrac{n(\theta + \phi)}{2}\right]$.

6. $(a)\ \dfrac{2^n + 2 \cos \dfrac{n\pi}{3}}{3}$, $(b)\ \dfrac{2^n - 2 \cos \dfrac{(n + 1)\pi}{3}}{3}$,

$(c)\ \dfrac{2^n - 2 \cos \dfrac{(n - 1)\pi}{3}}{3}$

9. $(a)\ \cos 3\phi = 4 \cos^3 \phi - 3 \cos \phi$, $\sin 3\phi = 3 \sin \phi - 4 \sin^3 \phi$.

$(b)\ \cos 5\phi = 16 \cos^5 \phi - 20 \cos^3 \phi + 5 \cos \phi$,

$\sin 5\phi = 16 \sin^5 \phi - 20 \sin^3 \phi + 5 \sin \phi$.

$(c)\ \cos 4\phi = 8 \cos^4 \phi - 8 \cos^2 \phi + 1$, $\dfrac{\sin 4\phi}{\cos \phi} = 4 \sin \phi - 8 \sin^3 \phi$.

10. $(a)\ 16 \sin^5 \phi = 10 \sin \phi - 5 \sin 3\phi + \sin 5\phi$.

$(b)\ 8 \sin^4 \phi = 3 - 4 \cos 2\phi + \cos 4\phi$.

§15.   **1.** $x = 2[\cos (67°30' + 90°k) + i \sin (67°30' + 90°k)]$, $k = 0, 1, 2, 3$.

2. $x = \sqrt{2}[\cos (11°15' + 90°k) + i \sin (11°15' + 90°k)]$, $k = 0, 1, 2, 3$.

3. $x = -\sqrt[3]{2}\ [\sin (120°k) + i \cos (120°k)]$, $k = 0, \pm 1$.

4. $x = \sqrt[4]{2}\ [\cos (120°k - 15°) + i \sin (120°k - 15°)]$, $k = 0, \pm 1$.

5. $x = \cos (30° + 90°k) + i \sin (30° + 90°k)$, $k = 0, 1, 2, 3$.

6. $x = \cos (40° + 120°k) + i \sin (40° + 120°k)$, $k = 0, 1, 2$.

7. $x = \sqrt[3]{2}\ [\cos (30° + 60°k) + i \sin (30° + 60°k)]$, $k = 0, 1, 2, 3, 4, 5$.

8. $x = \sqrt[4]{2}\ [\cos (60°k - 2°30') + i \sin (60°k - 2°30')]$, $k = 0, 1, 2, 3, 4, 5$.

9. $\cos 15° = \dfrac{\sqrt{6} + \sqrt{2}}{4}$, $\sin 15° = \dfrac{\sqrt{6} - \sqrt{2}}{4}$.

§16.   **1.** $(\cos 15° + i \sin 15°)^k = \left(\dfrac{\sqrt{6} + \sqrt{2}}{4} + i\ \dfrac{\sqrt{6} - \sqrt{2}}{4}\right)^k$, $k = \pm 1, \pm 5, \pm 7$,

± 11.

Chapter II

§2.   **1.** $x^8 + x^6 + x^4 + x^2 + 1$.   **2.** $2x^7 + 3x^6 - 9x^5 - x^4 + 5x^3 - 3x^2 - x + 1$.

3. $x^8 - 26x^6 + 21x^4 + 20x^2 + 4$.

4. $3x^8 - 2x^7 - 19x^6 + 11x^5 - 7x^4 - 3x^3 + 8x^2 - 2x + 1$.

§3. 1. $x^3 + 3x^2 + 1$.   **2.** $x^4 - x + 1$.

3. $x^3 - x^2 - 2$ quotient, $8x + 1$ remainder.

4. $x^8 - x^7 + x^5 - x^4 + x^3 - x + 1$.   **5.** $7x^2 + 7x$.

§5. 1. $q = 2x^3 - 10x^2 + 27x - 59$, $r = 119$.

2. $q = -x^3 + 4x^2 + 8x + 24$, $r = 72$.

3. $q = 6x^2 - 2.8x + 1.64$, $r = 4.968$.

4. $q = 5x^2 + 6.5x + 11.25$, $r = 32.75$.

5. $q = \tfrac{1}{3}x^3 + \tfrac{1}{9}x^2 - 1\tfrac{1}{27}x + 2\tfrac{2}{81}$, $r = \tfrac{37}{81}$.

6. $q = 5x^5 - 5x^4 - x^3 + x^2 - x + 1$, $r = 0$.

7. $q = 1 + 2x + 3x^2 + \cdots + (n-1)x^{n-2}$, $r = 0$.

8. 2.359375.   **9.** 1.7318.

§6. 1. $5(x-1) + 10(x-1)^2 + 10(x-1)^3 + 5(x-1)^4 + (x-1)^5$.

2. $5 - 15(x+1) + 9(x+1)^2 + 4(x+1)^3 - 5(x+1)^4 + (x+1)^5$.

3. $-317 + 927(x+2) - 1120(x+2)^2 + 720(x+2)^3 - 260(x+2)^4$
$$+ 50(x+2)^5 - 4(x+2)^6.$$

4. $3(x-0.3)^4 + 9.6(x-0.3)^3 + 8.02(x-0.3)^2 + 2.544(x-0.3) - 0.7237$.

§7. 1. $5, 15, 24, 12, -24$.   **2.** $-13, 11, -18, 78, -240, 240$.

3. $-67, 81, -62, 24$.   **4.** $\tfrac{247}{192}, -1, \tfrac{31}{12}, -6, 6$.

§8. 1. $x + 1$.   **2.** $(x+1)^2$.   **3.** $2x^2 - 2x + 1$.

4. $x^2 + 1$.   **5.** $x^2 - x - 1$.

Chapter III

§1. 1. $x^3 - 3x^2 + 2x = 0$.   **2.** $x^3 - 3x^2 + 4x - 2 = 0$.

3. $x^4 - 2x^3 + 3x^2 - 2x + 2 = 0$.

4. Roots $\dfrac{1}{2}, \dfrac{5 \pm \sqrt{5}}{10}$.   **5.** Roots $a + 1, \dfrac{a \pm \sqrt{a^2 - 4a}}{2}$.

6. Roots $\tfrac{3}{2}, -3, 1 + \sqrt{2}, 1 - \sqrt{2}$.   **7.** $\dfrac{x^3 - x}{6}$.

8. $-\dfrac{2x^4 - 5x^3 - 2x^2 + 15x}{10}$.   **9.** Roots $-1, 2, 1 + 2i$.

10. Roots $i, \sqrt{3}, -\sqrt{3}, 1 + i$.

§3. 1. $-\tfrac{1}{108}(x^2-1)(x-2)^2(x+3)^3$.   **2.** $-\tfrac{1}{108}x^2(x-1)^2(x+1)^3$.

3. $(x-1)\left(x + \dfrac{1}{2} + i\,\dfrac{\sqrt{3}}{2}\right)\left(x + \dfrac{1}{2} - i\,\dfrac{\sqrt{3}}{2}\right)$.

4. $(x+1)(x-1)(x+i)(x-i)$.

5. Factors $x + 1, x - 1, x + \dfrac{1}{2} \pm i\,\dfrac{\sqrt{3}}{2}, x - \dfrac{1}{2} \pm i\,\dfrac{\sqrt{3}}{2}$.

6. Factors $\left(x \pm \dfrac{1+i}{\sqrt{2}}\right), \left(x \pm \dfrac{1-i}{\sqrt{2}}\right)$.

7. $(x+i)\left(x + \dfrac{\sqrt{3}-i}{2}\right)\left(x - \dfrac{\sqrt{3}+i}{2}\right)$.

8. $\left(x + \dfrac{1-i}{\sqrt{2}}\right)\left(x + \dfrac{\sqrt{6}-\sqrt{2}}{4} + i\,\dfrac{\sqrt{6}+\sqrt{2}}{4}\right)\left(x - \dfrac{\sqrt{6}+\sqrt{2}}{4} - i\,\dfrac{\sqrt{6}-\sqrt{2}}{4}\right)$.

9. $\left(x + \dfrac{1}{2} + i\,\dfrac{\sqrt{3}}{2}\right)\left(x + \dfrac{1}{2} - i\,\dfrac{\sqrt{3}}{2}\right)\left(x - \dfrac{1}{2} + i\,\dfrac{\sqrt{3}}{2}\right)\left(x - \dfrac{1}{2} - i\,\dfrac{\sqrt{3}}{2}\right)$.

10. $\left(x + \dfrac{\sqrt{3} - i}{2}\right)\left(x + \dfrac{\sqrt{3} + i}{2}\right)\left(x - \dfrac{\sqrt{3} + i}{2}\right)\left(x - \dfrac{\sqrt{3} - i}{2}\right).$

11. Eight factors

$$x + \frac{\sqrt{6} + \sqrt{2}}{4} \pm i\,\frac{\sqrt{6} - \sqrt{2}}{4},\ x + \frac{\sqrt{6} - \sqrt{2}}{4} \pm i\,\frac{\sqrt{6} + \sqrt{2}}{4},$$

$$x - \frac{\sqrt{6} + \sqrt{2}}{4} \pm i\,\frac{\sqrt{6} - \sqrt{2}}{4},\ x - \frac{\sqrt{6} - \sqrt{2}}{4} \pm i\,\frac{\sqrt{6} + \sqrt{2}}{4}.$$

12. $2(x + \sqrt{3 + \sqrt{8}})(x - \sqrt{3 + \sqrt{8}})(x + \sqrt{3 - \sqrt{8}})(x - \sqrt{3 - \sqrt{8}}).$

13. $(x - 1)(x + 1)^2(x + 2)$　　　　**14.** $2(x - 1)^3(x + 1)(x + \frac{1}{2}).$

15. $7x(x + 1)\left(x + \dfrac{1}{2} + i\,\dfrac{\sqrt{3}}{2}\right)^2\left(x + \dfrac{1}{2} - i\,\dfrac{\sqrt{3}}{2}\right)^2.$

20. $\dfrac{(1 + xi)^{2m} + (1 - xi)^{2m}}{2} = \left(1 - \dfrac{x^2}{\tan^2 \dfrac{\pi}{4m}}\right)\left(1 - \dfrac{x^2}{\tan^2 \dfrac{3\pi}{4m}}\right) \cdots$

$$\left(1 - \frac{x^2}{\tan^2 \dfrac{(2m - 1)\pi}{4m}}\right),$$

$$\frac{(1 + xi)^{2m+1} + (1 - xi)^{2m+1}}{2} = \left(1 - \frac{x^2}{\tan^2 \dfrac{\pi}{4m + 2}}\right)\left(1 - \frac{x^2}{\tan^2 \dfrac{3\pi}{4m + 2}}\right) \cdots$$

$$\left(1 - \frac{x^2}{\tan^2 \dfrac{(2m - 1)\pi}{4m + 2}}\right)$$

§4.　**1.** $(x^2 + 2x + 2)(x^2 - 2x + 2).$　　　**2.** $(x^2 + x + 1)(x^2 - x + 1).$

3. $(x^2 + x\sqrt{3} + 1)(x^2 - x\sqrt{3} + 1).$

4. $(x - 1)(x + 1)(x^2 + x + 1)(x^2 - x + 1).$

5. $(x^2 + 1)(x^2 + x\sqrt{3} + 1)(x^2 - x\sqrt{3} + 1).$

6. $\left(x^2 + \dfrac{1 + \sqrt{5}}{2}\,x + 1\right)\left(x^2 + \dfrac{1 - \sqrt{5}}{2}\,x + 1\right).$

7. Roots $- 1, - 1, 2 + 3i, 2 - 3i.$

8. $- 1, 1 + i, 1 + i, 1 - i, 1 - i.$

9. Roots $1, 2, i, i, - i, - i.$

10. $1, i, i, - i, - i, - \dfrac{1}{2} + i\,\dfrac{\sqrt{3}}{2}, - \dfrac{1}{2} - i\,\dfrac{\sqrt{3}}{2}.$

11. Roots $- 2, 3, \sqrt{2} + i, \sqrt{2} - i, - \sqrt{2} + i, - \sqrt{2} - i.$

12. Roots $- 2, \dfrac{1 \pm i\sqrt{35}}{3}.$　　　　**14.** Roots $\pm \sqrt{2}, \dfrac{1 \pm i\sqrt{31}}{4}.$

15. Roots $\dfrac{1 \pm i\sqrt{8}}{3}, \dfrac{- 1 \pm i\sqrt{15}}{4}.$

§5.　**1.** $- 1, \dfrac{- 1 \pm i\sqrt{7}}{2}.$　　**2.** $- 3, 3, \frac{1}{2}.$　　**3.** $- 3, 2, \frac{1}{3}.$

4. $2, - \frac{1}{2}, - 1.$　　　**5.** $9, 4, - 6.$　　　**6.** $2, - 4, - 7.$

7. $\frac{2}{3}, \frac{4}{3}, 2.$　　　**8.** $\frac{2}{3}, 2, 6.$　　　**9.** $2, \dfrac{1 \pm \sqrt{3}}{2}; k = 2.$

10. $-\,23\!\!\tfrac{2}{3},\ \tfrac{2}{3},\ 9;\ k = -\,613\!\!\tfrac{2}{3}.$

11. $q^3 = rp^3.$

12. $4p^3 + 27q^2 = 0.$

14. $\dfrac{1 \pm \sqrt{5}}{2},\ \dfrac{1 \pm i\sqrt{7}}{2}.$

15. $1,\ \tfrac{1}{2},\ \pm\,\sqrt{5}.$

16. $3,\ 2,\ 1 \pm i.$

17. $-\,1 \pm \sqrt{2},\ \dfrac{1 \pm i\sqrt{3}}{2}.$

18. $-\,3,\ -\,3,\ \dfrac{-\,1 \pm i\sqrt{7}}{4}.$

19. $\tfrac{1}{3},\ \tfrac{2}{3},\ -\,1 \pm i.$

20. $-\,2,\ -\,\tfrac{1}{2},\ 1,\ 5\!\!\tfrac{1}{2}.$

21. $\tfrac{1}{2},\ 1,\ 2,\ 4;\ k = 35.$

22. (a) 4, (b) 2.

23. (a) 7, 39; (b) 1, 5.

§6. **1.** $X_1 = x - 3,\ X_2 = x - 2.$ **2.** $X_1 = x + 3,\ X_2 = x - 3.$

3. $X_1 = x + 3,\ X_2 = x - 2.$ **4.** $X_1 = x + 3,\ X_2 = x - 4.$

5. $X_1 = x - \dfrac{2}{\sqrt{3}},\ X_2 = x + \dfrac{1}{\sqrt{3}}.$ **6.** $X_1 = 4x^2 + 4x + 3,\ X_2 = 2x - 1.$

7. $X_1 = (x - 2)(x + 4),\ X_2 = x - 1.$

8. $X_1 = x^2 - 2x + 3,\ X_2 = x - 2.$

9. $X_1 = x^2 + 2x - 3,\ X_2 = x - 3.$ **10.** $X_1 = 2x^2 + 1,\ X_2 = x - 3.$

11. $X_1 = 1,\ X_2 = x + 1,\ X_3 = x - 1.$

12. $X_1 = x^2 - 5x + 6,\ X_2 = 1,\ X_3 = x + 1.$

13. $X_1 = x - 2,\ X_2 = x + 1,\ X_3 = x - 1.$

14. $X_1 = 1,\ X_2 = x - 2,\ X_3 = x + 2.$

15. $X_1 = 1,\ X_2 = 3x - 2,\ X_3 = x + 4.$

16. $X_1 = 1,\ X_2 = x^2 + 1,\ X_3 = x - 1.$

17. $X_1 = x - 2,\ X_2 = x^2 + 3x + 2,\ X_3 = x - 1.$

Chapter IV

§2. **1.** $-\,2,\ 7.$ **2.** $-\,2,\ 8.$ **3.** $-\,1,\ 4.$

4. $-\,2,\ 4.$ **5.** $-\,2,\ 2.$ **6.** $-\,10,\ 3.$

7. $-\,2,\ 7.$ **8.** $-\,3,\ 3.$ **9.** $-\,2,\ 5.$

10. $-\,7,\ 3.$

§3. **1.** 5. **2.** 3. **3.** 6.

4. 2.

§4. **1.** $2,\ 5,\ -\,5.$ **2.** 6. **3.** $-\,6.$

4. $-\,4,\ 4.$ **5.** $-\,3,\ 2.$ **6.** $-\,8.$

7. $-\,3,\ 2,\ 4.$ **8.** $-\,2,\ 3,\ 5.$ **9.** 2, 5.

10. 1 (double), $-\,2$ (double).

§5. **1.** $\tfrac{2}{3}.$ **2.** $-\,5.$ **3.** $\tfrac{2}{3}.$

4. $-\,3,\ \tfrac{1}{2},\ \tfrac{3}{5}.$ **5.** No rational roots. **6.** No rational roots.

7. $-\,\tfrac{2}{3},\ 2.$ **8.** $-\,2,\ 1,\ \tfrac{1}{2}$ (double). **9.** 2.

10. $1,\ \tfrac{1}{2},\ -\,\tfrac{2}{3}.$ **11.** $\tfrac{2}{3},\ -\,\tfrac{3}{2}.$ **12.** $-\,2,\ \tfrac{1}{2}.$

14. (a) $2,\ 3,\ -\,1$ (triple). (b) $\tfrac{1}{3},\ i$ and $-\,i$ (double).

Chapter V

§3. **1.** $\sqrt[3]{2} + \sqrt[3]{4}.$ **2.** $\sqrt[3]{2} + 2\sqrt[3]{4}.$ **3.** $\sqrt[3]{9} - \sqrt[3]{3}.$

4. $\sqrt[3]{18} - \sqrt[3]{12}.$ **5.** $\tfrac{1}{2}\sqrt[3]{4} - \sqrt[3]{2}.$

6. $-\sqrt[3]{\dfrac{5 + \sqrt{23}}{4}} - \sqrt[3]{\dfrac{5 - \sqrt{23}}{4}}.$ **7.** $\dfrac{2 + \sqrt[3]{32} - \sqrt[3]{2}}{3}.$

8. $-2 + \sqrt[3]{4} + \sqrt[3]{16}.$ **9.** $-2.$ **10.** $-5.$

11. $5 + \sqrt[3]{20} - \sqrt[3]{50}.$ **12.** $-\frac{1}{2} + 2\sqrt[3]{2} - \sqrt[3]{4}.$ **13.** $-1.769292.$

14. $0.5960716.$ **15.** $-4.3553013.$ **16.** $0.0960717.$

21. 4.847322 in.   **22.** 7.910170 in.   **23.** 1.2256 in.

§5. 1. $-0.53209,$ **2.** $-2.53209,$ **3.** $1.24698,$
$0.65270,$ $0.87939,$ $-1.80194,$
$2.87939.$ $-1.34730.$ $-0.44504.$

4. $-2.60168,$ **5.** $-3,$ **6.** $-2.65109,$
$2.26181,$ $-2.61803,$ $0.27389,$
$0.33988.$ $-0.38197.$ $1.37720.$

7. $-3.33006,$ **8.** $-3.86080,$
$-0.79836,$ $-2.25411,$
$1.12842.$ $0.11491.$

9. Distance of the plane from the base is 0.34730 of the radius.

10. Distances of the planes from the base are 0.22607 and 0.48170 of the radius, respectively.

11. 6.5270 in.

12. $2.87939; -1.87939; 1.53209; -0.53209; 0.65270; 0.34730.$

13. $1.61803; -0.61803; 1.19718; -0.19718; 0.5 \pm 1.90965i.$

§6. 1. $-1, 3, -1 \pm \sqrt{2}.$ **2.** $-2, 1, \dfrac{1 \pm \sqrt{5}}{2}.$ **3.** $-1 \pm \sqrt{3}, \dfrac{1 \pm \sqrt{5}}{2}.$

4. $-3, -1, -\frac{1}{2}, 2.$ **5.** $-1 \pm \sqrt{2}, 1 \pm i.$ **6.** $\dfrac{1 \pm \sqrt{5}}{2}, \dfrac{-1 \pm i\sqrt{3}}{2}.$

7. $2, -3, -1 \pm i.$ **8.** $1, -3, \dfrac{-3 \pm \sqrt{17}}{2}.$ **9.** $\dfrac{-1 \pm i\sqrt{3}}{2}, \dfrac{1 \pm i\sqrt{7}}{2}.$

10. $\dfrac{1 \pm i\sqrt{15}}{2}, \dfrac{-1 \pm i\sqrt{7}}{2}.$ **11.** $-\dfrac{1}{2}, 1, \dfrac{-1 \pm i\sqrt{7}}{2}.$

12. $-1 \pm i\sqrt{2}, \dfrac{1 \pm i\sqrt{15}}{2}.$

14. $2 \cos \dfrac{2\pi}{15} = \dfrac{1 + \sqrt{5}}{4} + \dfrac{\sqrt{30 - 6\sqrt{5}}}{4}, \; 2 \cos \dfrac{4\pi}{15} = \dfrac{1 - \sqrt{5}}{4} + \dfrac{\sqrt{30 + 6\sqrt{5}}}{4},$

$2 \cos \dfrac{8\pi}{15} = \dfrac{1 + \sqrt{5}}{4} - \dfrac{\sqrt{30 - 6\sqrt{5}}}{4}, \; 2 \cos \dfrac{16\pi}{15} = \dfrac{1 - \sqrt{5}}{4} - \dfrac{\sqrt{30 + 6\sqrt{5}}}{4}.$

15. $-1, \dfrac{-3 + \sqrt{5}}{2} \pm i \sqrt{\dfrac{\sqrt{5} - 1}{2}}, \dfrac{-3 - \sqrt{5}}{2} \pm \sqrt{\dfrac{\sqrt{5} + 1}{2}}.$

Chapter VI

§5. 5. $(-4, -3).$

6. $(3, 6); (6, 8); (8, 10); (-\infty, 3)$ if $\lambda > -1, (10, +\infty)$ if $\lambda < -1.$

7. $(-\infty, -\frac{1}{2}); (-\frac{1}{2}, \frac{1}{3}); (\frac{2}{3}, \frac{3}{2}); (\frac{3}{2}, +\infty).$

10. $(-3, -2); (-2, -1); (-1, 0); (0, +\infty).$

§7. 1. $(-2, -1); (-1, 1).$ **2.** $(-2, -1); (-1, 0);$ one double root.

3. $(-2, -1); (1, 2).$ **4.** $(-1, 0); (4, 5).$

5. $(-1, 0).$ **6.** $(0, 1).$

7. $(0, 1); (1, 2).$ **8.** $(-4, -3); (0, 1); (2, 3).$

9. $(0, 1); (5, 6).$ **10.** $(0, 1); (1, 2).$ **11.** $A \geqq 108.$ **12.** $A \geqq 27.$

§10. **1.** One positive, one negative root, and four imaginary.

2. One positive, one negative root, and two imaginary.

3. One positive root, four imaginary.

4. One positive root, four imaginary.

5. All roots imaginary. **6.** One positive root, four imaginary.

7. All roots imaginary. **8.** One positive root, the other roots imaginary.

§11. **1.** One root in $(0,1)$; one root in $(4,5)$.

2. One root in $(0,1)$; one root in $(1,2)$.

3. No roots in $(4,5)$. **4.** No root in $(1,2)$; no root in $(2,3)$.

5. One root in $(-3,-1)$; one root in $(0,1)$.

6. One root in $(0,1)$; two roots in $(-1,0)$.

§12. Roots in intervals

1. $(0,1)$. **2.** $(-1,0)$. **3.** $(-1,0)$; $(4,4\frac{1}{2})$; $(4\frac{1}{2},5)$.

4. $(-18,-17)$; $(1\frac{6}{5},2\frac{6}{9})$; $(2\frac{6}{9},1\frac{3}{4})$.

5. $(0,1)$; $(3,4)$; two imaginaries.

6. $(0,1)$; $(2,3)$; two imaginaries.

7. Four imaginaries.

8. $(-7,-6)$; -1; two imaginaries.

9. $(-5,-4)$; $(-2,-1)$; $(-1,0)$; $(0,1)$.

10. $(-7,-6)$; $(-3,-2)$; $(-1,0)$; $(0,1)$.

11. $(-1,0)$; four imaginaries.

12. $(1,2)$; four imaginaries.

13. $(-3,-2)$; four imaginaries.

14. $(-2,-1)$; $(0,1)$; $(2,3)$; two imaginaries.

15. $(-3,-2)$; $(1,2)$; four imaginaries.

16. $(-2,-1)$; $(1,2)$; four imaginaries.

17. Six imaginaries.

18. Six imaginaries.

19. $(-1,0)$; $(0,1)$; $(1,2)$; four imaginaries.

20. $(-1,0)$; $(0,\frac{1}{2})$; $(\frac{1}{2},1)$; four imaginaries.

21. $(-1,0)$; $(0,1)$; six imaginaries.

22. $(1,2)$; $(2,3)$; six imaginaries.

Chapter VII

§4. **1.** $V = x^3 - 3x + 1$, $V_1 = x^2 - 1$, $V_2 = 2x - 1$, $V_3 = +1$; $(-2,-1)$; $(0,1)$; $(1,2)$

2. $V = x^3 + 6x^2 + 10x - 1$, $V_1 = 3x^2 + 12x + 10$, $V_2 = 4x + 23$, $V_3 = -1$; $(0,1)$.

3. $V = x^3 - 4x + 2$, $V_1 = 3x^2 - 4$, $V_2 = 4x - 3$, $V_3 = +1$; $(-3,-2)$; $(0,1)$; $(1,2)$.

4. $V = x^3 - 6x^2 + 8x + 40$, $V_1 = 3x^2 - 12x + 8$, $V_2 = x - 17$, $V_3 = -1$; $(-2,-1)$.

5. $V = x^3 + x^2 - 2x - 1$, $V_1 = 3x^2 + 2x - 2$, $V_2 = 2x + 1$, $V_3 = +1$; $(-2,-1)$; $(-1,0)$; $(1,2)$.

6. $V = x^3 - 4x^2 - 4x + 20$, $V_1 = 3x^2 - 8x - 4$, $V_2 = 14x - 41$, $V_3 = +1$; $(-3,-2)$; $(2,3)$; $(3,4)$.

7. $V = 6x^4 - 24x^3 + 42x^2 - 32x + 11$, $V_1 = 6x^3 - 18x^2 + 21x - 8$,
$V_2 = -x^2 + x - 1$. All roots imaginary.

8. $V = 16x^4 - 32x^3 + 88x^2 - 8x + 17$, $V_1 = 8x^3 - 12x^2 + 22x - 1$,
$V_2 = -2x^2 - x - 1$. All roots imaginary.

9. $V = x^4 - 4x^3 + 10x^2 - 8x + 3$, $V_1 = x^3 - 3x^2 + 5x - 2$,
$V_2 = -2x^2 + x - 1$. All roots imaginary.

10. $V = x^4 - 4x^3 + 12x^2 - 12x + 5$, $V_1 = x^3 - 3x^2 + 6x - 3$,
$V_2 = -3x^2 + 3x - 2$. All roots imaginary.

11. $V = x^4 - 4x^3 + x^2 + 6x + 2$, $V_1 = 2x^3 - 6x^2 + x + 3$,
$V_2 = 5x^2 - 10x - 7$, $V_3 = x - 1$, $V_4 = +1$; $(-1, -\frac{1}{2})$; $(-\frac{1}{2}, 0)$;
$(2, \frac{5}{2})$; $(\frac{5}{2}, 3)$.

12. $V = x^4 - 4x^3 + x^2 - 1$, $V_1 = 2x^3 - 6x^2 + x$, $V_2 = 5x^2 - x + 2$;
$(-1, 0)$; $(3, 4)$.

13. $V = x^4 + x^3 + x - 1$, $V_1 = 4x^3 + 3x^2 + 1$, $V_2 = 3x^2 - 12x + 17$;
$(-2, -1)$; $(0, 1)$.

14. $V = x^4 + 2x^2 - 4x + 10$, $V_1 = x^3 + x - 1$, $V_2 = -x^2 + 3x - 10$.
All roots imaginary.

15. $V = x^5 - 5x^4 + 10x^3 - 5x^2 + 1$, $V_1 = x^4 - 4x^3 + 6x^2 - 2x$,
$V_2 = -3x^2 + 2x - 1$; $(-1, 0)$

16. $V = x^5 + 5x^4 - 20x^2 - 10x + 2$, $V_1 = x^4 + 4x^3 - 8x - 2$,
$V_2 = x^3 + 3x^2 - 1$, $V_3 = 3x^2 + 7x + 1$, $V_4 = 17x + 11$, $V_5 = +1$;
$(-4, -3)$; $(-3, -2)$; $(-1, 0)$; $(0, 1)$; $(1, 2)$.

17. $V = x^5 - 2x^4 + x^3 - 8x + 6$, $V_1 = 5x^4 - 8x^3 + 3x^2 - 8$,
$V_2 = 3x^3 - 3x^2 + 80x - 67$, $V_3 = 16x^2 - 23x + 9$; $(-2, -1)$; $(0, 1)$;
$(2, 3)$.

18. $V = x^6 - 6x^5 + 15x^4 - 20x^3 + 30x^2 - 24x + 14$,
$V_1 = x^5 - 5x^4 + 10x^3 - 10x^2 + 10x - 4$, $V_2 = -x^2 + x - 1$.
All roots imaginary.

19. $V = x^6 - 6x^5 + 16x^4 - 24x^3 + 22x^2 - 12x + 4$,
$V_1 = 3x^5 - 15x^4 + 32x^3 - 36x^2 + 22x - 6$,
$V_2 = -x^4 + 4x^3 - 8x^2 + 8x - 6$. All roots imaginary.

20. $V = 5x^6 - 30x^5 + 75x^4 - 90x^3 + 60x^2 - 18x - 2$,
$V_1 = 5x^5 - 25x^4 + 50x^3 - 45x^2 + 20x - 3$,
$V_2 = -x^3 + x^2 - x + 1 = -(x - 1)(x^2 + 1)$; $(-1, 0)$ and root 1.

22. $V = 1 - 3x^2 + x^3$, $V_1 = -2 + x$, $V_2 = +1$ for $x > 0$.
$V = 1 - 3x^2 + x^3$, $V_1 = 2 - x$ for $x < 0$.

23. $V = -1 + x^2 - 4x^3 + x^4$, $V_1 = 1 - 6x + 2x^2$, $V_2 = 190 - 67x$,
$V_3 = +1$ for $x > 0$.
$V = -1 + x^2 - 4x^3 + x^4$, $V_1 = -1 + 6x - 2x^2$ for $x < 0$.

24. $V = -1 + x^2 + x^3 + x^4$, $V_1 = 1 + 3x^2 + 4x^3$ for $x > 0$.
$V = -1 + x^2 + x^3 + x^4$, $V_1 = 1 + 3x^2 + 4x^3$, $V_2 = -3 - 2x + 3x^2$,
$V_3 = -4 - 3x$, $V_4 = -1$ for $x < 0$.

25. $V = -1 + 2x^3 + x^4 + x^5$, $V = 6 + 4x + 5x^2$ both for $x > 0$ and $x < 0$.

26. $4p^3 + 27q^2 < 0$.   27. $p^5 - q^2 > 0$.

Chapter VIII

§2. **1.** 1.912 **2.** 2.6207. **3.** 1.573. **4.** 2.1147.

5. 2.511. **6.** 4.264. **7.** $-$ 1.663. **8.** 1.532, 0.347.

9. 1.493, 0.250 **10.** $x = 1.345$, $y = -0.809$; $x = -1.711$, $y = -1.926$.

11. $x = 1$, $y = 2$; $x = -1.771$, $y = 1.365$.

12. 14.7. **13.** 47.46. **14.** 122.9.

15. 123.4. **16.** 43.98 in. **17.** 21.47 in.

18. 73.4 in. **19.** (a) 179 in., (b) 21.5 in. **20.** 1.347 radius.

§3. **1.** 1.442249. **2.** 1.357208. **3.** 1.324717. **4.** $-$ 0.644513.

5. $-$ 0.37213778. **6.** 1.164247. **7.** 3.41147412.

8. 1.36523001. **9.** $-$ 3.04891734. **10.** 0.250992.

11. 1.688697. **12.** 1.90785326. **13.** 2.31725562.

14. $-$ 1.117535. **15.** 0.295865; 1.449080.

16. 0.568396; 2.378596. **17.** $-$ 1.734691.

18. 0.228467; 3.232396; $-$ 0.197445.

§4. **1.** 7.062569 +. **2.** 2.57455046 +. **3.** 0.009477370187+.

4. 57.29577957. **5.** 1.47817455 +. **6.** 58.86620452 +.

7. 0.17728226 +. **8.** 0.64359442 +.

§6. **1.** Error negative and in absolute value $< 3 \times 10^{-8}$; $x = 1.5320888$ +.

2. Error in absolute value $< 8 \times 10^{-8}$; $x = 1.35689195$.

3. Error negative and in absolute value $< 10^{-10}$; $x = -0.372137785$.

4. Error negative and in absolute value $< 1.2 \times 10^{-10}$; $x = 1.493359197$ $\div$.

5. Error negative and in absolute value $< 10^{-10}$; $x = 1.907853262$ +.

6. Error negative and in absolute value 3.6×10^{-13}; $x = 1.734691345692$ +.

§7. **3.** 1.7693. **4.** 3.3876. **5.** 3.1038. **6.** 4.3311.

7. 4.2644. **8.** 0.16744. **9.** 0.33765. **10.** 0.20097.

11. 0.9216. **12.** 0.099115. **13.** 1.49336. **14.** 4.06443.

15. 4.33106. **16.** 0.055567. **17.** 0.052167. **18.** 0.35173.

19. 0.56714. **20.** $-$ 0.86287. **21.** 3.59728. **22.** 0.92042.

23. 0.51097. **24.** 0.73908. **25.** 0.32470. **26.** 4.49341.

27. 0.86033.

§8. **1.** 1.53209, **2.** 1.16425, **3.** 1.69202,
    0.34730, $-$ 1.77287,   1.35690,
    $-$ 1.87939. $-$ 3.39138. $-$ 3.04892.

4. 0.25099, **5.** 0.309906. **6.** 2.309881.
    1.49336.

7. 1.338483. **8.** 0.73908. **9.** 4.55553.

10. 217°12′0.5″. **11.** 108°36′14″. **12.** 149°16′27″.

13. 1.265 of a radius. **4.** 0.804744.

Chapter IX

§1. **1.** $x = {}^{25}\!/_4$, $y = -{}^{55}\!/_4$. **2.** $x = {}^{52}\!/_{47}$, $y = {}^{71}\!/_{47}$.

3. $x = -{}^{3}\!/_5$, $y = -1$. **4.** $x = 5$, $y = -8$.

5. $x = \dfrac{1}{a}$, $y = 0$. **6.** $x = (a + b)^2$, $y = (a - b)^2$.

7. $x = 5$, $y = -5$. **8.** $x = 1$, $y = \frac{1}{2}$.

§2. **1.** (a) 3; (b) **5**; (c) 5; (d) 6. **2.** (a) and (d) homogeneous.

4. No such polynomial exists.

§5. **11.** (a) -3; (b) -18.   **12.** (a) -187; (b) 470.

§6. **1.** (a) 19; (b) 16; (c) 13.   **2.** (a) even; (b) even; (c) odd.

§9. **1.** 0.   **2.** 0.   **3.** 0.   **4.** 0.   **5.** 0.   **6.** 0.

§11. **1.** 2.   **2.** 1.   **3.** 1.   **4.** -15.

  5. 0.   **6.** 160.   **7.** 18.   **8.** -5.

  9. -15.   **10.** 0.   **11.** $2(a + b + c)(a - b)(b - c)$.

  12. $-2abc$.   **13.** $4abc$.   **14.** $-2(a^3 + b^3)$.

  15. $(c - a)(c - b)(b - a)$.

  16. $(a - b)(a - c)(a - d)(b - c)(b - d)(d - c)$.

  17. $(c - a)(c - b)(b - a)(a + b + c)$.

  18. $(c - a)(c - b)(b - a)(ab + ac + bc)$.

  19. $(c - a)(c - b)(b - a)(ab + ac + bc)$.

  20. $(c - a)(c - b)(b - a)[a^2 + b^2 + c^2 + ab + ac + bc]$.

  21. $(a - b)(a - c)(b - c)(a + b + c)(a^2 + b^2 + c^2)$.

  22. $2(c - a)(c - b)(b - a)(z - x)(z - y)(y - x)$.

  23. $4(a + c)(a + b)(b + c)$.   **24.** $2abc(a + b + c)^3$.

  25. $a(b - a)(c - b)(d - c)$.

  26. $(a + b + c + d)(a + c - b - d)(a + b - c - d)(a + d - b - c)$.

  27. $(x - a_1)(x - a_2)(x - a_3)(x - a_4)$.

  28. $(x_1 - 1)(x_2 - 1)(x_3 - 1)(x_4 - 1)$.

  29. bcd.   **30.** $abcd\left(1 + \dfrac{1}{a} + \dfrac{1}{b} + \dfrac{1}{c} + \dfrac{1}{d}\right)$.

  31. $(x + n - 1)(x - 1)^{n-1}$.   **32.** $\phi(x) - x\phi'(x)$;
  $$\phi(x) = (a_1 - x)(a_2 - x) \cdots (a_n - x).$$

  33. $(c - a)(c - b)(b - a)(d - a)(d - b)(d - c)(a + b + c + d)$.

  34. $(c - a)(c - b)(b - a)(d - a)(d - b)(d - c)$
  $$[a^2 + b^2 + c^2 + d^2 + ab + ac + ad + bc + bd + cd].$$

Chapter X

§1. **1.** $x = 1, y = 0, z = 1$.   **2.** $x = a - 1, y = 1, z = 0$.

  3. $x = t = 1, y = z = -1$.   **4.** $x = -\frac{1}{5}, y = \frac{2}{5}, z = -\frac{24}{5}, t = -\frac{3}{5}$.

  5. $x = -2, y = -1, z = 1, t = 2$.

  6. $x_1 = \dfrac{a + d}{2}, x_2 = \dfrac{c - d}{2}, x_3 = \dfrac{b - c}{2}, x_4 = \dfrac{a - b}{2}$.

  7. $x_1 = x_3 = \frac{3}{5}, x_2 = x_4 = -\frac{2}{5}$.

  8. $x_1 = 7, x_2 = x_3 = x_4 = -1$.   **9.** $x_1 = x_2 = x_3 = x_4 = x_5 = 1$.

  10. $x_1 = x_2 = x_4 = -1, x_3 = 2, x_5 = 0$.

§4. **1.** $\rho = 2$.   **2.** $\rho = 2$.   **3.** $\rho = 3$.   **4.** $\rho = 2$.

  5. $f_3 = 3f_1 - 2f_2$.   **6.** $f_3 = -f_1 - f_2$.

  7. $f_3 = -3f_1 + 2f_2, f_4 = -f_1$.   **8.** $f_3 = -2f_1 + f_2, f_4 = -\frac{3}{2}f_1 + \frac{1}{2}f_2$.

§5. **1.** Incompatible.   **2.** $x = -3z - 4, y = 2z + 3$.

  3. $x = 1, z = -1, t = -y$.

  4. $x = -\frac{3}{2} + 3z + t, y = 1 - 2z$.   **5.** $x = 2 + t, y = -2t, z = 1$.

  6. Incompatible.

Chapter XI

§1. **1.** $\Sigma x_1^2 x_2 + 2\Sigma x_1 x_2 x_3$.   **2.** $\Sigma x_1^2 x_2 + 6\Sigma x_1 x_2 x_3$.

  3. $\Sigma x_1^2 x_2 - 6\Sigma x_1 x_2 x_3$.   **4.** $2\Sigma x_1^2 x_2^2 + 2\Sigma x_1^2 x_2 x_3$.

5. $\Sigma x_1^4 - 2\Sigma x_1^3 x_2 + 3\Sigma x_1^2 x_2^2.$

6. $\Sigma x_1^4 x_2^2 - 2\Sigma x_1^4 x_2 x_3 + 2\Sigma x_1^3 x_2^2 x_3 - 6\Sigma x_1^2 x_2^2 x_3^2 - 2\Sigma x_1^3 x_2^3.$

7. $\Sigma x_1^2 x_2^2 + 6\Sigma x_1 x_2 x_3 x_4.$ **8.** $\Sigma x_1^2 x_2^2 x_3^2 + \Sigma x_1^3 x_2 x_3 x_4.$

9. $3\Sigma x_1^2 - 2\Sigma x_1 x_2.$ **10.** $3\Sigma x_1^4 - 4\Sigma x_1^3 x_2 + 2\Sigma x_1^2 x_2^2 + 4\Sigma x_1^2 x_2 x_3 - 24\Sigma x_1 x_2 x_3 x_4.$

§2. **1.** $s_0 = 3,\ s_1 = 0,\ s_2 = 6,\ s_3 = -3,\ s_4 = 18,\ s_5 = -15,\ s_6 = 57.$

2. $s_0 = 3,\ s_1 = 3,\ s_2 = 5,\ s_3 = 12,\ s_4 = 29.$

3. $s_0 = 4,\ s_1 = 0,\ s_2 = 0,\ s_3 = 12,\ s_4 = 4.$

4. $s_{-2} = 13,\ s_{-3} = -19.$ **5.** $s_{-1} = 0,\ s_{-2} = -2,\ s_{-3} = 3,\ s_{-4} = 2.$

6. $s_1 = 0,\ s_2 = -1,\ s_3 = \frac{3}{2}.$ **7.** $2x^3 - 5x - 4 = 0.$

8. $12x^4 - 4x - 3 = 0.$ **9.** $x^3 - 18x^2 + 81x - 81 = 0.$

10. $x^5 - x^4 + x^3 - x^2 + x - 1 = 0.$ **11.** $x^3 - 3x^2 + 6x + 6 = 0.$

12. $5x^4 + 4x^2 - 8 = 0.$ **14.** (a) 1.6170 (1.618034),

(b) 6.372281318 (6.372281323).

15. 0.3247181 (0.3247179). **17.** $y^3 - 3y - 1 = 0.$

19. $y^3 - 18y^2 + 81y - 81 = 0.$ **20.** $y^3 - 42y^2 + 441y - 49 = 0.$

21. (a) 1, (b) $\frac{1}{3}$.

§3. **1.** $f_1 f_2 - 3f_3.$ **2.** $f_1^4 - 6f_1^2 f_2 + 9f_2^2.$ **3.** $f_1 f_5.$

4. $f_2^2 - 2f_1 f_3 + 2f_4.$

§4. **1.** $3p_3 - p_1 p_2.$ **2.** $p_2^2 - 2p_1 p_3 + 2p_4.$

3. $p_1^2 p_2 - p_1 p_3 - 2p_2^2 + 4p_4.$

4. $2p_1^2 p_3 - p_1 p_2^2 + p_2 p_3 - 5p_1 p_4 + 5p_5.$

5. $p_1 p_3 - 4p_4.$ **6.** $3p_1 p_4 - p_2 p_3 - 5p_5.$

7. $7p_1 p_5 + 4p_2 p_4 - 3p_3^2 - 3p_1^2 p_4 + p_1 p_2 p_3 - 12p_6.$

8. $p_2 p_4 - 4p_1 p_5 + 9p_6.$

9. $9p_3 - p_1 p_2.$ **10.** $2p_2^2 - 2p_1 p_3 - 4p_4.$

11. $4p_6 + p_1 p_5 - 4p_2 p_4 - p_3^2 + p_1^2 p_4.$

12. $y^3 - 3y^2 - 6y + 17 = 0.$ **13.** $y^3 - 12y^2 + 45y - 53 = 0.$

14. $y^3 - qy^2 + pry - r^2 = 0.$ **15.** $y^3 - 9y^2 + 26y - 23 = 0.$

16. (a) 3.24698, 1.55495; (b) $-3.04727 \pm 1.13594i.$

19. $-3.$ **20.** 3. **21.** 39. **22.** $-\ 2^{19}\!\frac{9}{104}.$

23. 9. **24.** 12. **25.** 4. **26.** 0.

Chapter XII

§2. **1.** $x = 1, 0, 1, -1;$ **2.** $3y^4 + 10y^2 - 1 = 0,\ (y + 1)x = y - 1.$

$y = 0, 1, 1, -1.$

3. $(3y^2 + 4y - 1)(y^2 + 4y - 3) = 0,\ x(1 + y) = 1 - y.$

4. $x = 0, -2;$ **5.** $(y - 1)(25y^3 - 45y^2 - 171y + 243) = 0,$

$y = 0, 2.$ $(7y - 9)x = 5y^2 - 7y.$

6. $(y - 1)(y^3 + 13y^2 + 100y - 100) = 0,\ (3y - 10)x + y^2 + 6y = 0.$

7. $x = -1,\ y = -1.$

8. $x = 0, 1, -1 + \dfrac{i}{\sqrt{2}}, -1 - \dfrac{i}{\sqrt{2}}.$

$y = 1, 0, -1 - \dfrac{i}{\sqrt{2}}, -1 + \dfrac{i}{\sqrt{2}}.$

9. $x^5 - 2x^4 - x^3 + 2x^2 + x - 1 = 0.$

10. $x^3 + 3x^2 - 24x + 1 = 0.$ **11.** (a) 1, (b) $y^2 - y.$

§3. **1.** $x = 0, -2,$ **2.** $x = 0, -4, -5,$ **3.** $x = 0, 0, 0, 1, 1, -1, -1, 2, -2,$

$y = 0, 1.$ $y = 2, 2, 3.$ $y = 1, -1, 3, 0, 2, 0, 2, 1, 1.$

4. (*a*) $y = x$, x arbitrary; (*b*) $x = 1, 2$,
$$y = -2, -1.$$

5. No solution.   **6.** $x = 0, 2, 1, \, {}^{42}\!\!/_{25}$,
$$y = 0, -1, 2, \, {}^{63}\!\!/_{25}.$$

7. $x = 1, 0, \dfrac{1 + i\sqrt{3}}{4}, \dfrac{1 - i\sqrt{3}}{4}, \dfrac{-3 + i\sqrt{3}}{4}, \dfrac{-3 - i\sqrt{3}}{4}$,

$y = 0, 1, \dfrac{-3 + i\sqrt{3}}{4}, \dfrac{-3 - i\sqrt{3}}{4}, \dfrac{1 + i\sqrt{3}}{4}, \dfrac{1 - i\sqrt{3}}{4}$.

8. $y^4 + 9y^2 + 54 = 0$, $(y^2 + y + 6)x + 2y + 6 = 0$.

9. $28y^3 + 713y^2 - 100y = 0$, $(119y^2 - 55y + 5)x = 14y^2 - 32y + 5$.

10. $\lambda = 10$, $\mu = -5$.

§6.   **1.** $1 \pm 2i$.   **2.** $-0.884646 \pm 0.589743i$.

      3. $-1.047276 \pm 1.135940i$.   **4.** $1 \pm i$, $-1 \pm i\sqrt{2}$.

      5. $\pm i$, $1 \pm 2i$.   **6.** $1 \pm i$.

      7. $0.72714 \pm 0.43001i$, $-0.72714 \pm 0.93409i$.

      8. $0.304877 \pm 0.754528i$.

INDEX

A

Abel, 83–84
Abridged divisor, 159–163
Absolute value, 8
  inequality for, 31
  of products of complex numbers, 10
  of sum of complex numbers, 12–14
Addition, of complex numbers, 3, 31
  definition of, 2
  laws for, 3
  of matrices, 217, 218
Algebra, fundamental theorem of, 55–58, 293–297
  of matrices, 217, 312
Algebraic equations, 50–69
  Horner's method of evalutaion for, 151–155
  solution of, 82–84
Algebraic solutions, impossibility of, 272–274
Algorithm, Euclid, 47
Amplitudes, 20
  normalization of, 313
  orthogonality of, 311
Angles between directed lines, 18–19
Approximate methods, 104, 151–180
Arbitrary complex coefficients, 59
Area of triangle, use of determinants for, 249
Argument (amplitude), of complex number, 20
Arithmetic mean, 117
Associative law, for addition, 3
  for matrices, 218, 220
  for multiplication, 3
Augmented matrix, 246
Axis, of imaginaries, 18
  of reals, 18

B

Bernoulli's method of computing roots, 263, 316
Binomial coefficients, 23, 53
  identity for, 54, 55
Binomial equation, representation of roots of, 24
  trigonometric solution of, 24–26
Binomial theorem, 43
Binomials, algebraic, 6
Biquadratic equations, 83
  Lagrange's solution of, 274–275
  solution of, 94–97
Bremiker tables, 319

C

Cardan (1501–1576), 83
  formulas of, 84–89
Casus irreducibilis, 91, 90–94
Cauchy's inequality, 117
Change in sign of polynomials (*see* Variation)
Characteristic equation, 311
Circle, division of, 28–30
  equation of, 247
  power of a point with respect to, 251–252
Coefficients, 35
  arbitrary complex, 55
  computation by Newton's formula, 261–262
  detached, method of, 36, 38
  given numbers for, 99
  of quotient, 42
  real, 59–60
  relations to roots, 61–65
Cofactors, in expansion of determinants, 210–212
Collinear complex numbers, 33

Catalog

If you are interested in a list of fine Paperback
books, covering a wide range of subjects
and interests, send your name and address,
requesting your free catalog, to:

McGraw-Hill Paperbacks
330 West 42nd Street
New York, New York 10036